Lecture Notes in Computer Science　　11549

Commenced Publication in 1973
Founding and Former Series Editors:
Gerhard Goos, Juris Hartmanis, and Jan van Leeuwen

More information about this series at http://www.springer.com/series/7409

Robert Thomson · Halil Bisgin ·
Christopher Dancy · Ayaz Hyder (Eds.)

Social, Cultural, and Behavioral Modeling

12th International Conference, SBP-BRiMS 2019
Washington, DC, USA, July 9–12, 2019
Proceedings

 Springer

Editors
Robert Thomson
United States Military Academy
West Point, NY, USA

Christopher Dancy
Bucknell University
Lewisburg, PA, USA

Halil Bisgin
University of Michigan–Flint
Flint, MI, USA

Ayaz Hyder
The Ohio State University
Columbus, OH, USA

ISSN 0302-9743   ISSN 1611-3349   (electronic)
Lecture Notes in Computer Science
ISBN 978-3-030-21740-2   ISBN 978-3-030-21741-9   (eBook)
https://doi.org/10.1007/978-3-030-21741-9

LNCS Sublibrary: SL3 – Information Systems and Applications, incl. Internet/Web, and HCI

© Springer Nature Switzerland AG 2019
This work is subject to copyright. All rights are reserved by the Publisher, whether the whole or part of the material is concerned, specifically the rights of translation, reprinting, reuse of illustrations, recitation, broadcasting, reproduction on microfilms or in any other physical way, and transmission or information storage and retrieval, electronic adaptation, computer software, or by similar or dissimilar methodology now known or hereafter developed.
The use of general descriptive names, registered names, trademarks, service marks, etc. in this publication does not imply, even in the absence of a specific statement, that such names are exempt from the relevant protective laws and regulations and therefore free for general use.
The publisher, the authors and the editors are safe to assume that the advice and information in this book are believed to be true and accurate at the date of publication. Neither the publisher nor the authors or the editors give a warranty, expressed or implied, with respect to the material contained herein or for any errors or omissions that may have been made. The publisher remains neutral with regard to jurisdictional claims in published maps and institutional affiliations.

This Springer imprint is published by the registered company Springer Nature Switzerland AG
The registered company address is: Gewerbestrasse 11, 6330 Cham, Switzerland

Preface

Improving the human condition requires understanding, forecasting, and impacting sociocultural behavior both in the digital and nondigital world. Increasing amounts of digital data, embedded sensors collecting human information, rapidly changing communication media, changes in legislation concerning digital rights and privacy, spread of 4G technology to developing countries and so on are creating a new cyber-mediated world where the very precepts of why, when, and how people interact and make decisions are being called into question. For example, Uber took a deep understanding of human behavior vis-à-vis commuting, developed software to support this behavior, ended up saving human time (and so capital) and reducing stress, and so indirectly created the opportunity for humans with more time and less stress to evolve new behaviors. Scientific and industrial pioneers in this area are relying on both social science and computer science to help make sense of and impact this new frontier. To be successful, a true merger of social science and computer science is needed. Solutions that rely only on social science or only on computer science are doomed to failure. For example, Anonymous developed an approach for identifying members of terror groups such as ISIS on the social media platform Twitter using state-of-the-art computational techniques. These accounts were then suspended. This was a purely technical solution. The response was that those individuals with suspended accounts just moved to new platforms, and resurfaced on Twitter under new IDs. In this case, failure to understand basic social behavior resulted in an ineffective solution.

The goal of this conference is to build this new community of social cyber scholars by bringing together and fostering interaction between members of the scientific, corporate, government, and military communities interested in understanding, forecasting, and impacting human sociocultural behavior. It is the charge to this community to build this new science, its theories, methods, and its scientific culture in a way that does not give priority to either social science or computer science, and to embrace change as the cornerstone of the community. Despite decades of work in this area, this new scientific field is still in its infancy. To meet this charge, to move this science to the next level, this community must meet the following three challenges: deep understanding, sociocognitive reasoning, and re-usable computational technology. Fortunately, as the papers in this volume illustrate, this community is poised to answer these challenges. But what does meeting these challenges entail?

Deep understanding refers to the ability to make operational decisions and theoretical arguments on the basis of an empirically based deep and broad under-standing of the complex sociocultural phenomena of interest. Today, although more data are available digitally than ever before, we are still plagued by anecdotally based arguments. For example, in social media, despite the wealth of information available, most analysts focus on small samples, which are typically biased and cover only a small time period, and use that to explain all events and make future predictions. The analyst finds the magic tweet or the unusual tweeter and uses that to prove their point.

Tools that can help the analyst to reason using more data or less biased data are not widely used, are often more complex than the average analyst wants to use, or take more time than the analyst wants to spend to generate results. Not only are more scalable technologies needed, but so too is a better understanding of the biases in the data and ways to overcome them, and a cultural change to not accept anecdotes as evidence.

Sociocognitive reasoning refers to the ability of individuals to make sense of the world and to interact with it in terms of groups and not just individuals. Today most social–behavioral models either focus on (1) strong cognitive models of individuals engaged in tasks and so model a small number of agents with high levels of cognitive accuracy but with little if any social context, or (2) light cognitive models and strong interaction models and so model massive numbers of agents with high levels of social realisms and little cognitive realism. In both cases, as realism is increased in the other dimension, the scalability of the models fails, and their predictive accuracy on one of the two dimensions remains low. By contrast, as agent models are built where the agents are not just cognitive by socially cognitive, we find that the scalability increases and the predictive accuracy increases. Not only are agent models with sociocognitive reasoning capabilities needed, but so too is a better understanding of how individuals form and use these social cognitions.

More software solutions that support behavioral representation, modeling, data collection, bias identification, analysis, and visualization support human sociocultural behavioral modeling and prediction than ever before. However, this software is generally just piling up in giant black holes on the Web. Part of the problem is the fallacy of open source; the idea that if you simply make code open source others will use it. By contrast, most of the tools and methods available in Git or R are only used by the developer, if at all. Reasons for lack of use include lack of documentation, lack of interfaces, lack of interoperability with other tools, difficulty of linking to data, and increased demands on the analyst's time due to a lack of tool-chain and workflow optimization. Part of the problem is the not-invented-here syndrome. For social scientists and computer scientists alike, it is simply more fun to build a quick and dirty tool for one's own use than to study and learn tools built by others. And, part of the problem is the insensitivity of people from one scientific or corporate culture to the reward and demand structures of the other cultures that impact what information can or should be shared and when. A related problem is double standards in sharing where universities are expected to share and companies are not, but increasingly universities are relying on that intellectual property as a source of funding just like other companies. While common standards and representations would help, a cultural shift from a focus on sharing to a focus on re-use is as, or more, critical for moving this area to the next scientific level.

In this volume, and in all the work presented at the SBP-BRiMS 2019 conference, you will see suggestions of how to address the challenges just described. SBP-BRiMS 2019 carried on the scholarly tradition of the past conferences out of which it has emerged like a phoenix: the Social Computing, Behavioral–Cultural Modeling, and Prediction (SBP) Conference and the Behavioral Representation in Modeling and Simulation (BRiMS) Society's conference. A total of 72 papers were submitted as regular track submissions. Of these, 28 were accepted as full papers for an acceptance rate of 38%. Additionally, there were a large number of papers describing emergent ideas and late-breaking results. This was an international group with papers submitted with authors from many countries.

The conference has a strong multidisciplinary heritage. As the papers in this volume show, people, theories, methods, and data from a wide number of disciplines are represented including computer science, psychology, sociology, communication science, public health, bioinformatics, political science, and organizational science. Numerous types of computational methods are used that include, but are not limited to, machine learning, language technology, social network analysis and visualization, agent-based simulation, and statistics.

This exciting program could not have been put together without the hard work of a number of dedicated and forward-thinking researchers serving as the Organizing Committee, listed on the following pages. Members of the Program Committee, the Scholarship Committee, publication, advertising and local arrangements chairs worked tirelessly to put together this event. They were supported by the government sponsors, the area chairs, and the reviewers. We thank them for their efforts on behalf of the community. In addition, we gratefully acknowledge the support of our sponsors – the Army Research Office (W911NF-17-1-0138), the National Science Foundation (IIS-1926691), and the Artificial Intelligence Journal. Enjoy the conference proceedings.

April 2019   Kathleen M. Carley
                                                                              Nitin Agarwal

Organization

Conference Co-chairs

Kathleen M. Carley	Carnegie Mellon University, USA
Nitin Agarwal	University of Arkansas – Little Rock

Program Co-chairs

Halil Bisgin	University of Michigan–Flint, USA
Christopher Dancy II	Bucknell University, USA
Ayaz Hyder	The Ohio State University, USA
Robert Thomson	United States Military Academy, USA

Advisory Committee

Fahmida N. Chowdhury	National Science Foundation, USA
Rebecca Goolsby	Office of Naval Research, USA
Stephen Marcus	National Institutes of Health, USA
Paul Tandy	Defense Threat Reduction Agency, USA
Edward T. Palazzolo	Army Research Office, USA

Advisory Committee Emeritus

Patricia Mabry	Indiana University, USA
John Lavery	Army Research Office, USA
Tisha Wiley	National Institutes of Health, USA

Scholarship and Sponsorship Committee

Nitin Agarwal	University of Arkansas – Little Rock, USA
Christopher Dancy II	Bucknell University, USA

Industry Sponsorship Committee

Jiliang Tang	Michigan State University, USA

Publicity Chair

Donald Adjeroh	West Virginia University, USA
Katrin Kania Galeano	University of Arkansas – Little Rock, USA

Local Area Coordination

David Broniatowski   The George Washington University, USA

Proceedings Chair

Robert Thomson   United States Military Academy, USA

Agenda Co-chairs

Robert Thomson   United States Military Academy, USA
Kathleen M. Carley   Carnegie Mellon University, USA

Journal Special Issue Chair

Kathleen M. Carley   Carnegie Mellon University, USA

Tutorial Chair

Kathleen M. Carley   Carnegie Mellon University, USA

Graduate Program Chair

Yu-Ru Lin   University of Pittsburgh, USA

Challenge Problem Committee

Kathleen M. Carley   Carnegie Mellon University, USA
Nitin Agarwal   University of Arkansas – Little Rock, USA
Sumeet Kumar   Massachusetts Institute of Technology, USA
Brandon Oselio   University of Michigan, USA
Justin Sampson   Arizona State University, USA

BRiMS Society Chair

Christopher Dancy II   Bucknell University, USA

BRiMS Steering Committee

Christopher Dancy II   Bucknell University, USA
William G. Kennedy   George Mason University, USA
David Reitter   The Pennsylvania State University, USA
Dan Cassenti   US Army Research Laboratory, USA

SBP Steering Committee

Nitin Agarwal	University of Arkansas – Little Rock, USA
Sun Ki Chai	University of Hawaii, USA
Ariel Greenberg	Johns Hopkins University/Applied Physics Laboratory, USA
Huan Liu	Arizona State University, USA
John Salerno	Exelis, USA
Shanchieh (Jay) Yang	Rochester Institute of Technology, USA

BRiMS Executive Committee

Brad Best	Adaptive Cognitive Systems, USA
Brad Cain	Defense Research and Development, Canada
Daniel N. Cassenti	US Army Research Laboratory, USA
Bruno Emond	National Research Council, USA
Coty Gonzalez	Carnegie Mellon University, USA
Brian Gore	NASA, USA
Kristen Greene	National Institute of Standards and Technology, USA
Jeff Hansberger	US Army Research Laboratory, USA
Tiffany Jastrzembski	Air Force Research Laboratory, USA
Randolph M. Jones	SoarTech, USA
Troy Kelly	US Army Research Laboratory, USA
William G. Kennedy	George Mason University, USA
Christian Lebiere	Carnegie Mellon University, USA
Elizabeth Mezzacappa	Defence Science and Technology Laboratory, UK
Michael Qin	Naval Submarine Medical Research Laboratory, USA
Frank E. Ritter	The Pennsylvania State University, USA
Tracy Sanders	University of Central Florida, USA
Venkat Sastry	University of Cranfield, USA
Barry Silverman	University of Pennsylvania, USA
David Stracuzzi	Sandia National Laboratories, USA
Robert Thomson	Unites States Military Academy, USA
Robert E. Wray	SoarTech, USA

SBP Steering Committee Emeritus

Nathan D. Bos	Johns Hopkins University/Applied Physics Lab, USA
Claudio Cioffi-Revilla	George Mason University, USA
V. S. Subrahmanian	University of Maryland, USA
Dana Nau	University of Maryland, USA

SBP-BRIMS Steering Committee Emeritus

Jeffrey Johnson	University of Florida, USA

Technical Program Committee

Kalin Agrawal
Shah Jamal Alam
Elie Alhajjar
Scott Batson
Jeffrey Bolkhovsky
Lashon Booker
David Broniatowski
Magdalena Bugajska
Jose Cadena
Subhadeep Chakraborty
Rumi Chunara
Andrew Crooks
Peng Dai
Hasan Davulcu
Jana Diesner
Wen Dong
Koji Eguchi
Bruno Emond
William Ferng
Michael Fire
Ariel Greenberg
Kristen Greene
Kyungsik Han
Walter Hill
Shen-Shyang Ho
Tuan-Anh Hoang
Yuheng Hu
Robert Hubal
Terresa Jackson
Aruna Jammalamadaka
Bill Kennedy
Shamanth Kumar
Huan Liu
Yu-Ru Lin
Deryle W. Lonsdale

Stephen Marcus
Venkata Swamy Martha
Elizabeth Mezzacappa
Allen Mclean
Sai Moturu
Keisuke Nakao
Radoslaw Nielek
Kouzou Ohara
Byung Won On
Alexander Outkin
Hemant Purohit
Aryn Pyke
Weicheng Qian
S. S. Ravi
Travis Russell
Amit Saha
Samira Shaikh
Narjes Shojaati
David Stracuzzi
Zhijian Wang
Changzhou Wang
Yafei Wang
Xiaofeng Wang
Changzhou Wang
Rik Warren
Elizabeth Whitaker
Paul Whitney
Kevin S. Xu
Xiaoran Yan
Laurence Yang
Yong Yang
Mo Yu
Reza Zafarani
Rifat Zahan
Kang Zhao

Contents

Analyzing the Dabiq Magazine: The Language and the Propaganda Structure of ISIS

Halil Bisgin[1]($^{(\boxtimes)}$) (ORCID), Hasan Arslan[2], and Yusuf Korkmaz[1]

[1] University of Michigan-Flint, Flint 48502, USA
bisgin@umich.edu
[2] Western Connecticut State University, Danbury, CT 06810, USA

Abstract. The Islamic State of Iraq and Sham (ISIS) still poses a significant concern worldwide due to its brutal attacks and unconventional recruitment strategy despite its recent defeat and loss of territory. ISIS distinguished itself from other notorious terrorist organizations regarding Techniques, Tactics, and Procedures (TTP). It has been observed that ISIS is highly capable of attracting foreign fighters through its improved "netwar" skills. Whereas its propaganda videos and images have been extensively analyzed, a systematic analysis of textual content is still lacking. Therefore, we examine the Dabig magazine to discover propagandist elements by performing natural language processing (NLP) and text mining methods. Namely, we first automatically detect three types of entities (person, location, organization) in each article for fifteen Dabiq issues. Then we build entity networks based on co-occurrence of entities to observe the entity relationships over time. We further employ topic modeling on all articles and calculate statistics for entities. We observe entities revolve around the term "jihad," and the ISIS consistently seems to exploit the sources of Islam in their propaganda. The analysis also revealed that ISIS primarily targets Shiites by using derogatory language about their belief system and try to justify their attacks against them.

Keywords: ISIS · Entity Recognition · Network analysis · Topic modeling

1 Introduction

The Islamic State of Iraq and Sham or Levant (ISIS/ISIL) is a significant concern worldwide due to its brutal attacks and unconventional recruitment strategy. ISIS distinguished itself from other notorious terrorist organizations like Al Qaeda and Boko Haram regarding Techniques, Tactics, and Procedures (TTP). It's death cult mentality, and violent modus operandi aimed to attract young Muslims joining them in their delusional world of a rudimentary state. Instead, "it presents itself as the avant-garde of a mass movement, like the Khmer Rouge" [1]. Like the *Thing*, it assimilates the other life forms to survive in the Muslim Lands. "They use violent images to attract alienated, young people to their cause and have gained international attention by posting their execution and beheading videos online for all to see" [2]. Certainly, the ISIS members used the available technology and the cyber world not only to recruit new soldiers but also disseminate their propaganda worldwide. They are "crafted not just to

R. Thomson et al. (Eds.): SBP-BRiMS 2019, LNCS 11549, pp. 1–11, 2019.
https://doi.org/10.1007/978-3-030-21741-9_1

stir the hearts of potential recruits but also to boost the organization's ghastly brand—to reinforce Westerners' perceptions of the Islamic State and its devotees as ruthless beyond comprehension" [3]. The ISIS videos distributed via social media often depict brutal acts of violence.

The Islamic State brands itself and spreads its message through a comprehensive social media strategy, including video elements designed to appeal to a young generation and a professional-quality magazine publication. This article does a text data mining on the words ISIS magazine *Dabiq*. The Islamic State uses social media outlets and the publication *Dabiq* as part of comprehensive marketing, branding, and recruitment strategy [4, 5]. Therefore, one of the best ways to visualize the mental map of ISIS is to look at its online publication, *Dabiq* and to derive high-quality information from its text content. Previously whereas propaganda videos and images have been analyzed to account for general propaganda techniques, a systematic analysis of textual content is still lacking.

Nevertheless, this research examines ISIS texts to discover propagandist elements that can have an impact on recruitment. More specifically, we perform text mining and natural language processing methods on the ISIS periodical, Dabiq magazine. We present our results that include word frequencies and their network representations to illustrate the relationship of entities mentioned in the magazine. Since ISIS emerged from the cadres of Al Qaeda, introductory information about them is also necessary for the reader.

2   Background

2.1   Before ISIS, There Is Al Qaeda

Al Qaeda is a Salafi-jihadi Islamic extremist organization currently headquartered in an area somewhere along the Pakistan-Afghanistan border. Salafi-jihadi movements like Al Qaeda seek the "destruction of current Muslim societies through the use of force and creation of what they regard as a true Islamic society" [6]. Al Qaeda was founded by Osama bin Laden and came out of the early 1980s-era fight against the Soviet Union in Afghanistan.

Al Qaeda makes use of both new and old platforms and techniques to spread its message. As Thompson points out, Al-Qaida and its members understand the vulnerabilities of the Western world's reliance on information sharing and the use of technology to communicate effectively [7]. Al Qaeda's messaging strategy combines a mix of proselytizing through older Internet technologies and a more modern attempt at branding and messaging that targets a younger audience. One example of a more modern messaging and branding strategy from Al Qaeda comes from Al Qaeda in the Arabian Peninsula (AQAP). They publish a magazine called *Inspire*, which focuses the narrative in a few key areas and distributed online.

"Al-Qaida's use of media strategy give it the means to infiltrate Muslim communities around the world for easier access to potential recruits. Al-Qaida uses a combination of written and audiovisual messages that transcends both technology and literacy barriers... The Inspire magazine encourages young Muslims in the West to commit terrorist attacks and publishes step by step directions for 'homegrown' terrorists" [8].

The magazine promotes successful operations and martyrs, a "DIY guide for weapon assembly" including information about bomb-making and other tools, and it promotes "lone wolf" terrorism. However, *Inspire* seems to lack any references to Muslim charity or specific discussion of the Islamic faith [9]. Also, the majority of their messaging consists of "long videos featuring senior Al Qaeda ideologues pontificating about various aspects of jihad and quoting extensively from the Koran," which may not appeal to younger generations of potential recruits [10]. In term of recruiting women, Al-Qaida sought women in operational terrorist roles. *Al-Shamikha*, known as the majestic woman magazine was launched in March 2011 and is designed to target and attract the female jihadi audience.

2.2   The Islamic State and Dabiq Magazine

The Islamic State of Iraq and Syria (ISIS) (aka. DAESH in Arabic) is a Salafi-jihadi organization operating predominantly in Syria and Iraq. The group emerged out of the 'Al Qaeda in Iraq' organization which had several disputes with Al Qaeda core leadership over tactics, strategy, and targets [11]. As a Salafi-jihadi organization, they see themselves as the only true Muslims and seek to purge others considered as apostates or "deserters of the religion," including "the Shia and, for many Salafis, democrats, or those participating in a democratic system" [12]. As being one of the deadliest terrorist organizations in recent history, ISIS was very effective to inspire the "soldiers" carrying out terror attacks throughout the Western world in their sophisticated propaganda efforts. For more than four years, they captured and controlled large areas in Iraq and Syria, including some major cities and resources, then declared themselves the Islamic State [10]. ISIS, indeed, was the long-term vision of its infamous Jordanian leader, Abu Musab Al Zarqawi, an Al Qaeda commander, who was killed in a targeted killing attack by U.S. government in 2006. The terrorist organization is currently led by Abu Bakr al-Baghdadi, an educated Iraqi who declared himself as the caliph of the alleged Islamic State of Iraq and Levant in 2013.

In contrast to Al Qaeda's "far enemy" strategy, ISIS sees its most urgent enemy as apostate Shi'ite regimes in Syria and Iraq that impede the creation of a 'pure,' radically sectarian Islamic state" (Bertrand, 2015, May 21). The group leadership strongly believes that others who practice "major idolatry" are outside the Muslim faith, including "those worshipping - or perceived to be worshipping - stones, saints, tombs, etc." [13] such as the destruction of the ancient ruins in Palmyra. ISIS also deploys other brutal tactics to terrorize, maintain control, and purge apostates and non-believers from their territory, including public executions, rape, and crucifixion. George Packer in the New Yorker magazine [1] describes the degree of ISIS's fallacy:

"The Islamic State doesn't behave according to recognizable cost-benefit analyses. It doesn't cut its losses or scale down its ambitions. The very name of the self-proclaimed caliphate strikes most people, not least other Muslims, as ridiculous, if not delusional. But it's the vaulting ambition of an actual Islamic State that inspires ISIS recruits. The group uses surprise and shock to achieve goals that are more readily grasped by the apocalyptic imagination than by military or political theory."

The Islamic State also takes advantage of other social media outlets and strategies. They regularly communicate between leaders and followers via social media and are "creating competition within the jihadi world in cyberspace as well as in the arts of terrorism and atrocity" [14]. Female members of ISIS often use Twitter, Facebook, and Instagram to recruit others, and those converts also appear in other propaganda videos [15].

The name *Dabiq* refers to a town close to the Turkish border in Syria. ISIS believes that the place has significance in the sense of Islamic eschatology, where a final battle would take place between the armies of the East and the West, which would lead to the Apocalypse. Dabiq magazine is published in five different languages (Arabic, English, French, Russian and Turkish), and serves as an online propaganda arm for the terrorist organization. They have issued 15 issues in pdf format and distributed them all online. The magazine was printed in a non-periodical style following particularly the Muslim calendar called *Hijri calendar*, which is a lunar calendar consisting of 12 months in a year of 354 or 355 days. Regarding the Gregorian calendar, all issues came out between June 2014 and July 2016.

Many articles in Dabiq discussed the "social services, security, and dignity" ISIS claims can be found within their caliphate [9]. Each issue reported news about the state of their alleged caliphate and had crafty images of their utopian statehood. Every issue included a lot of threats against the West and sent intimidating messages to certain Muslim and non-Muslim figures mostly from the politics, media, and academics. ISIS is cautious when they mention the names and the identities of its soldiers and commanders in Dabiq. There is very little information that can assist the intelligence agencies to reveal the true identities of those mentioned members. The magazine constantly glorified the suicide bombers as heroic 'Istishhad' (death of a martyr) operations. Indeed, Dabiq consistently put side by side pictures and stories of ISIS social support for its people (i.e., medical care to children, repairing bridges and roads, etc.) with profiles of fighters who were killed, allegedly in defense of such projects [16]. Interestingly, the feature 'In the words of the Enemy' displayed the statement and opinions of others who portrayed ISIS as dangerous, strong and scary as a positive thing for its members. Unlike its predecessor Al Qaeda, ISIS leadership encouraged its supporters to wage "jihad at home" if they lacked the means of joining them in the front lines. Indeed, the terrorist organization asks Muslims who have been living in non-Muslim lands to "attack, kill, and terrorize the crusaders on their own streets and in their own homes" [17].

3   Methodology

All of the fifteen issues of the magazine were included in this research. The digital copies in PDF file format were downloaded from the Clarion project website [18], which defines itself as "a non-profit organization that educates the public about the dangers of radical Islam." The Clarion project being biased and Islamophobic are beyond the scope of this article. Figure 1 below displays the following steps before the analysis of the content concerning identifying specific patterns and methods ISIS's use of language and terminology. First, we split each issue into articles, which varied in size, and fixed non-English characters. Then, we utilized natural language processing (NLP) and text mining tools to discover entity types for which we computed frequencies and looked for co-occurrences to determine their relationship.

For each article, we ran Entity Recognition (ER) task by using Stanford Named Entity Recognizer (NER) [19], which probabilistically determines if a word represents a *person* (P), an *organization* (O), or a *location* (L). As a result of this step, we identified 1,486 persons, 285 organizations, and 651 locations, within all 15 Dabiq magazines. There were also 621 entities that could not be categorized by the tool and labeled as miscellaneous. We further manually checked and curated all entity names including miscellaneous ones, which resulted in 1,786 entities. For this final list of entities in any category above, we performed searched within the articles across all issues and computed their frequencies. Besides, we sought patterns by inspecting their co-occurrences. Namely, if any two entities were mentioned in the same article, we not only built a link in between but also counted how often they appeared together. This resulted in a dynamic network analysis which evolved throughout fifteen issues. Finally, we took a more holistic approach, which mainly relies on the "bag of words" assumption to extract a group of keywords that would help us infer concepts. We, therefore, adapted topic-modeling approach by using Latent Dirichlet Allocation (LDA) [20] for concept discovery, which does not rely on word sequence in contrast to NER. Instead, it assumes words are used together by following a probability distribution which is imposed by a given topic, and each document (article) can discuss multiple topics which are not known a priori.

Fig. 1. Procedures for methodological analysis.

4 Results and Discussion

We started our text analysis by merely counting the occurrences of each entity that they curated and verified via external resources. Figure 2 below demonstrates the distributions frequency of the words in the texts. Different 'cutting points' were established to view the distribution of words in the articles; clearly, not all entities within the texts were equally used; as expected, the majority of the entities (∼ 77%) were repeated at most five times throughout all issues following a significant decrease. These entities consist of many individual names of head of the states, politicians, religious figures, military commanders, journalists, various geographical locations and organizations. For example, names like Putin (*Person*), Sisi (*Person*), and Senegal (*Location*) have been mentioned more than the others. Understandably, the word '*Islam*' was the most frequently used entity with 6,568 times in the Dabiq magazine.

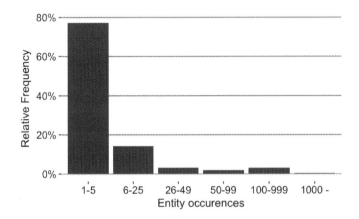

Fig. 2. Distribution of words (entities) based on their occurrences.

While the skewed nature of the entity distribution gave an idea about the overall selection or use of the entities, a more detailed analysis that included the pattern of words over the time was also necessary. Figure 3 presents the entity frequencies that appears the most in a given issue and indicates that certain entities such as *Islam* and *Jihad* were stressed continuously in the texts followed by the words, *caliphate*, *Iraq*, and *Syria*. Since the terrorist organization attempted to legitimize itself by taking the name of "the Islamic State of Iraq and Syria (ISIS)" and declared itself as a Caliphate, the initial findings were not unusual. However, Abu Bakr Al Baghdadi, who has been the group's leader and so-called "Caliph of Islam" appeared more in the very early issues, but eventually did not become a central figure like the entities mentioned above. The data also revealed that ISIS consistently seemed to exploit the primary sources of Islam; selected verses from the Qur'an were used to justify their positions; Hadith books written by prominent Muslim scholars like Sahih Bukhari and Sahih Muslim along with Fqih (Islamic Jurisprudence) were also used to support their level of violence against certain.

We also noticed that the term "Rafd" (Arabic origin: Rafd), which means 'the rejecter' refers explicitly to Shiites, appeared frequently in some of the issues. The term has been interchangeably used by ISIS magazine to dehumanize the Shiites in many Muslim lands. Indeed, ISIS attacks them due to their rejection of the authority of the Caliphate in the Islamic history.

Religious groups like ISIS have been using the media to satisfy many objectives related to their activities such as radicalizing their purpose, funding initiatives, spreading fear, gaining publicity, and strengthening recruitment domestically and internationally. They seek media coverage for their acts of terrorism to communicate with the public and to propagate their cause. Therefore, as part of their recruitment strategy along with the de-legitimization and de-humanization efforts of their enemies, specific derogatory terms, ethnic slurs, and religious slangs were interchangeably used in the Dabiq magazine. More specifically, ISIS views *Kurds, Yazidis, Christians* and other minorities as heretics; but makes the Shiites as their primary and constant target of a genocidal campaign [21], which also explains the frequent use of "Rafd".

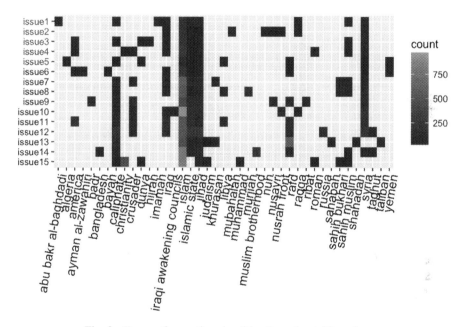

Fig. 3. Frequently mentioned entities throughout fifteen issues

We investigated the co-occurrences of entities in the same articles. Namely, they counted the number of times that a pair of entities is mentioned in the same text to infer the strength of a potential relationship. Then, a co-occurrence graph was built to explain the connections in a given issue topologically. An analysis of all fifteen issues revealed that the word 'Islam' is the most common entity that has been mentioned with various entities in the articles (Table 1). We picked the top ten co-occurrences, which not only revealed expected associations such as *Jihad* and *Islam* but also retrieved entity pairs that are perceived rival and opposites by members of the ISIS such as *Christianity vs. Islam*.

Our temporal networks that were constructed for each issue included different entities specific to that particular issue. However, a network of six entities, which were strongly connected in each issue, was detected in the analysis: Islamic State, Islam,

Table 1. Entity pairs mentioned most frequently together in the same text.

Entity 1	Entity 2	Freq.
Islamic state	Islam	1581
Jihad	Islam	1134
Jihad	Islamic State	869
Caliphate	Islamic State	594
Christianity	Islam	590
Syria	Islam	589
Syria	Islamic State	572
Sahih Muslim	Islam	568
Caliphate	Islam	567
Crusader	Islam	523

Jihad, Iraq, Syria, and Caliphate. Figure 4 below displays that each entity is represented with a circle in a big network web; the larger and the bigger the oval, the higher the frequency is; the bolder the link between the entities, the stronger the connection is. As depicted in Fig. 4 below, the language of the ISIS distinguishably shows strong relationships between the five constructs (Islam, Jihad, Iraq, Syria, Caliphate) where all five also tied to the organization "Islamic State."

As a part of our entity analysis, we also mainly looked into location entity type for which we identified 437 names. Of these, 177 locations were found with their latitudes and longitudes on at Nominatim[1] by using R. It was discovered that Raqqa, a city in

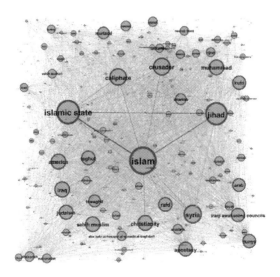

Fig. 4. Entity network for issue 1

---

[1] http://nominatim.openstreetmap.org.

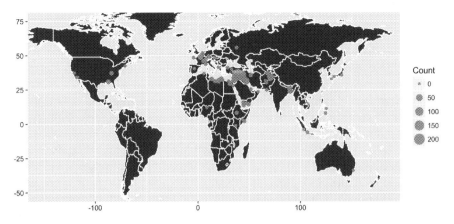

Fig. 5. Locations found in Dabiq (Color figure online)

Syria, was ranked the first due to the capital city of the Islamic State followed by Aleppo, Dimashq, Al-Khayr, and Baghdad.

In further visualization of those 177 locations on a map (Fig. 5), orange dots with varying sizes represent the locations with their frequencies and intensified in the Middle East region.

In our final analysis on the corpus, we built a topic model by implementing LDA with Mallet [22] to find out a bag of words, which followed a probabilistic pattern that could lead us to concepts (topics) used in writing. We ran the model for ten topics after removing the stop words from each article each of which represents a single document. Then, we manually went through the topics that were initially represented with topic numbers and listed words that are in decreasing order of association to the given topics. Among ten topics, three were selected for which we were able to infer concepts (topics) by looking at their top words listed in Table 2. While the first topic was talking about politics-related words including the countries in the Middle East, we noticed that Topic 2 mostly focused on Jihadi discourse and Topic 3 drew attention to administrative perspective. Since topic modeling allows using the same word in different contexts, we saw that some words such as "War" appeared in multiple topics. However, their order in the list would give an idea about their importance, weight or relatedness to a particular topic.

Table 2. Selected topics

Topic 1 (middle east/politics)	Topic 2 (Jihadi discourse)	Topic 3 (administrative)
Iraq	Allah	Iraq
Military	State	Hijrah
Syrian	Islamic	Dabiq
Iran	Muslims	Land
Political	War	Authority
Syria	Muslim	Remaining
Council	Soldiers	Tribes
War	People	Ummah

5 Conclusion

This paper demonstrated that the ISIS used the Dabiq magazine as an effective propaganda document to recruit young and naïve Muslims worldwide. Applying the various text analysis models, we showed that ISIS ideology is solely based on war, destruction and the politicization of Islamic sources and concepts like jihad and selecting weak Hadiths justifying their policies. Overall, the members of the terrorist organization always view the Shiism as the primary enemy that need to be rooted out of the Middle Eastern region. The magazine content used the words and language as an elegant and deceptive method of recruiting potential Jihadists worldwide. However, not only ISIS was very careful to reveal information about the true identity of its' members, but also exposed little about its leader Al Baghdadi in the content of the magazine.

As technology continues to evolve and advance, social media has helped terrorist organizations to expand and enlarge their numbers. The Internet has extended media's capabilities in delivering information to the public on a global level. It should be noted that military campaigns can only neutralize some level of hardcore violence committed by the ISIS members. There should be strong narratives that would counter the messages of the Dabiq magazine as well as its online communications. Both using the stories of ISIS defectors and the Syrian refugees against the terrorist group would definitely "delegitimize their claims of creating an Islamist utopia when they are nothing but a ruthless terrorist organization" [23].

References

1. Packer, G. Why did ISIS murdered Kenji Goto. The New Yorker (2015). https://www.newyorker.com/news/daily-comment/isis-murdered-kenji-goto. Accessed 3 Aug 2018
2. Wong, K.: Five ways ISIS, al Qaeda differ (2016). http://thehill.com/policy/defense/218387-five-ways-isis-is-different-than-al-qaeda. Accessed 3 Aug 2018
3. Koerner, B.I.: Why ISIS is winning the social media war? Wired (2016). https://www.wired.com/2016/03/isis-winning-social-media-war-heres-beat/. Accessed 3 Aug 2018
4. Brandon, C.: What does Dabiq do? ISIS hermeneutics and organizational fractures within Dabiq magazine. Studies in Conflict & Terrorism **0**(0), 1–18 (2016)
5. Ingram, H.J.: An analysis of Islamic State's Dabiq magazine. Aust. J. Polit. Sci. **51**(3), 458–477 (2016)
6. Zimmerman, K.: America's Real Enemy: The Salafi-Jihadi Movement. American Enterprise Institute (2017). https://www.criticalthreats.org/wp-content/uploads/2017/07/Zimmerman_Americas-Real-Enemy-The-Salafi-Jihadi-Movement.pdf. Accessed 13 Aug 2018
7. Thompson, R.L.: Radicalization and the use of social media. J. Strat. Secur. **4**(4), 9 (2011)
8. Al-Tabaa, E.S.: Targeting a female audience: american muslim women's perceptions of Al-Qaida propaganda. J. Strat. Secur. **6**(5), 4 (2013)
9. Fink, N.C., Sugg, B.: A tale of two jihads: comparing the Al-Qaeda and ISIS narratives. IPI Glob. Obs. **9** (2015)
10. Byman, D., Al Qaeda, the islamic state, and the global jihadist movement: what everyone needs to know. What Everyone Needs To Know (2015)

11. Groll, E., Francis, D.: Osama bin Laden Would Not Have Taken Ramadi (2015). http://foreignpolicy.com/2015/05/20/osama-bin-laden-would-not-have-taken-ramadi/. Accessed 3 Aug 2018
12. Bunzel, C.: From Paper State to Caliphate: The Ideology of Islamic State. The Brookings Project on US Relations with the Islamic World. Analysis Paper No 19 (2015)
13. Bertrand, N. We're getting to know just how different ISIS is from al Qaeda (2015). http://www.businessinsider.com/difference-between-isis-and-al-qaeda-2015-5. Accessed 3 Aug 2018
14. Liu, E.: Al Qaeda Electronic: A Sleeping Dog? The Critical Threats Project of the American Enterprise Institute, no. 12 (2015). https://www.criticalthreats.org/wp-content/uploads/2016/07/Al_Qaeda_Electronic-1.pdf. Accessed 3 Aug 2018
15. Faiola, A., Mekhennet, S.: From hip-hop to jihad. How the Islamic State became a magnet for converts. The Washington Post **6** (2015)
16. Ford, T.: How Daesh uses language in the domain of religion. Mil. Rev. **96**(2), 16 (2016)
17. From Hypocrisy to Apostasy. The Extinction of the Grayzone, in Dabiq
18. Clarion Project. http://www.clarionproject.org/news/islamic-state-isis-isil-propaganda-magazine-dabiq. Accessed 20 Feb 2016
19. Finkel, J.R., Grenager, T., Manning, C.: Incorporating non-local information into information extraction systems by Gibbs sampling. In: Proceedings of the 43rd Annual Meeting on Association for Computational Linguistics. Association for Computational Linguistics (2005)
20. Blei, D.M., Ng, A.Y., Jordan, M.I.: Latent Dirichlet allocation. J. Mach. Learn. Res. **3**, 993–1022 (2003)
21. Alaaldin, R.: The ISIS campaign against Iraq's Shia Muslims is not politics. It's genocide. Guardian, 5 January 2017
22. McCallum, A.K., Mallet: a machine learning for language toolkit (2002)
23. Coaty, P.: Understanding the War on Terror, 3rd edn. Kendall Hunt Publishing, Dubuque (2012)

Characterizing Organizational Micro-climates in Structural Groups

Geoffrey P. Morgan$^{(\boxtimes)}$ and Kathleen M. Carley

Carnegie Mellon University, Pittsburgh, PA 15217, USA
gmorgan@cs.cmu.edu

Abstract. We use text to characterize micro-climates in a specific organizational context. We use Louvain clustering to create trees, and identify communities and smaller groups, which we call leaves. In comparing these structural groups, we see that most structural groups within this organization share locations and organizational functions. Our analyses show that location, gender, and minority status dominance tend to increase the strength of the textual distinction of the found structural groups.

Keywords: Organizational climate · Organization Micro-climates · Structural analysis · Text analysis

1 Introduction

Micro-climates, group-level organizational climates [1] within an organization, are important to understanding the organization as a whole. Many interacting organizational climates exist and influence reaction to interventions, framing of work, and response to outside stimulus. Strong micro-climates indicate areas of the organization where identification may be primarily to the group, not to the organization as a whole. This identification with sub-groups may lead to significant organizational dysfunction and ultimately contribute to the organization's failure.

In this work, we offer a new method for identifying organizational micro-climates through textual analysis of sub-groups within the organization. We have developed this textual analysis method to circumvent many of the difficulties with traditional surveys to evaluate micro-climates, such as survey demand characteristics [2]. We evaluate organizational micro-climates using functional, structural, and locational groups, along with all available combinations (structural and functional, functional and locational, structural and functional and locational). We use functional, location, and structural groups, along with their overlaid combinations to aggregate texts and evaluate these groups against the entire organization. Functional groups are collections of people who serve a similar function within the organization, such as Human Resources or Legal – functional specialization often informs the language people use. Location groups are collections of people who serve in the same place – location often informs language use. Structural groups are people who frequently interact – we use Louvain Clustering [3] to identify structural groups. Structural groups are distinct from formal group identification – the distinction can be thought of as "who actually works together"

© Springer Nature Switzerland AG 2019
R. Thomson et al. (Eds.): SBP-BRiMS 2019, LNCS 11549, pp. 12–20, 2019.
https://doi.org/10.1007/978-3-030-21741-9_2

versus "who is supposed to work together". Functional, locational and structural groups all can, and should, influence the language being used – we use textual analysis to identify the influence can be attributed to functional, locational, and structural factors along with their combinations.

Often, functional and formal groups overlap, depending on the organizational structure adopted by the organization [4]. Structural groups and formal groups should often overlap, given that formal groups are intended to represent coherent groups within the organization, and structural groups are identified as coherent communities inside the network via Louvain Clustering [3].

2 Organizational Climate Research and Identifying Climate Within Groups

Organizational climate research is heavily influenced by field theory [5, 6], which offers the useful perspective that a person's behavior is moderated by the organizational context in which that person operates. Later studies of organizational climate have begun to tease apart the 'organizational context', suggesting that an individual operates within multiple climates at the same time [7–9]. These efforts, however, seem to have universally relied on survey measures, and the principal goal was to identify whether a behavioral practice was more moderated by the group or organizational climate. Instead, we are interested in delineating the boundaries of where there are strong micro-climates within the organization. We consider structural, functional, and formal groupings.

The problem of evaluating whether a given group should be attributed as having a climate has been an open question in prior survey studies. Early work in the field acknowledged that a given set of individuals must have some level of agreement in outcomes in order to attribute the measure to the larger unit at all, although the amount of agreement varied [10]. There are a variety of measurements of agreement, including the $r_{WG(j)}$ [11], and inter-class correlations. Other researchers argue that inter-class correlations are misleading, since the issue is agreement within groups, not the amount of explainability the units offer [12]. Other work has also suggested that the survey measures frequently used to assess organizational culture are too binary [13].

Because of these and similar issues, we wanted to take an alternative approach: use the language of the organization at-work to identify micro-climates. This approach resonates with modern organizational communication scholarship that identifies organizational culture similarly as to how earlier work defined organization climate, as "the set(s) of artifacts, values, and assumptions that emerges from the interactions of organizational members" [14]. This implicitly includes language [15].

3 Data

Our data is from a particular multi-national organization, MergedCo, going through a horizontal merger, the two elements merging are called, for the purpose of this paper, LuxuryCo and StandardCo. MergedCo provided both email data and survey data. The email data includes both meta-data and text-data for approximately 2.5M emails. The survey data included over 1600 participants, and we use it in this work to identify functions and locations for employees in the email data.

4 Method

In this work, we are principally interested in identifying group climates that differ substantially from parent and peer climates. Such climates, as suggested by field theory, are likely to generate behavior detrimental to the organization. Because climates emerge through the interactions of individuals [16, 17], and their perceptions of those interactions [18], locally dense clustering of interactions may be conducive towards the development of local climates substantially different from the larger organizational climate. Viewing the organization through a network [19] formalism, we use Louvain clustering [3] to identify structurally coherent communities within the larger organizational structure. All network analyses are supported by ORA [20].

Unlike prior work in group climate research, we consider the language of the organizations' members. Language shapes our perceptions of others, and thus should be correlated with organizational climate outcomes. In this we subscribe to structural symbolic interactionism [21, 22], where our interactions with others and the words we use themselves contribute to our understanding of those others and ourselves.

Our method subsections delineate the answer to several questions, (1) how do we identify structural groups of interest, and (2) how do we characterize the strength of a given micro-climate?

4.1 Louvain Trees to Identify Groups of Interest

Using a network formalism, we hierarchically cluster the network using Louvain Clustering to form what we call a Louvain Tree. As shown in Fig. 1, there are two important areas of the Louvain Tree. The first is what we call Louvain Communities, the resulting clusters from the initial partition of the organization. The second are the Louvain Leaves, where we ran Louvain Clustering on each resulting community until the groups found were thirty (30) or less people. For the purposes of this analysis, we ignore the meso-communities between the Louvain Leaves and the Louvain Communities.

4.2 Using Text to Find and Characterize Micro-climates

We use the language, the tokens people use to convey meaning, to characterize the strength of a micro-climate through differentiation from the larger organizational climate. In this section we detail this method through a set of equations. This method is distinct, in particular, from the family of methods that embody TF-IDF (Term Frequency, Inverse Document Frequency) [23] because it jettisons, for the purposes of cultural comparison, the notion of inverse document frequency. While TF-IDF is designed to identify terms most likely to be both exhaustively (through Term Frequency) and specifically (through being heavily used in specific documents) discussed

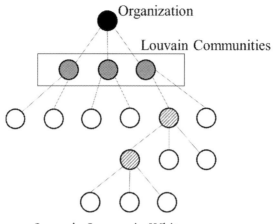

Fig. 1. Notional Louvain Tree - Louvain Clustering identifies the communities, then further hierarchical Louvain Clustering is done on resulting structures until group size is ≤ 30

in a given corpus, this approach instead compares term use between two corpora, controlling for total terms in each corpus, and identifies the terms that are most likely to distinguish the two corpora. From a cultural perspective, the notion that a term is important because it appears often but only in a few documents is not useful and may even be harmful. For more information, see [24]. Both TF-IDF and our method perform similarly in identifying stop-words and similar phenomena as being low-value.

We illustrate the mis-use of TF-IDF in this context with an example. Imagine a new joiner to the organization, J, who always closes their emails with the valediction, "Cheers". J becomes influential, impressing their colleagues with their insight and perspective. Many of their frequent email correspondents also begin to close their emails with "Cheers". Gradually, "Cheers" as a valediction dominates J's work-group and can be used to readily differentiate whether an email comes from J's group or another group in the organization. Our method would score this term highly when comparing J's group to other groups, while TF-IDF would score this term, since it appears in every email, as low as stop words like "the", or "and".

Each token, t, is a unique symbol among all tokens, T. Group membership and document authorship collectively generate two distinct corpora, A and G, which we then compare. Equation 1 identifies what we call the normalized odds ratio, ranges from -0.5 to 0.5. Scores near 0 indicate that a particular token does not offer distinction between the two corpora. Scores below 0 indicate that the token is more common in Corpora A than G, scores above 0 indicate that the token is more common in Corpora G than A.

To calculate distinction between two climates, we need to define a series of values. Our first value is the normalized odds ratio. The normalized odds ratio is the occurrence of term t in Corpus A normalized by all terms in Corpus A, divided by the number of times term t appears in Corpus G divided by all terms in Corpus G

$$odds(t, AG) = \left(1 - \left(\frac{1}{\left(\frac{|t_A|}{|T_A|} \middle/ \frac{|t_G|}{|T_G|} \right)} \right) \right) - .5 \tag{1}$$

'We use a cutoff value, c, to avoid assigning importance to marginal cases – if the absolute value from Eq. 1's $odds(t)$ is less than c, than $odds(t)$ is reassigned to 0. We do this to avoid assigning importance to marginal cases. We call this the flattened (normalized) odds ratio. We set c to 0.1 for this paper.

$$\begin{aligned} fOdds(t, AG, c) &= abs(odds(t, AG)) \\ &> c, odds(t, AG) \end{aligned} \tag{2}$$

After identifying whether a term is more strongly associated with A or with G, we calculate a contextualized frequency term that compares the prominence of the term in a particular corpora against that term in an appropriate prior, P.

$$\begin{aligned} freq(t, AGP, c) &= fOdds(t, AG, c) > 0, max(odds(t, AP, 0)) \\ &\quad fOdds(t, AG, c) < 0, max(odds(t, GP), 0) \end{aligned} \tag{3}$$

Equation 4 combines the frequency term calculated in Eq. 3 with the flattened-odds of Eq. 2 to score the token's impact. Flattened terms are canceled out no matter how prominent their terms compared to the prior. Thus, high scoring terms:

– Distinguish A and G
– Are more prominent in either A or G than P

$$S(t, AGP, c) = fOdds(t, AG, c) * freq(t, AGP, c) \tag{4}$$

Equation 5 summarizes the scores for all terms to produce a relative comparison of Corpora A and G against Prior P with cutoff value c for a given set of tokens T. Unless noted otherwise, all tokens found in A and G are used to inform T.

$$Score(T, AGP, c) = \sum_t abs(s(t, AGP, c)) \tag{5}$$

In this section, we have identified how we score tokens and corpora. High scoring corpora informed by group memberships are likely, to us, to indicate substantial group micro-climates.

5   Results

We use the structural and textual approach documented in Sect. 4 against our data, briefly discussed in Sect. 3.

5.1   Structural Analysis

MergedCo has 16 communities and 172 leaves (with the average leaf size thus being 12 members). The initial Louvain Clustering that identified the communities had a found modularity value of 0.64, indicating substantial non-random structure in the grouping of nodes. We can also place these structures on a continuum between the legacy organizations LuxuryCo and StandardCo, as shown in Fig. 2. In this figure, communities and leaves gravitated towards the legacy organization to which they have the most relationship. Eight (8) of the sixteen communities identify completely or strongly with LuxuryCo, four (4) identify completely or strongly with StandardCo, and the remaining four (4) are balanced between LuxuryCo and StandardCo.

Fig. 2. Louvain Communities (Grey) and Louvain Leaves (White) are more or less associated with the two legacy organizations, LuxuryCo and StandardCo. The more left-ward, the more LuxuryCo is dominant, the more right-ward, the more StandardCo is dominant. There are four communities which are highly balanced between LuxuryCo and StandardCo actors.

For each Louvain Community (Sc) and Louvain Leaf (Sl), we can classify the structural group as:

- Structural (Sc or Sl)
- Structural and Functional (ScF or SlF)
- Structural and Locational (ScL or SlL)
- Or Structural, Functional, and Locational (ScFL or SlFL)

A structural group gains either the functional or locational tag when 60% or more of its members are all of one functional or locational group, which is not the "no answer provided" group. Groups that overlap in structure, function, and location should, we expect, have more unique language and a stronger micro-climate. Figure 3 shows the composition of the various classifications for both Louvain Communities and Louvain Leaves.

As a network theorist might hope, the majority of our found structural groups have dominant functions and locations.

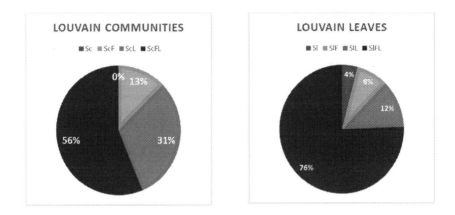

Fig. 3. The distribution of the different structural types for both Louvain Communities (Left) and Louvain Leaves (Right)

5.2   The Impact of Categorical Dominance on Micro-climates

When looking at the Louvain Leaves, the small groups, we can ask ourselves, do functional, locational, minority, and gender homogeneity influence the overall distinctiveness of the text of the group? We believe that shared factors should tend to increase the distinctive-ness score of the group, as language tends to descend into an argot built on many shared similarities. Our results bear this out.

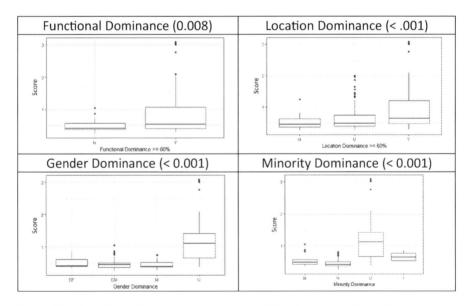

Fig. 4. Score boxplots, with group types as a categorical value and our distinctiveness score as the Y-axis

In Fig. 4, the top row results for functional and locational dominance suggest that users were willing (or were required if they participated at all) to provide their function and location – information the organization possessed and would be unlikely to identify them as individuals. For both function and location, a homogenous (single dominant function/location) indicated a stronger, higher-scoring, climate. The bottom row, showing gender and minority results, indicates that actors unwilling to provide potentially identifying information are frequently actors who participate in small-group climates where language is often very different from the organization as a whole. This non-response is suggestive that surveys, which even when mandatory are almost never mandatory on a question-by-question basis, are least likely to offer useful input in places where there is the greatest need.

6   Conclusion

Few who have worked in an organization can deny the power of an organization's culture and climate to define what it means to be a member of the organization. An organization is culture [15], as much as it is a collection of people and resources organized to some purpose. Yet classifying and measuring such an ever-present yet usually invisible phenomenon has been difficult to do with survey approaches.

This work takes a novel approach to measuring organizational climate by using the textual artifacts of the organization to measure distinctiveness. This approach identifies not only key symbols of note, which TF-IDF [23] also provides, but also provides the capability of aggregating these term differences together into a useful metric for comparing two groups to another. We also explain how TF-IDF, as an algorithm is inappropriate to this particular context. Further, these groups may be explicitly and usefully subsets of each other (e.g., MergedCo compared to a specific community), or they may be explicitly distinct (e.g., LuxuryCo compared to StandardCo), allowing support for explorations suggested by field-theory [6] and later micro-culture research without limiting the method's scope.

This method thus allows us to both consider the specifics, symbols and authors that particularly stand out, while also considering the macro-view, what groups are most distinct from each other, and what sub-groups are most distinct from their parent organization.

References

1. Hoy, W.K.: Organizational climate and culture: a conceptual analysis of the school workplace. J. Educ. Psychol. Consult. **1**(2), 149–168 (1990)
2. Orne, M.T.: On the social psychology of the psychological experiment: with particular reference to demand characteristics and their implications. Am. Psychol. **17**, 776 (1962)
3. Blondel, V.D., Guillaume, J.-L., Lambiotte, R., Lefebvre, E.: Fast unfolding of communities in large networks. J. Stat. Mech: Theory Exp. **2008**, P10008 (2008)
4. Burton, R.M., Obel, B., DeSanctis, G.: Organizational Design: A Step-by-Step Approach. Cambridge University Press, Cambridge (2011)

5. Lewin, K.: Field theory and experiment in social psychology: concepts and methods. Am. J. Sociol. **44**, 868–896 (1939)
6. Lewin, K.: Field Theory in Social Science. Harper & Brothers, New York (1951)
7. Zohar, D., Luria, G.: A multilevel model of safety climate: cross-level relationships between organization and group-level climates. J. Appl. Psychol. **90**, 616 (2005)
8. Brondino, M., Silva, S.A., Pasini, M.: Multilevel approach to organizational and group safety climate and safety performance: co-workers as the missing link. Saf. Sci. **50**, 1847–1856 (2012)
9. Rodrigues, M.A., Arezes, P.M., Leão, C.P.: Multilevel model of safety climate for furniture industries. Work **51**, 557–570 (2015)
10. Erhrhart, M.G., Schneider, B., Macey, W.H.: Organizational Climate and Culture: An Introduction to Theory, Research, and Practice. Routledge, New York (2014)
11. James, L.R., Demaree, R.G., Wolf, G.: Estimating within-group interrater reliability with and without response bias. J. Appl. Psychol. **69**, 85 (1984)
12. George, J.M., James, L.R.: Personality, affect, and behavior in groups revisited: comment on aggregation, levels of analysis, and a recent application of within and between analysis. J. Appl. Psychol. **78**, 798–804 (1993)
13. LeBreton, J.M., Senter, J.L.: Answers to 20 questions about interrater reliability and interrater agreement. Organ. Res. Methods **11**, 815–852 (2008)
14. Keyton, J.: Communication and Organizational Culture: A Key to Understanding Work Experiences. SAGE, Thousand Oaks, CA (2011)
15. Keyton, J.: Organizational Culture: Creating Meaning and Influence. In: Putnam, L.L., Mumby, D.K. (eds.) The SAGE Handbook of Organizational Communication: Advances in Theory, Research, and Methods. SAGE, Thousand Oaks (2014)
16. Schneider, B., Reichers, A.E.: On the etiology of climates. Pers. Psychol. **36**, 19–39 (1983)
17. Glick, W.H.: Response: Organizations are not central tendencies: shadowboxing in the dark, round 2. Acad. Manag. Rev. **13**, 133–137 (1988)
18. Schneider, B.: Organizational behavior. Annu. Rev. Psychol. **36**, 573–611 (1985)
19. Brandes, U., Robins, G., McCranie, A., Wasserman, S.: What is network science? Netw. Sci. **1**, 1–15 (2013)
20. Carley, K.M., Pfeffer, J., Reminga, J., Storrick, J., Columbus, D.: ORA User's Guide 2013. Carnegie Mellon University, School of Computer Science, Institute for Software Research (2013)
21. Serpe, R.T., Stryker, S.: The construction of self and reconstruction of social relationships. Adv. Group Process. **4**, 41–66 (1987)
22. Stryker, S.: Symbolic Interactionism: A Social Structural Version. Benjamin/Cummings Publishing Company, Menlo Park (1980)
23. Sparck Jones, K.: A statistical interpretation of term specificity and its application in retrieval. J. Doc. **28**, 11–21 (1972)
24. Tietze, S., Cohen, L., Musson, G.: Understanding Organizations Through Language. Sage, Thousand Oaks (2003)

Pro/Con: Neural Detection of Stance in Argumentative Opinions

Marjan Hosseinia[1]([✉]), Eduard Dragut[2], and Arjun Mukherjee[1]

[1] University of Houston, Houston, TX, USA
mhosseinia@uh.edu, arjun@cs.uh.edu
[2] Temple University, Philadelphia, PA, USA
edragut@temple.edu

Abstract. Accurate information from both sides of the contemporary issues is known to be an 'antidote in confirmation bias'. While these types of information help the educators to improve their vital skills including critical thinking and open-mindedness, they are relatively rare and hard to find online. With the well-researched argumentative opinions (arguments) on controversial issues shared by Procon.org in a non-partisan format, detecting the stance of arguments is a crucial step to automate organizing such resources. We use a universal pre-trained language model with weight-dropped LSTM neural network to leverage the context of an argument for stance detection on the proposed dataset. Experimental results show that the dataset is challenging, however, utilizing the pretrained language model fine-tuned on context information yields a general model that beats the competitive baselines. We also provide analysis to find the informative segments of an argument to our stance detection model and investigate the relationship between the sentiment of an argument with its stance.

Keywords: Stance detection · Universal language model fine-tuning · AWD-LSTM

1 Introduction

The problem of stance detection is to identify whether a given opinion supports an idea or contradicts it. It is relatively new in the area of opinion mining and is recently being explored by more researchers [1–4, 7, 12]. Table 1 provides two arguments. The arguments answer a question while taking a stance of the two possible sides against a controversial issue. A stance that supports an issue is a *pro*, and the other side that is against it is a *con*.

In opinion mining identifying a stance of an opinion is a more challenging task than sentiment analysis [4] and naturally differs from it. Here, the problem is no longer finding the whole polarity of an opinion but is to identify its polarity against an issue. Recently, the argumentative opinions of controversial issues have attracted more people who want to take a stance after seeking enough information about the reason behind opinions from both sides. For example, one might wonder *"Should Marijuana Be a Medical Option?"*. This question might be found in many online debate forums

© Springer Nature Switzerland AG 2019
R. Thomson et al. (Eds.): SBP-BRiMS 2019, LNCS 11549, pp. 21–30, 2019.
https://doi.org/10.1007/978-3-030-21741-9_3

Table 1. Tow arguments, a pro and a con, for the issue *medical marijuana*. The question is: *"Should Marijuana Be a Medical Option?"*. Each example is a tuple of type (issue, question, context, argument).

Issue: Medical marijuana
Question: Should Marijuana Be a Medical Option?
Context: In 1970, the US Congress placed marijuana in Schedule I of the Controlled Substances… Proponents of medical marijuana argue that it can be a safe and effective treatment for the symptoms of cancer, AIDS, multiple sclerosis, pain, glaucoma, epilepsy, and other conditions… Opponents of medical marijuana argue that it is too dangerous to use, lacks FDA-approval, and that various legal drugs make marijuana use unnecessary

Argument (Pro): Ultimately, the issue is not about laws, science or politics, but sick patients. Making no distinction between individuals circumstances of use, the war on drugs has also become a war on suffering people. Legislators are not health care professionals and patients are not criminals, yet health and law become entwined in a needlessly cruel and sometimes deadly dance… I sincerely hope our work will illuminate the irrational injustice of medical marijuana prohibition

Argument (Con): We can't really call marijuana medicine. It's not a legitimate medicine. The brain is not fully developed until we're about 25. That's just the way it is, and using any kind of mind-altering substance impacts that development. It needs to go through the FDA process…

and people who like to consume marijuana or the ones who hate it take a stance without bringing an acceptable justification. These types of opinions are usually short and express the stance directly (e.g. tweets). However, argumentative opinions are generally long, more complex, contain high-level ideas, and take a stance while bringing some reasons. Finding the stance of an argument is not straightforward compared to opinions with spontaneous language (e.g. tweets). See Table 1-pro as an example. We study the problem of stance detection in argumentative opinions of 46 different controversial issues. The arguments are collected and represented in a nonpartisan way which means that they are not biased specifically towards any party.

We make the following contributions in this paper. First, we propose a new stance detection dataset[1] from ProCon[2], a collection of critical controversial issues. Each entity of our ProCon dataset is a tuple of type (issue, question, context, argument) where an *issue* refers to the underlying domain, a *question* asks for an opinion, *context* brings a summary of proponent and opponent view- points about the *issue*, and an *argument* is a reason-based opinion for or against the *issue*. Table 2 shows how people justify/condemn "legalization of abortion" while bringing some reasons.

We, also, propose a model that leverages the context of an issue to predict the stance of the given opinion. In ProCon dataset, the average number of opinions per issue per class is 24. This size of data may not be large enough for training a neural network. To compensate for this small size of data we build our model on top of the Universal pretrained Language Model, ULMFiT, [8] and fine tune it to our stance detection task. The pretrained language model is a "counterpart of ImageNet for NLP"

---

[1] https://github.com/marjanhs/stance.

[2] https://www.procon.org/.

Table 2. Two arguments about "legalization of abortion".

Pro	Con
The US Supreme Court has declared abortion to be a "fundamental right" guaranteed by the US Constitution.... decision stated that the Constitution gives "a guarantee of certain areas or zones of privacy," and that "This right of privacy... is broad enough to encompass a woman's decision whether or not to terminate her pregnancy"	Unborn babies are considered human beings by the US government. The federal Unborn Victims of Violence Act, which was enacted "to protect unborn children from assault and murder", states that under federal law, anybody intentionally killing or attempting to kill an unborn child should be punished... for intentionally killing or attempting to kill a human being

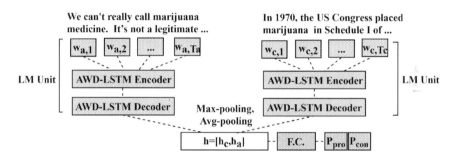

Fig. 1. The model; $w_{a,i}, w_{c,i}$ ith word of argument and context sequence; F.C.: Fully connected layer; p_{pro}, p_{con}: class probabilities; dark boxes use pretrained weights.

and can be used in various tasks independent of document size and label as well as the number of in-domain documents [8].

The model is detailed in Sect. 4. We now discuss the related works. Then, we describe our dataset, the proposed model, and the experimental study.

2   Related Works

Most works on stance classification focus on online debates including posts and tweets while typically restricting the underlying data to a few (up to 8) targets (issues). A probabilistic approach that models stance, the target of stance and sentiment of a tweet is trained on a large dataset of around 3 K and evaluated on more than 1.2 K tweets toward five targets [5]. Hasan et al. explore a new task of reason classification by modeling stances and reasons on a corpus of ideological debate posts from five domains [7]. They show that developed models of stances and reasons provide better reason and stance classification results than their simpler models. SemEval2016 provides two different stance detection frameworks for tweets [4, 11]: one supervised framework with five targets containing 4 K tweets and another weakly supervised with one target [1]. Unlike others, Bar et al. propose a contrast detection algorithm for 55

Table 3. ProCon dataset statistics. Docs refer to argumentative opinions (arguments).

	Train		Dev		Test		Train	
	Docs	Docs/issue	Docs	Docs/issue	Docs	Docs/issue	Words/arg	Words/cntx
Size	1517	33	178	4	530	12	166 ± 65	177 ± 34

different topics [2]. It is designed to detect the stance of *claims* that are defined as *"general, concise statements that directly support or contest the given topic"*. Actually, claims are *"often only a small part of a single Wikipedia sentence"* in their dataset [9]. While all these works have made important contributions, they do not address the problem of detecting stance in fluid and long arguments, which is the focus of this work.

3   Dataset

We collect the information of 46 controversial issues from ProCon, a top-rated non-profit organization that provides professionally-researched pros and cons to create our dataset (Table 3)[3]. We define each instance as a tuple of type I = (issue, context, question, argument) where the *issue* is a general topic, the *context* introduces the issue and brings a summary of proponent and opponent opinions and an *argument* is a reason-based opinion taking a stance on or against the given *issue* (Table 1). The issues cover various topics from health and medicine, education, politics, science and technology to entertainment and sports. We will use the words *target* and *issue* interchangeably in this paper as *target* convey same meaning in other research. *Argument* supports a position with powerful and compelling statements. The dataset is divided into 1, 517 train, 178 dev, and 530 test samples (Table 3).

4   Model

Inspired by ULMFiT [8], we propose a model to handle both diverse and small training data per issue (Fig. 1). Our model has three units: (a) parallel Language Model (LM) units to learn an argument and the context of its underlying issue. (b) one fusion unit that summarizes all elements of the data and (c) the classification unit that predicts the stance. We describe them below.

4.1   Parallel LM Units

We let the model jointly learn an argument with its corresponding context using two LM units. A context usually covers a few sentences introducing the issue and two summaries of proponent and opponent arguments (Table 1-context). We hypothesize

---

[3] For more details visit https://www.procon.org/faqs.php.

that pro-arguments and con-arguments are related to disjoint parts of the context because of the intrinsic contradiction of pro- and con-arguments.

Let $P = ([w_{1,a}, \ldots, w_{T_a,a}], [w_{1,c}, \ldots, w_{T_c,c}])$ be input pair where $w_{i,a}$ and $w_{i,c}$ are the ith word of an argument and its context sequence respectively and T_a, T_c are the last time steps. Each LM unit is a three-layer neural network (Fig. 1). First, words are represented as vectors of size $d_e = 400$ using the embedding matrix W_e. The matrix is the result of pretraining the Language Model on Wikitext data with more than 103 M words [10]. Then, a weight-dropped LSTM (AWD-LSTM) encodes word embedding to a higher dimension (1, 150), and another AWD-LSTM decodes the hidden representation of words into the embedding dimension and predicts the next word of the sequence. AWD-LSTM applies recurrent regularization on the hidden-to-hidden weight matrices to prevent over-fitting across its connections. It adds Activation Regularization (AR) and Temporal Activation Regularization (TAR) to the loss function [10]. Later we provides more details of the two regularization techniques. The argument LM unit is the following:

$$
\begin{aligned}
x_{i,a} &= W_e w_{i,a}, \\
z_{i,a} &= \mathrm{lstm}_{enc,a}(x_{i,a}), \ i \in [1, \ T_a], \\
h_{i,a} &= \mathrm{lstm}_{dec,a}(z_{i,a}), \ i \in [1, \ T_a]
\end{aligned}
\tag{1}
$$

where $z_{i,a}$, $h_{i,a}$ are the hidden state of LSTM encoder and decoder respectively. Similarly, $h_{i,c}$ is the output of context LM unit.

4.2   Fusion and Classification

The fusion layer leverages the information of both LM outputs. Most information of an argument is hidden in the last hidden state of the LSTM decoder of the LM unit. However, important information might be hidden anywhere in a *long* document. We use max-pooling and average-pooling of both inputs (argument, context) along with the last hidden state of LSTM decoder for fusion.

$$
\begin{aligned}
h_a &= [h_{T_a,a}, \ \text{max - pool}(h_{T_a}, \ a), \ \text{avg - pool}(h_{T_a}, \ a)], \\
h_c &= [h_{T_c,c}, \ \text{max - pool}(h_{T_c}, \ c), \ \text{avg - pool}(h_{T_c}, \ c)]
\end{aligned}
\tag{2}
$$

where $h_{T_a,a}$, $h_{T_c,c}$ are the hidden state of LSTM decoder of argument and context LM units at time T_a, T_c and [,] is concatenation. Finally, the pooled information, $h = [h_a, h_c]$, builds the fusion layer and connects an argument with any significant parts of the context. We feed h through a fully connected layer with $d_r = 50$ hidden neurons activated with a rectifier. The second fully-connected layer but with the linear activation gives us 2d vectors to be used by a softmax function for classification. We apply batch-normalization and dropout to both fully-connected layers to avoid over-fitting. As we mentioned earlier, AWD- LSTM adds TAR (l_{tar}) and AR (l_{ar}) to the final loss. AR is an L2-regularization that controls the norm of the weights to reduce over-fitting. And TAR acts as L2 decay and is used on individual activations. It considers the difference of the outputs of the LSTM decoder at consecutive time steps:

$$l_{ar} = \alpha * \left\| [h_{T_a,a}, \; h_{T_c,c}] \right\|2, \; l_{tar} = \beta * \left\| [h'_{T_a,a}, \; h'_{T_c,c}] - [h'_{T_{a-1},a}, \; h'_{T_{c-1},c}] \right\|_2$$
$$L = - \sum_d \log h_{s,j} \; + \; l_{ar} \; + \; l_{tar} \tag{3}$$

where j is the label of the document and $\alpha = 2$, $\beta = 1$ are the scaling coefficients. $h'_{T_a,a}$, $h'_{T_c,c}$ are the last hidden states of the two LSTM decoders without dropout.

5   Evaluation

We compare our model with state-of-the-art methods in stance detection. The methods are as follows:

- BoW-s: is a Bag of Words model that gains the best performance in TaskA of SemEval2016 [12] with SVM classifier. The features are boolean representation (0/1) of word uni-, bi- and tri-grams as well as character 2, 3, 4 and 5-grams. The presence/absence of any manually selected keywords of the underlying issue is also added to the feature vector. For example, for the issue of 'Hillary Clinton' the presence of *Hillary* or *Clinton* sets this feature to true. We manually select at least three keywords per issue in Procon dataset. Unlike [12], we do not build an individual classifier for each issue separately. We create one general classifier trained on the whole dataset. We examine the BoW-s feature vectors with SVM, Gaussian Naive Bayes (GNB), Logistic Regression (LR), and Random Forest (RF).[4]
- Independent Encoding (IE): is one of the baselines reported in [1]. It learns the representation of a document and its target independently using two parellel LSTMs. Then, the last hidden states of the two LSTMs are concatenated and projected with the *tanh* function. Finally, a softmax predicts the class distribution over the non-linear projection.
- ULMFiT: is the backbone of our model [8]. We keep all settings intact and train the model by applying the discriminative fine-tuning technique.
- Bidirectional Conditional Encoding (BiCoEn): outperforms the existing methods of SemEval 2016-TaskB. In TaskB the goal is to predict the stance of a tweet over one single unseen target, 'Donald Trump' [1]. The model takes a *tweet* and its underlying *target* and initializes the state of the bidirectional LSTM of tweets with the last hidden state of the forward and backward encoding of the target. In this way the model builds target-dependent representations of a tweet while both the left and right sides of a word are considered. This model takes a document (tweet) and its target (e.g. 'Climate Change is a Real Concern' or 'Atheism') as input for training. To make the comparison more reliable we examine BiConEn for both types of input: (*argument, context*) and (*argument, issue*). Here, *issue* is the *target* as in [1].[5]

---

[4] We use scikit-learn with default settings.

[5] We use their code shared on https://github.com/sheffieldnlp/stance-conditional.

Table 4. Procon dataset results. arg: argument, cntx: context, P: Precision, R: Recall

Method	Input	Pro			Con			Macro-F1	Acc
		P	R	F1	P	R	F1		
BoW-s+SVM	arg	61	63	62	62	61	61	62	62
BoW-s+RF	arg	64	67	65	66	62	64	65	65
BoW-s+LR	arg	61	63	62	62	60	61	61	61
BoW-s+GNB	arg	58	65	62	61	54	57	59	59
ULMFiT	(arg, cntx)	65.8	61.2	63.4	64	**68.5**	**66.2**	64.8	64.9
IE	(arg, issue)	55.4	60.5	57.5	56.7	51.1	53.2	55.4	56.7
IE	(arg, cntx)	56.9	52.7	54.5	56.1	60.1	57.8	56.2	57.3
BiCoEn	(arg, issue)	56.5	57.7	57	57.2	55.9	56.4	56.7	56.9
BiCoEn	(arg, cntx)	55.9	57.6	56.4	56.7	54.6	55.2	55.8	56.6
Our model	(arg, cntx)	**65.9**	**82.6**	**73.3**	**77**	57.7	65.9	**69.6**	**70.1**

6   Results and Analysis

We apply discriminative fine-tuning for ULMFiT and our model. We execute the evaluation 5 times and report the average results for all methods. Table 4 provides the experimental results. The largest values are highlighted in bold and the second largest are underlined. According to the table, both accuracy and Macro-F1 of all baselines do not exceed 65%, showing that the presence of diverse issues makes the problem hard to solve. It is expected that the Neural Network (NN) baselines give weak results compared to BoW for ProCon data. With 1, 517 training samples and 46 different issues, the average number of arguments per issue is 33 which is not enough for fitting NN models unless we provide some external knowledge for them such as what we do for our model (Pre-trained Language Model). It notes that stance detection is not a pure binary classification problem, because detecting the underlying issue is required for identifying the polarity of opinion against it. Aside from the above notes, BiCoEn is designed for detecting the stance of tweets for one unseen single target (issue), however, in ProCon the size of input argument is much longer than a tweet (166 compared to 20 words) and belongs to a diverse number of issues. We set the maximum length of an input to be 20 words for IE and BiConEn, as recommended by the authors of [1]. However, we find that by increasing this threshold, accuracy decreases. The reason is that the sequence length of both LSTMs must be equal, because the initial weights of argument-LSTM are the output of issue-LSTM. When we increase the maximum length, issue-LSTM takes no new information but padding indices (the average length of argument sequence is much greater than average length of issue sequence, $166 \gg 3$).

Ultimately, our model achieves an accuracy increase of more than 5% compared to BoW+RF. It indicates that leveraging the context information along with LM Fine-Tuning helps the model identify the issue and the stance against it more accurately. We provide more analysis in the following sections.

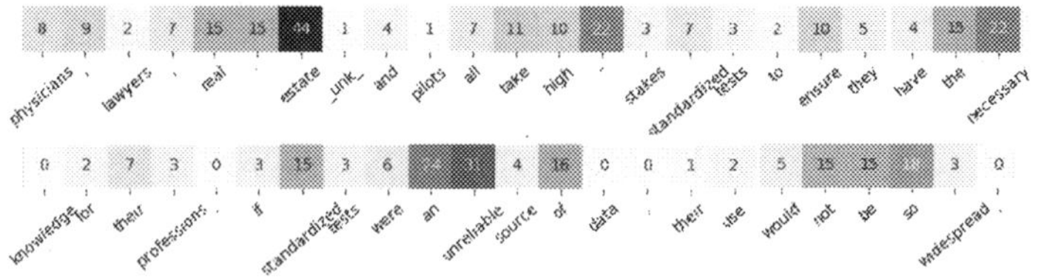

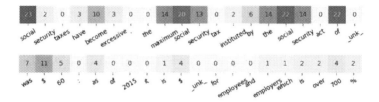

Fig. 2. Heatmap of max-pooling matrix of one argument. The underlying question: *"Is the Use of Standardized Tests Improving Education in America?"*. Darker colors show larger scores.

Fig. 3. Heatmap of max-pooling matrix of the first half of an argument (the second half scores are mostly zero). The underlying question: *"Should Social Security Be Pri- vatized?"*. Darker colors show larger scores.

6.1 Effect of Max-pooling

The fusion layer merges the information from previous layers for prediction. To understand what the model learns in this layer we plot the word scores in the max-pooling matrix of an argument. We define the score of word w at time t, to be the index frequency of the embedding vector of w in pooling operation. The larger the score, the more important that word is to the model, because more embedding dimensions of that word appear in the max-pooled matrix (same word in different time steps may have different scores). Figures 2 and 3 show the heatmaps of a short and the first half of a longer argument respectively that are correctly classified. We cannot provide more plots due to space constraints. However, we find that the words at the beginning of long documents are more informative (Fig. 3). One reason is that the first sentence of long arguments is *usually* the topic sentence that conveys the stance. Moreover, for shorter arguments, the model finds the information across all parts of the argument almost evenly.

6.2 Effect of Pre-trained LM

To assess the impact of pre-trained LM, we examine our model without utilizing the pre-trained LM. We do not fine-tune the LM units over the training data, too. The experiment are represented in Table 5. The dramatic drop in all metrics shows the effect of the ablated techniques. Pre-training helps generalization and prevents our model from overfitting the relatively small training data.

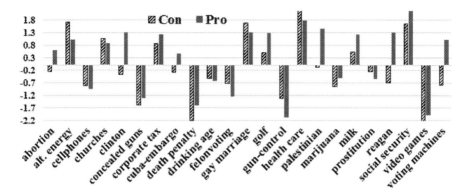

Fig. 4. Average sentiment score per issue per class.

Table 5. Effect of LM fine-tuning

Method	LM-FT	Pro			Con			Macro-F1	Acc
		P	R	F1	P	R	F1		
Our model	No	53.2	68.6	59.9	56.3	40.2	46.9	53.4	54.3
Our model	Yes	65.9	82.6	73.3	77	57.7	65.9	69.6	70.1

6.3 Sentiment Analysis

How does sentiment relate to stance? Are pro-opinions often positive while cons are negative? To answer this question and find the relation between the stance and sentiment we define the sentiment score s_d of document d as $s_d = \sum_{s \in d} s_v$ where s_v is the VADER sentiment score of sentence s [6]. We compare the average sentiment score of the 23 issues from training set arguments between two classes (Fig. 4). According to the plot, in some cases such as *abortion* and *voting machines* the score of pro is positive while con has negative overall score, indicating that proponents and opponents have different sentiments in their arguments about the issue. For some other cases, such as *health care*, both classes have a positive sentiment score. We identify as a key reason the concept of *'the right to health'*, has a positive sentiment. That makes opponents use this concept and its synonyms frequently making their arguments statistically positive. For "churches" where the underlying question is "should churches remain tax-exempt?" con has a larger positive score than pro. We find that some supporters (pro class) bring negative justifications by predicting the unsatisfactory situation after withdrawing the tax-exempt for churches. This unsatisfactory situation is explained while having negative sentiment.

7 Conclusion

We propose a general model for stance detection of arguments. Unlike most models, our documents are long (with the average size of 166 words) and come from a large number of different domains. Experiments show promising results compared to the baselines. We also find our proposed model relies on the beginning of long arguments for stance detection. And depending on the discussed issue, sentiment of an argument varies in pro or con class. Namely, pro-arguments express negative while con-argument have positive sentiment.

Acknowledgement. This work is supported in part by the U.S. NSF grants 1838145, 1527364, and 1838147. We also thank anonymous reviewers for their helpful feedback.

References

1. Augenstein, I., Rocktäschel, T., Vlachos, A., Bontcheva, K.: Stance detection with bidirectional conditional encoding. arXiv preprint arXiv:1606.05464 (2016)
2. Bar-Haim, R., Bhattacharya, I., Dinuzzo, F., Saha, A., Slonim, N.: Stance classification of context-dependent claims. In: Proceedings of the 15th Conference of the European Chapter of the Association for Computational Linguistics: Volume 1, Long Papers, vol. 1, pp. 251–261 (2017)
3. Chen, W.F., Ku, L.W.: UTCNN: a deep learning model of stance classification on social media text. arXiv preprint arXiv:1611.03599 (2016)
4. Du, J., Xu, R., He, Y., Gui, L.: Stance classification with target-specific neural attention networks. In: International Joint Conferences on Artificial Intelligence (2017)
5. Ebrahimi, J., Dou, D., Lowd, D.: A joint sentiment-target-stance model for stance classification in Tweets. In: Proceedings of COLING 2016, the 26th International Conference on Computational Linguistics: Technical Papers, pp. 2656–2665 (2016)
6. Gilbert, C.H.E.: Vader: a parsimonious rule-based model for sentiment analysis of social media text. In: Eighth International Conference on Weblogs and Social Media (ICWSM-2014) (2014) http://comp.social.gatech.edu/papers/icwsm14.vader.hutto.pdf. Accessed 20 Apr 16
7. Hasan, K.S., Ng, V.: Why are you taking this stance? Identifying and classifying reasons in ideological debates. In: Proceedings of the 2014 Conference on Empirical Methods in Natural Language Processing (EMNLP), pp. 751–762 (2014)
8. Howard, J., Ruder, S.: Universal language model fine-tuning for text classification. In: Proceedings of the 56th Annual Meeting of the Association for Computational Linguistics (Volume 1: Long Papers), vol. 1, pp. 328–339 (2018)
9. Levy, R., Bilu, Y., Hershcovich, D., Aharoni, E., Slonim, N.: Context dependent claim detection. In: Proceedings of COLING 2014, the 25th International Conference on Computational Linguistics: Technical Papers, pp. 1489–1500 (2014)
10. Merity, S., Keskar, N.S., Socher, R.: Regularizing and optimizing LSTM language models. arXiv preprint arXiv:1708.02182 (2017)
11. Mohammad, S., Kiritchenko, S., Sobhani, P., Zhu, X., Cherry, C.: SemEval-2016 task 6: detecting stance in Tweets. In: Proceedings of the 10th International Workshop on Semantic Evaluation (SemEval-2016), pp. 31–41 (2016)
12. Mohammad, S.M., Sobhani, P., Kiritchenko, S.: Stance and sentiment in Tweets. ACM Trans. Internet Technol. (TOIT) **17**(3), 26 (2017)

Modeling Gender Inequity in Household Decision-Making

Allegra A. Beal Cohen[1(✉)], Paul R. Cohen[2], and Gregory Kiker[1]

[1] Department of Agricultural and Biological Engineering,
University of Florida, Gainesville, FL 32603, USA
aa.cohen@ufl.edu
[2] School of Computing and Information,
University of Pittsburgh, Pittsburgh, PA 15213, USA

Abstract. The Food and Agriculture Organization (FAO) estimates that if female farmers in developing countries had access to the same resources as men, the number of undernourished people would decrease by 12%–17% [9]. Clearly, gender equity is a vital part of increasing agricultural production to feed the world's projected 9.7 billion people by 2050. However, programs designed to empower women in agricultural systems are expensive, and no quantitative model exists to define and explore the efficacy of policies in cultural contexts. We introduce a formal model of household decisions embedded in an agent-based model of community gender dynamics and show how the explicit definition of gender inequity can help inform decision-making about programs intended to empower women.

1 Introduction

As the global population increases, the world's demand on agricultural production is predicted to rise by 70% to 100% [10]. With almost ten billion people expected by mid century, those concerned with agricultural productivity must also be concerned with gender inequity. The FAO estimates that providing female farmers access to resources could increase agricultural productivity in developing countries by 2.5% to 4%, which translates to a 12% to 17% reduction in the number of undernourished people [9]. Also, women tend to allocate more resources in favor of their children, which can lead to more productive households in the future [9,12]. Thus, any efforts to increase the world's food security should include programs to empower the women who make up nearly half of the world's farmers [7].

Programs designed to empower women in agricultural systems can focus on improving women's personal assets, farming inputs, land rights, education, or access to health care. However, interventions to reduce gender inequity do not always account for power imbalances within households. For example, as women's activities become more profitable, their husbands often usurp them [8].

Supported by the NSF Graduate Research Fellowship Program.

R. Thomson et al. (Eds.): SBP-BRiMS 2019, LNCS 11549, pp. 31–38, 2019.
https://doi.org/10.1007/978-3-030-21741-9_4

In other cases, men become more violent as women's personal assets increase [15]. Accounting for the cultural context of interventions is vital to their success, as culture heavily influences decision-making within households.

This paper provides a quantitative model of gender dynamics in households. Our model is the first to quantify gender inequity across multiple households and to consider the impact of community on intra-household bargaining. Given the expense of program pilot studies, the goal of our model is to make program design and evaluation cheaper and faster for non-governmental organizations (NGOs), government agencies, and other entities invested in gender equity.

The remainder of this paper is organized as follows: In Sect. 2, we introduce a formal model of intra-household bargaining. In Sect. 3, we describe how households interact and how their members learn new bargaining policies. In Sect. 3.1, we show how the model can be used to explore and evaluate strategies for improving women's outcomes. We conclude with a discussion of future work.

2   Intra-household Bargaining

Early models of agricultural households assumed that household decisions are "unitary" [6], despite the reality that husbands and wives sometimes disagree. Recent attempts to quantify household decision-making emphasize bargaining power and intra-household resource allocation. The "collective" framework accommodates scenarios in which household members have different preferences about how to allocate time, land and capital, where choices are made through Pareto-efficient cooperative bargaining [3–5]. However, theories of intra-household allocation are often applied in field studies to show that heterogeneity of preference can lead to inefficient choices [14]. Smith and Chavas model household decision-making as a two-stage game where household members optimize their utility functions and make final choices based on whether they would be better off with a divorce. A model is used to show that Pareto-efficiency does not hold [16]. Basu maximizes household members' utility as well, but allows members' bargaining power to be affected by choices made in the previous timestep [2]. In her summary of intra-household bargaining frameworks, Agarwal notes that household bargains do not occur in a vacuum, and that models must account for social norms and gender differences [1]. With this in mind, we base our model of intra-household decision-making on two dynamics derived from the literature: Men generally have more power than women, and women generally allocate more money to the household and their children [11,12].

2.1   Portfolios and Payoffs

Here we present a bargain between Alice and Bob (A and B), members of a household (H). Alice and Bob must choose one of two portfolios, π_1 and π_2, which represent the allocation of land, labor and capital to activities. Let V_i^A, V_i^B, and V_i^H be the payoffs of $\pi_{i=1,2}$ to Alice, Bob and the household, respectively. To simplify, in this example we assume that V_i is known by Alice and Bob and that they only care about payoffs, not about the activities themselves. In general, Alice and Bob must estimate the return of π_i based on their preferences and past experience.

We assume $V_2 = \delta V_1$ for $0 < \delta < 1$; that is, π_1 has a greater total payoff than π_2. We also assume that $V_2^B > V_1^B$; that is, π_2, though smaller in total, gives more payoff to Bob. Thus, Bob prefers π_2 and Alice prefers π_1. Let p_1^B and p_2^B be Bob's proportion of V_1 and V_2, respectively:

$$V_1^B = V_1 p_1^B \tag{1}$$
$$V_2^B = \delta V_1 p_2^B \tag{2}$$

The proportions of V_1 and V_2 that remain for Alice and the Household are therefore:

$$V_1^{A,H} = V_1(1 - p_1^B) \tag{3}$$
$$V_2^{A,H} = \delta V_1(1 - p_2^B) \tag{4}$$

Alice prefers π_1 because it returns more to her and the household. Alice and Bob can negotiate the choice between π_1 and π_2 in three ways:

1. A wants π_1 and B agrees to select π_1.
2. A wants π_1 and B agrees, subject to A paying B a *penalty* from V_1^A;
3. A wants π_1 but agrees to select π_2, with no penalty.

Choice 1 is best for Bob iff $p_1^B V_1 \geq p_2^B \delta V_1$, or $p_1^B \geq \delta p_2^B$. Choices 2 and 3 can yield the same return to Bob if Alice pays a *penalty* by increasing Bob's proportion of V_1. To make Bob indifferent between Choices 2 and 3, Alice must offer a p_1^{B*} that makes $p_1^{B*} V_1 = p_2^B \delta V_1$:

$$p_1^{B*} = \frac{(p_2^B \delta V_1)}{V_1} = p_2^B \delta. \tag{5}$$

If Choice 2 yields $p_1^{B*} V_1$ to Bob, then Alice and the Household get

$$p_1^{A,H} = (1 - p_1^{B*})V_1 \tag{6}$$
$$= (1 - p_2^B \delta)V_1 \tag{7}$$

In contrast, Choice 3 yields:

$$p_2^{A,H} = (1 - p_2^B)V_2 \tag{8}$$
$$= (1 - p_2^B)\delta V_1 \tag{9}$$
$$= (\delta - p_2^B \delta)V_1 \tag{10}$$

For $0 < \delta < 1.0$, Alice prefers Choice 2 to Choice 3. In this example, the penalty Alice pays is in units of the portfolio's payoff. Depending on where Alice and Bob live, the penalty might be in terms of money, crops, labor, or things of less certain value like intimacy.

3   External Interactions

Until now, we have assumed that Bob is indifferent between portfolios that return the same to him, but this is not the case in the real world. Agarwal (1997) states that social norms can affect household bargaining in several ways, including moving the point of compromise to fit what is "acceptable" [1]. Even if Alice offers a penalty to make Bob's $p_1^{B*}V_1 = p_2^B V_2$, Bob may be unsatisfied with the deal if it does not conform to the social norms of his culture. With the following example, we extend our formal model of the household bargain to include a p_{min} for Bob that represents the minimum proportion of payoff he accepts based on what he perceives to be the norm.

Suppose Bob lives in a town where men always get more than half of portfolio payoffs. Based on this norm, Bob has $p_{min} = 0.51$, where he rejects any deal that does not give him more than half of any portfolio's returns. Suppose he and Alice must choose between π_1 with $V_1 = 100$ and $p_1^B = 0.4$ and π_2 with $V_2 = 80$ and $p_2^B = 0.6$. Following the logic of the previous section, Alice offers Bob $p_1^{B*} = 0.48$ of π_1, accepting a penalty of 8 portfolio units to get her way. Unfortunately, $p_1^{B*} \leq p_{min}$, so Bob is not happy with her offer. However, Alice can "sweeten the deal" by allowing him a greater fraction of V_1; she has between $(1 - p_2^B \delta)V_1$ (Eq. 7) and $(\delta - p_2^B \delta)V_1$ (Eq. 10) to offer him and increases her penalty by 3 units to make him happy. She still does better than if she and Bob had agreed on π_2, but Bob's attention to social norms has decreased her payoff.

Social norms must change if women are to achieve equity in agricultural systems. We extend our model to analyze the effects of changes in social norms on Alice and Bob. We allow Bob to adjust his p_{min} in response to his social network in a manner consistent with the theory of social influence learning [17]. In our model, there are N households, each with one male (Bob) and female (Alice) agent. Male agents have a "neighborhood" of M randomly assigned male neighbors. At each time step, after each Bob has bargained with his Alice, he talks to his M neighbors, and with probability ρ talks to an additional m random male agents. Bob assesses the mean household wealth of those agents he talks to, and he adjusts his p_{min} based on the p_{min}^Rs of the agents who are richer than average:

$$p_{min,t+1}^{Bob} = \alpha \cdot (p_{min,t}^R - p_{min,t}^{Bob}) \tag{11}$$

where α is Bob's learning rate. Equation 11 updates Bob's p_{min} at time $t+1$ given his p_{min} at time t and the p_{min}s of the richest Bobs in his neighborhood. Currently Bob's estimates of his neighbors' wealth and p_{min}s are accurate; in future we will explore the effects of biased estimates.

3.1   An Example

Here we show the behavior of our model with the following parameters: $N = 20$; $M = 4$; $m = 1$; $\rho = 0.5$; $\alpha = 0.2$.

We ran a simulation of $N = 20$ households over 30 time steps. During each time step, each household's male and female agent choose between portfolios

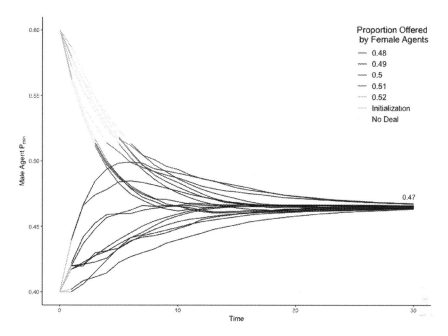

Fig. 1. Basic behavior of the model. Each line represents one male agent's p_{min} over time. The color of the line shows how much the female agent offered the male agent to choose her preferred portfolio at that time step. Black, "No Deal", means that male agents' p_{min}s are higher than the female agents can offer and the male agent's preference is chosen. (Color figure online)

π_1 and π_2. Agents do not care about the activities in the portfolio, only the portfolio's returns. π_1 returns 10 units of payoff with $p_1^B = 0.4$; π_2 favors male agents with a p_2^B of 0.6 and a total return of 8. Thus, female agents have between 3.2 and 5.2 of π_1 to bribe their husbands to select that portfolio, and male agents must have $p_{min} \leq 0.52$ for the bargain to work. 10 of the male agents begin with an initial p_{min} of 0.6, and the other 10 begin with a p_{min} of 0.4. In this example, portfolios are communal, so when a male agent updates his p_{min}, he judges his neighbors based on their household's total wealth.

Figure 1 shows the p_{min} of the 20 male agents over 30 time periods. Each line represents the p_{min} of one male agent at time t. The color of the line indicates how much the household's female agent offered to get her preferred portfolio, rounded to the hundredth place; "No Deal" indicates the male agent's p_{min} was too high to make her preferred portfolio worth it, so the male agent's preferred portfolio is selected. As expected, higher p_{min}s in male agents reduce the payoffs for their female partners, who must "sweeten the deal" further. Also as expected, male agents' p_{min}s converge to slightly below the initial average, with an average $p_{min} = 0.47$ at $t = 30$. Agents' p_{min}s converge to slightly below both the average of the initial p_{min}s (0.5) and the threshold for choosing π_1 (0.48) because the male agents with initial p_{min}s of 0.4 have slightly more influence.

Male agents with p_{min}s that start at 0.4 and increase (we'll call these agents M_{low}) consistently agree to choose π_1, which has a higher payoff. As a result, the households of M_{low} add to their coffers faster than the households of the agents that start with the higher p_{min} (M_{high}). Agents with richer households tend to be part of the "richer than average" (R) group in Eq. 11; thus, male agents adjusting their p_{min}s are more often influenced by M_{low} to reduce their p_{min}s. As all agents increase or decrease their p_{min}s towards the threshold for choosing π_1, the difference between any given pair's p_{min}s gets smaller and the rate of adjustment decreases.

Although our results are preliminary, it is clear how they might be used to plan or improve a program intended to empower women. If new social norms can be formed in a community simply by increasing the wealth of those who choose the preferred behavior, then an NGO might successfully increase women's share of household payoffs by giving cash to households that allocate funds fairly.

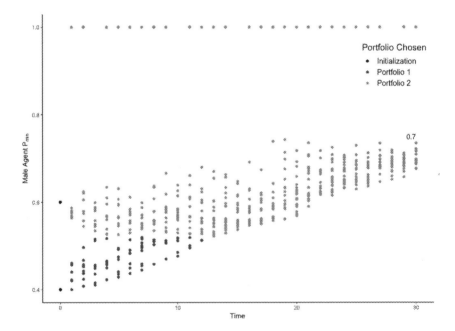

Fig. 2. The model with insistence. Each point is a male agent's p_{min} at time t; its color indicates which portfolio that agent's household chose.

However, suppose male agents *insist* on their preference some fraction of the time. In our model, insistence means that Bob unexpectedly increases his p_{min} to 1.0; that is, he demands the entire payoff of his preferred portfolio, regardless of what he usually does. At the next time step, Bob's p_{min} returns to what it was before he became insistent. Figure 2 shows that even when the probability of insistence in a bargain is just 10%, the initial conditions of the first experiment lead to a different outcome. Lines for each agent are reduced to points to preserve clarity. Each point represents the p_{min} of one male agent at time t; the color of the point indicates which portfolio the household chose at that time step. The presence of insistence in the model results in an upward trend where male agents increase their $p_{min}s$ to an average of 0.7 at $t = 30$. When the model is run out further, the average p_{min} converges to 1.0. As expected, without any counteracting efforts, even small upward perturbations in the model eventually lead to the maximization of p_{min}.

These results can be compared to situations in which husbands unexpectedly enforce their authority when they feel it is being challenged. For example, in a community where men are traditionally wealthier than women, a woman's husband may feel threatened if she receives a microcredit loan to build her own business. He may resort to violence, theft or isolation to keep her from increasing her power in the household, which might in turn leave a community's gender inequity worse off than it was before the intervention. Modeling gender inequity within a community can help identify potential conflicts before a program has launched. The responsibility to know whether these factors matter or not in a culture lies with the area expert; but defining the problem space can provide valuable insights. In this simple example, an NGO made aware of the possibility of male retaliation might budget resources to reduce it, perhaps by educating men in the community or using their program to reduce women's isolation [13].

4   Future Work

Future implementations will make the model more realistic by increasing the range of actions available to households, and revising some simplifying assumptions. Most importantly, Alice has no bargaining power outside of the penalty she offers Bob, whereas in reality, Alice's options may include threatening divorce or publicly shaming Bob. How the bargain changes when Alice's bargaining power increases is a complicated problem. Many factors can either increase Alice's bargaining power or lead to violence, from her personal wealth to how connected she is to other women in the community. Second, the model assumes that all portfolios are communal; that is, Bob and Alice do not discriminate between their individual inputs, and payoffs from all activities are summed and divided up. In many places, intra-household allocation depends on spheres of influences: Personal activities, such as vegetable gardens, and communal activities, such as cereal crops, have separate inputs and outputs. With this addition, we will be able to better model time poverty (the idea that if Bob requires Alice to labor in the fields, she has less time to tend her children and her own plots) and intervention strategies that increase Alice's personal assets. Third, Bob and Alice do not learn to better estimate the payoffs of portfolios. The next iteration of our model will allow agents to adjust their preferences for activities based on the perceived results of their neighbors' chosen portfolios. Finally, we will "close the loop" by making the available portfolios depend on earlier decisions; for example, if Bob and Alice amass enough capital, they will be able to invest in higher-quality seeds or education for their children.

Acknowledgments. Thanks to the National Science Foundation Graduate Research Fellowship for funding this research. Thank you also to Lacey Harris-Coble and Luca Mantegazza who discussed gender dynamics and bargaining with us.

References

1. Agarwal, B.: "Bargaining" and gender relations: within and beyond the household. Feminist Econ, **3**(1), 1–51 (1997)
2. Basu, K.: Gender and say: a model of household behaviour with endogenously determined balance of power. Econ. J. **116**(511), 558–580 (2006)
3. Browning, M., Chiappori, P.A.: Efficient intra-household allocations: a general characterization and empirical tests. Econometrica **66**(6), 1241–1278 (1998)
4. Chiappori, P.A., Fortin, B., Lacroix, G.: Marriage market, divorce legislation, and household labor supply. J. Polit. Econ. **110**(1), 37–72 (2002)
5. Chiappori, P.A., et al.: Rational household labor supply. Econometrica **56**(1), 63–90 (1988)
6. Deaton, A., Muellbauer, J., et al.: Economics and Consumer Behavior. Cambridge University Press, Cambridge (1980)
7. Doss, C.: If women hold up half the sky, how much of the world's food do they produce? In: Quisumbing, A., Meinzen-Dick, R., Raney, T., Croppenstedt, A., Behrman, J., Peterman, A. (eds.) Gender in agriculture, pp. 69–88. Springer, Dordrecht (2014). https://doi.org/10.1007/978-94-017-8616-4_4
8. Doss, C.R.: Designing agricultural technology for African women farmers: lessons from 25 years of experience. World Devel. **29**(12), 2075–2092 (2001)
9. FAO: The State of Food and Agriculture: Women in Agriculture: Closing the gender gap for development. Food and Agriculture Organization (FAO) of the United Nations (2010–2011)
10. Godfray, H.C.J., et al.: Food security: the challenge of feeding 9 billion people. Science **327**(5967), 812–818 (2010)
11. Lopez-Claros, A., Zahidi, S.: Women's empowerment: measuring the global gender gap. Geneva Switzerland World Economic Forum 2005 (2005)
12. Meinzen-Dick, R., Behrman, J., Menon, P., Quisumbing, A.: Gender: A key dimension linking agricultural programs to improved nutrition and health. In: Reshaping Agriculture for Nutrition and Health, pp. 135–44 (2012)
13. Rosenberg, M.L., Butchart, A., Mercy, J., Narasimhan, V., Waters, H., Marshall, M.S.: Interpersonal violence. Dis. Control Priorities Dev. Countries **2**, 755–70 (2006)
14. Schaner, S.: Do opposites detract? Intrahousehold preference heterogeneity and inefficient strategic savings. Am. Econ. J.: Appl. Econ. **7**(2), 135–74 (2015)
15. Schuler, S.R., Hashemi, S.M., Badal, S.H.: Men's violence against women in rural Bangladesh: undermined or exacerbated by microcredit programmes? Dev. Pract. **8**(2), 148–157 (1998)
16. Smith, L.C., Chavas, J.P.: Supply response of West African agricultural households: Implications of intrahousehold preference heterogeneity. IFPRI Food Consumption and Nutrition Division, discussion paper 69 (1999)
17. Young, H.P.: Innovation diffusion in heterogeneous populations: contagion, social influence, and social learning. Am. Econ. Rev. **99**(5), 1899–1924 (2009)

Bot Detection: Will Focusing on Recall Cause Overall Performance Deterioration?

Tahora H. Nazer[1][✉], Matthew Davis[1][✉], Mansooreh Karami[1],
Leman Akoglu[2], David Koelle[3], and Huan Liu[1]

[1] Arizona State University, Tempe, AZ 85281, USA
{tahora.nazer,matt.davis,mkarami,huan.liu}@asu.edu
[2] Carnegie Mellon University, Pittsburgh, PA 15213, USA
lakoglu@andrew.cmu.edu
[3] Charles River Analytics, Cambridge, MA 02138, USA
dkoelle@cra.com

Abstract. Social bots are an effective tool in the arsenal of malicious actors who manipulate discussions on social media. Bots help spread misinformation, promote political propaganda, and inflate the popularity of users and content. Hence, it is necessary to differentiate bot accounts and human users. There are several bot detection methods that approach this problem. Conventional methods either focus on precision regardless of the overall performance or optimize overall performance, say F_1, without monitoring its effect on precision or recall. Focusing on precision means that those users marked as bots are more likely than not bots but a large portion of the bots could remain undetected. From a user's perspective, however, it is more desirable to have less interaction with bots, even if it would incur a loss in precision. This can be achieved by a detection method with higher recall. A trivial, but useless, solution for high recall is to classify every account (human or bot) as bot, hence, resulting in poor overall performance.

In this work, we investigate if it is feasible for a method to focus on recall without considerable loss in overall performance. Extensive experiments with recall and precision trade-off suggest that high recall can be achieved without much overall performance deterioration. This research leads to a recall-focused approach to bot detection, REFOCUS, with some lessons learned and future directions.

Keywords: Social media · Twitter · Social bots · Bot detection · Recall

1 Introduction

Bots are prevalent on social media and their malicious actions have been observed repeatedly. An example of this wide-spread activity of bots was seen during the

© Springer Nature Switzerland AG 2019
R. Thomson et al. (Eds.): SBP-BRiMS 2019, LNCS 11549, pp. 39–49, 2019.
https://doi.org/10.1007/978-3-030-21741-9_5

US Presidential Election in 2016. Allcott and Gentzkow [1] extensively studied this event and reported that millions of pro-Trump and pro-Clinton fake stories were shared on Facebook, in part, by bot accounts. They also provide evidence that more than 40% of fake news sources use social media to spread their content.

Thus, researchers have put great effort into understanding bots and developing methods to detect them. In supervised bot detection methods, which are the focus of this work, a labeled dataset of bots and human users is available prior to training a machine learning classifier. Using these labels, we can learn characteristics, also know as features (described in detail in Sect. 2), that discriminate bots from humans and use them to build classifiers that predict class labels (bot or human). The classifiers are then tested on unobserved datasets and evaluated using two prominent metrics: precision and recall.

A common theme among previous bot detection methods is attempting to maximize precision [4, 9, 16]. This is one extreme: the sole purpose is to minimize false positives and avoid mistakenly marking a human user as bot. By doing this, detection methods avoid removing human users from the site but leave many bots undetected. The other extreme is eliminating bots from social media at the price of removing human users. This approach is not preferable either. A method for finding a trade-off between precision and recall is optimizing for F_1 score which is the harmonic mean between precision and recall. Harmonic mean is dominated by the minimum of its arguments. Hence, F_1 cannot become arbitrarily large when either precision or recall is unchanged and the other metric is increased. This prevents bot detection algorithms from landing on trivial solutions (marking all users as bots or humans) to gain high F_1. However, considering the same weight for precision and recall in F_1 prevents us from having control over the final values of either precision or recall. In other words, two classifiers are considered equally good if they have the same F_1 regardless of the fact that one might result in higher recall and the other one a higher precision. The ideal case is finding a solution close to optimum F_1 that allows us to focus on precision or recall depending on the application.

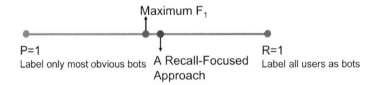

Fig. 1. Our goal is having a recall-focused approach close to the optimal F_1.

To align with corporate goals (having a large number of active users and retaining human users by avoiding accidentally suspending their accounts), bot detection models with high precision are preferable. However, from a user's perspective, both social media users and researchers alike, the preferable situation is encountering a minimum number of bots. So, in this case, high recall is preferred. In this work, we focus on developing a supervised algorithm aligned with a user's perspective: a REcall FOCUSed bot detection model, REFOCUS. We use multiple real-world datasets to show how we can find a sweet spot between blindly optimizing for F_1 or recall as shown in Fig. 1. We also compare REFOCUS with state-of-the-art bot detection models to show that focusing on recall does not necessarily result in overall performance deterioration in terms of F_1.

2 Supervised Bot Detection Methods

To use supervised bot detection models, one must identify differences among bot and human users in terms of features such as content or activity in a labeled dataset. Then, a classifier is trained on the features and labels to distinguish bots from humans in an unobserved dataset. Different classification methods can be used for this purpose such as Support Vector Machines [11], Random Forests [9], and Neural Networks [8]. We describe some common user features below:

- *Content*: the measures in this category focus on the content shared by users. Words, phrases [16], and topics [11] of social media posts can be a strong indicator of bot activity. Also, bots are motivated to persuade real users into visiting external sites operated by their controller, hence, share more URLs in comparison to human users [4,13,17]. Bots are observed to lack originality in their tweets and have large ratio of retweets/tweets [14].
- *Activity Patterns*: Bots tweet in a "bursty" nature [4,10], publishing many tweets in a short time and being inactive for a longer period of time. Bots also tend to have very regular (e.g. tweeting every 10 min) or highly irregular (randomized lapse) tweeting patterns over time [18].
- *Network Connections*: bots connect to a large number of users hoping to receive followers back but the majority of human users do not reciprocate. Hence, bots tend to follow more users than follow them back [4].

As bots become more complex and harder to detect [5], bot detection methods incorporate a larger number of features and datasets with more samples of humans and bots. One of the recent approaches is BotOrNot [6]. This method exploits 1,150 features from five categories: user-based, friends, network, temporal, language, and sentiment [16]. The initial model was trained on a dataset of ~40,000 bots and humans and has been updated multiple times using seven more datasets with total of ~87,000 samples. The strength of this method has encouraged researchers to use it to label ground-truth datasets [7]. We will compare our proposed method with BotOrNot in Sect. 5.

3 Data for Supervised Bot Detection

To show the robustness of our model with respect to the language, topic, time, and labeling mechanism, we use three datasets represented in Table 1.

Table 1. Statistics of the datasets used in this study.

Property	Arabic Honeypot	Social Spambot 1	Social Spambot 2
Tweets	637,435	4,449,395	4,257,918
Retweets	209,703	782,267	754,104
Human accounts	2,317	1,083	1,083
Bot accounts	1,978	991	464
Bot ratio	46.05%	47.78%	29.99%
Labeling approach	Honeypot	Manual	Manual

As seen in Table 1, in this work, we utilize three existing bot detection datasets. We describe how each of these raw datasets were collected and what they contain in Sect. 3.1. Then, we specify how we preprocessed the raw data using a content-based feature extraction method in Sect. 3.2.

3.1   Datasets

The first dataset is a honeypot dataset collected by Morstatter et al. [11] which we refer to as the Arabic Honeypot dataset. It was collected using a network of 9 honeypot accounts which tweeted Arabic phrases, as well as randomly following and retweeting each other. Any user who followed a honeypot was considered a bot, because bot behaviors are sporadic and provide no intelligent information to humans. For collecting a set of human users, the authors manually inspected users who tweeted same Arabic phrases as some of the bots, then crawled data for them and other users that the inspected users immediately followed; assuming that humans only follow other humans and not bots. In August 2018, we re-crawled this dataset using the tweet IDs shared by the authors.

Additionally, we employ two datasets introduced by Cresci et al. [5] in their previous work on detecting social bots on Twitter namely: `test set #1` and `test set #2`. We call these datasets Social Spambots 1 and 2, respectively, in our work. Each dataset is a combination of social spambots and human users on Twitter. To collect the human user accounts, Cresci et al. contacted random users, asked a natural language question, and manually evaluated if the user was a human. Social Spambots 1 contains these genuine accounts plus social bots that were discovered during the 2014 Mayoral election in Rome, Italy which were used to retweet a candidate within minutes of his original posting. Social Spambots 2 includes the genuine accounts and social bots that advertised products on *Amazon.com* by deceitfully spamming URLs which point to the products. We obtained these datasets directly from the BotOrNot Bot Repository [6].

3.2   Feature Extraction

Bot accounts are created by malicious actors to serve specific purposes. Thus, their content can be a strong indicator to expose such potentially automated accounts. The problem with using content for bot detection is that the raw text features are of high dimensionality and sparse. Inspired by the recent advances of topic modeling, we adopt latent Dirichlet allocation (LDA) [3] to obtain a topic representation of each user. LDA, which treats each document as a distribution over topics and each topic as a distribution over the vocabulary in the dataset, has been proven useful for extracting latent semantics of documents. As such, we use LDA to extract features from users' tweet content. In this work, each user is considered one document and the content of that document is his tweets. We trained separate LDA models on each of the three datasets and develop classifiers independently. We follow the assumption that, since bots are naturally more interested in certain topics, denoting each user as a distribution over different topics may help to better identify them from regular accounts [11].

4   A Recall-Focused Approach – REFOCUS

A bot detection classifier generates the probability of being a bot (belonging to the positive class) for each instance in the dataset. To assign a binary label to users, a classifier uses a threshold (commonly set to 0.5 [12]) to decide; if the probability of being a bot is more than the classification threshold then the user is labeled as a bot, otherwise as a human. Based on the assigned labels, classifiers can be evaluated using precision (P) and recall (R) illustrated in Fig. 2 and defined bellow:

$$P = \frac{tp}{tp + fp}, \quad R = \frac{tp}{tp + fn} \tag{1}$$

Precision and recall can be independently maximized easily. A trivial approach for increasing the recall is lowering the classification threshold and classifying more users as bots. Alternatively, increasing the classification threshold results in labeling most users as humans, with only the unquestionably obvious bot users labeled as bots, and causes a trivial increase in precision. However, precision and recall are not independent from each other: increasing one might result in decreasing the other. One approach for finding a trade-off between precision and recall is using the F_β score. F_β is the *weighted* harmonic mean of precision and recall [15] and is defined as follows:

$$F_\beta = \frac{(1 + \beta^2)PR}{\beta^2 P + R} \tag{2}$$

With β values greater than one, β times more weight is put on recall and for values less than one, β times more weight is associated with precision.

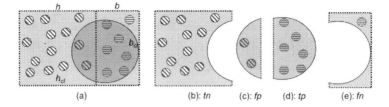

Fig. 2. Illustration of true negative - (b): tn, false positive - (c): fp, true positive - (d): tp, and false negative - (e): fn for a classifier trained on dataset (a) when the classifier labels a subset of users as bots (positive class) - b_{cl} - and the rest as humans (negative class) - h_{cl}.

4.1   Searching for a Trade-Off: Selecting β

Our goal is optimizing for recall, hence, we utilize F_β with $\beta > 1$ to find the best classification threshold: a sweet spot between where F_1 (overall performance) is maximized and where $R = 1$. The framework of our recall focused approach is presented in Fig. 3. We divide the dataset to 90% $Train$ and 10% $Test$. Then, for ten iterations, we divide Train to 90% $Train_i$ and 10% Val_i which are training and validation sets respectively; $Train_i$ is 81% of the whole data and Val_i is 9%. In each iteration, we train a classifier C_i on $Train_i$, change the classification threshold between 0.1 and 0.9 with 0.1 steps, and find the threshold that results in the highest F_β score on Val_i; we call this threshold t_i. After the tenth iteration, we get an average of the thresholds t_1 to t_{10} to find the average threshold t. Then we train a classifier, C, on $Train$ and using t, we find the precision, recall, and F_1 score. We repeat this process ten times and report the average of precision, recall, and F_1 scores as our final results.

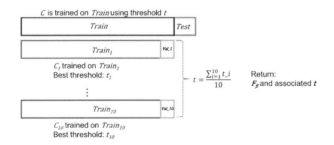

Fig. 3. Framework for the proposed bot detection model, REFOCUS.

We need to test different values of β in the training phase to find the best classification threshold using F_β. As we increase β, precision has a non-increasing trend and recall has a non-decreasing trend. This happens because as we increase β we put more weight on recall in comparison to precision. More formally

$$R_{\beta_i} \geq R_{\beta_j} \quad and \quad P_{\beta_i} \leq P_{\beta_j} \quad if \quad \beta_i > \beta_j \tag{3}$$

Due to this non-increasing pattern of precision with increase of β, we prefer to maintain a low β as long as we do not sacrifice the chance of achieving a higher recall with minor loss in precision. To find the right β, we start from $\beta = 1$ and in each step we choose the current β as β_{opt} if

$$(R^\beta - R^{\beta_{opt}}) > (F_1^{\beta_{opt}} - F_1^\beta) \tag{4}$$

Meaning that we choose a larger β if the gain in R is more than the loss in F_1.

5   Experiments

In this section we empirically investigate the performance of our proposed approach. First, we investigate the effect of β and then, we compare REFOCUS with baseline bot detection models in terms of P, R, and F_1.

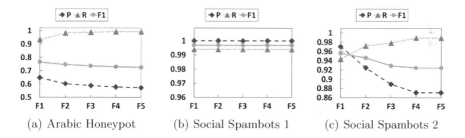

(a) Arabic Honeypot   (b) Social Spambots 1   (c) Social Spambots 2

Fig. 4. Effect of β on precision (P), recall (R), and overall performance (F_1). In each dataset, we change β from 1 to 5, use F_β for finding the best classification threshold in the training phase and report P, R, and F_1 on the test set.

5.1   Searching for the Right β

It is intuitive that using a F_β when $\beta > 1$ for training a classifier helps us find the classification threshold that results in higher recall as compared to when $\beta = 1$. However, it raises two questions: (1) what is best value of β and can we increase it indefinitely to reach the highest recall possible? (2) Does the model trained using $\beta > 1$ still perform well in terms of F_1 or we will drastically lose precision? We answer the first question here and the second one in Sect. 5.2.

We test our model on three datasets: Arabic Honeypot, Social Spambot 1, and 2. The results are shown in Fig. 4. In Social Spambots 1, we do not observe any change in the overall performance in terms of F_1 as we change the β. This can be due to the way this dataset was collected resulting in humans and bots being quite distinct from each other. This distinction causes the classifier perform well no matter what the threshold is. Hence, any of the F_β scores can be used to find the best classification threshold. In Arabic Honeypot and Social Spambot 2, we see some variations in precision, recall, and overall performance. $\beta = 2$ gives us the best trade-off between precision and recall because the loss in the overall performance is smaller that the gain in recall; in other words, the slope of recall line is larger than the slope of F_1 line. Further increase in β does not provide enough gain on recall in comparison to the loss in the overall performance, hence, we stop at F_2.

5.2   Testing the Overall Performance

For comparing the overall performance of REFOCUS with other bot detection methods, we need to decide on the number of topics in LDA and the classification model. Due to the similarity between our feature extraction and the one by Morstatter et al. [11] we follow their observation that 200 topics generated the highest F_1 in the Arabic Honeypot dataset and set number of topics to 200.

We test multiple classification algorithms that are observed to have high performance in the problem of bot detection [2] to find the best fit for REFOCUS:

Table 2. Performance of REFOCUS when implemented using different classifiers.

Classifier	Arabic Honeypot			Social Spambot 1			Social Spambot 2		
	P	R	F_1	P	R	F_1	P	R	F_1
Decision Tree	0.738	0.746	0.741	0.995	0.996	0.996	0.933	0.956	0.944
Random Forest	0.460	1.0	0.630	0.992	0.996	0.994	0.854	0.947	0.897
Logistic Regression	0.603	0.984	0.748	1.0	0.996	0.998	0.984	0.919	0.950
SVM	0.601	0.983	0.746	1.0	0.993	0.996	0.924	0.971	0.945

Decision Tree, Random Forest, Logistic Regression, and SVM. We use Python Scikit-learn package [12] for implementation with default settings except $max\_depth = 1$ for Random Forest and $max\_iter = 1$ for Logistic Regression to avoid overfitting. As shown in Table 2, all classifiers achieve very similar (difference less than 0.5%) F_1 score except for Random Forest that has lower performance. We choose SVM for the rest of our experiments because it has similar or higher R and similar F_1. Worth mentioning that our method can be built on top of any classifier to help improve recall without sacrificing the overall performance.

We compare our proposed approach, REFOCUS, with two baselines:

- *SVM*: REFOCUS uses SVM to train multiple classifiers on subsamples of the dataset and learns the best recall-precision trade-off using F_β. Hence, we compare our method with SVM when its parameters are set to default and it generates the class labels (1 or -1) using 0.5 as threshold. Users are represented with 200 LDA topics and we use 10-fold cross validation.
- *BotOrNot* [6, 16]: this supervised bot detection model exploits 1150 features in six categories: user-based, friends, network, temporal, content and language, and sentiment. The model uses a Random Forest classifier and is trained on

multiple publicly available datasets. BotOrNot has been used for generating ground-truth due to its performance.

We perform two sets of experiments. In the first one, we use the Arabic Honeypot dataset. We use an LDA model with 200 topics to extract features from the dataset then we apply REFOCUS and report the results. However, using this dataset raises the concern that our approach might not perform as well on non-Arabic tweets. Hence we also perform the second experiment. We follow the same procedure but use the datasets that were collected by Cresci et al. [5]. These datasets (as explained in Sect. 3) have three advantages: they are among the most recent publicly available labeled datasets for bots and include newer bots, they use manual labeling which is different from the honeypot dataset, and a majority of the tweets are in English. Hence, by testing our approach on Cresci's datasets, we show that our model performs well regardless of the language of tweets and is resilient to new bots that emerge on social media.

The results are presented in Table 3. For the experiments on Cresci's datasets, we do not balance the classes due to small size of the data. Hence, we also include the ROC AUC in our results. The ROC AUC for a classifier that randomly assigns labels to instances is 0.5 regardless of the class balance and is a helpful metric to assess classifiers when the samples of one class are more than the other. Reserving the class imbalance is also helpful to mimic the real world scenario where bots are a small portion of all users on social media [16].

Table 3. Comparison between REFOCUS and baseline bot detection methods.

Dataset	Method	P	R	F_1	ROC
Arabic Honeypot	SVM	0.655	0.919	0.765	0.849
	BotOrNot	0.472	0.523	0.496	0.514
	REFOCUS	0.601	0.983	0.746	0.849
Social Spambot 1	SVM	1.0	0.991	0.995	0.997
	BotOrNot	0.963	0.961	0.962	0.969
	REFOCUS	1.0	0.993	0.996	0.997
Social Spambot 2	SVM	0.986	0.915	0.949	0.996
	BotOrNot	0.954	0.939	0.946	0.957
	REFOCUS	0.924	0.971	0.945	0.996

In the first experiment, on the Arabic Honeypot dataset, SVM has higher precision and lower recall in comparison to REFOCUS. The reason is that SVM only labels a user as bot if the predicted probability of being a bot for that user is over 0.5. However, our method learns the best threshold for optimizing recall while reaching a high F_1. Hence REFOCUS chooses a lower threshold (0.35 in this case). This choice results in 2% lower F_1, however, we are willing to tolerate this loss due to 6% gain in recall. BotOrNot performs considerably worse in this dataset in comparison to the Social Spambot datasets. The reason is that Social Spambot datasets have been used in training BotOrNot and it is expected for classifiers to have lower performance on unseen datasets (e.g. Arabic Honeypot).

In the second experiment, we test our method on two non-Arabic datasets which are obtained using a manual annotation method to show that our results are robust to variations in datasets such as language. In Social Spambots 1, SVM and our proposed approach perform almost identically with an slightly better recall in REFOCUS. The reason is that the differences between instances in human and bot classes are well captured by the classifiers to the extent that the classifier (either SVM or REFOCUS) are very confident in the labeling. Hence, each instance gets a high probability of being in its actual class and changing the threshold does not change the classification results much. We also observe that our approach outperforms BotOrNot. On Social Spambots 2, SVM and BotOrNot outperform our approach in precision and have lower recall, similar to the Arabic Honeypot dataset, because they are not is not designed to optimize on recall. F_1 of our approach is similar to the baselines.

6   Conclusion and Future Directions

The dominant trend among the previously proposed methods for bot detection is solely focusing on precision, making sure that no human user is marked as a bot, or optimizing for F_1. In this work, we showed that we can focus on recall of a bot detection model without sacrificing the overall performance. We tested our method on three real-word datasets and observed that using F_2 score in the training phase results in finding the best classification threshold for optimizing recall and having high overall performance in terms of F_1. In the future, we wish to explore the robustness of our method on translated datasets and also measure its effectiveness in discriminating different types of bots in a dataset.

Acknowledgements. Support was provided, in part, by NSF grant 1461886 on "Disaster Preparation and Response via Big Data Analysis and Robust Networking" and ONR grants N000141612257 (on "Intelligent Analysis of Big Social Media Data for Crisis Tracking") and N000141812108 (on "Bot Hunter"). We would like to thank anonymous reviewers for their valuable feedback.

References

1. Allcott, H., Gentzkow, M.: Social media and fake news in the 2016 election. J. Econ. Perspect. **31**(2), 211–36 (2017)
2. Alothali, E., Zaki, N., Mohamed, E.A., Alashwal, H.: Detecting social bots on Twitter: a literature review. In: IIT, pp. 175–180. IEEE (2018)
3. Blei, D.M., Ng, A.Y., Jordan, M.I.: Latent Dirichlet allocation. J. Mach. Learn. Rese. **3**(Jan), 993–1022 (2003)
4. Chu, Z., Gianvecchio, S., Wang, H., Jajodia, S.: Who is tweeting on Twitter: human, bot, or cyborg? In: ACSAC, pp. 21–30. ACM (2010)

5. Cresci, S., Di Pietro, R., Petrocchi, M., Spognardi, A., Tesconi, M.: The paradigm-shift of social spambots: evidence, theories, and tools for the arms race. In: The Web Conference, pp. 963–972 (2017)
6. Davis, C.A., Varol, O., Ferrara, E., Flammini, A., Menczer, F.: Botornot: a system to evaluate social bots. In: The Web Conference, pp. 273–274 (2016)
7. Khaund, T., Al-Khateeb, S., Tokdemir, S., Agarwal, N.: Analyzing social bots and their coordination during natural disasters. In: Thomson, R., Dancy, C., Hyder, A., Bisgin, H. (eds.) SBP-BRiMS 2018. LNCS, vol. 10899, pp. 207–212. Springer, Cham (2018). https://doi.org/10.1007/978-3-319-93372-6_23
8. Kudugunta, S., Ferrara, E.: Deep neural networks for bot detection. Inf. Sci. **467**, 312–322 (2018)
9. Lee, K., Eoff, B.D., Caverlee, J.: Seven months with the devils: a long-term study of content polluters on Twitter. In: ICWSM, pp. 185–192. AAAI (2011)
10. Lee, S., Kim, J.: Early filtering of ephemeral malicious accounts on Twitter. Comput. Commun. **54**, 48–57 (2014)
11. Morstatter, F., Wu, L., Nazer, T.H., Carley, K.M., Liu, H.: A new approach to bot detection: striking the balance between precision and recall. In: ASONAM, pp. 533–540. IEEE (2016)
12. Pedregosa, F., et al.: Scikit-learn: machine learning in Python. J. Mach. Learn. Res. **12**, 2825–2830 (2011)
13. Ratkiewicz, J., et al.: Truthy: mapping the spread of astroturf in microblog streams. In: The Web Conference, pp. 249–252. ACM (2011)
14. Ratkiewicz, J., Conover, M., Meiss, M., Gonçalves, B., Flammini, A., Menczer, F.: Detecting and tracking political abuse in social media. In: ICWSM, pp. 297–304. AAAI (2011)
15. Rijsbergen, C.J.V.: Information Retrieval, 2nd edn. Butterworth-Heinemann, Newton (1979)
16. Varol, O., Ferrara, E., Davis, C.A., Menczer, F., Flammini, A.: Online human-bot interactions: detection, estimation, and characterization. In: ICWSM, pp. 280–289. AAAI (2017)
17. Xie, Y., Yu, F., Achan, K., Panigrahy, R., Hulten, G., Osipkov, I.: Spamming botnets: signatures and characteristics. ACM SIGCOMM Comput. Commun. Rev. **38**(4), 171–182 (2008)
18. Zhang, C.M., Paxson, V.: Detecting and analyzing automated activity on Twitter. In: Spring, N., Riley, G.F. (eds.) PAM 2011. LNCS, vol. 6579, pp. 102–111. Springer, Heidelberg (2011). https://doi.org/10.1007/978-3-642-19260-9_11

A Quantitative Portrait of Legislative Change in Ukraine

Zachary K. Stine[(✉)] and Nitin Agarwal

University of Arkansas at Little Rock, Little Rock, AR 72204, USA
{zkstine, nxagarwal}@ualr.edu

Abstract. Over the past decade, Ukraine has undergone tremendous socio-political changes, which continue to this day. While such changes may be analyzed and interpreted from a variety of sources, we utilize recent advancements in the quantitative analysis of culture to identify how these changes are encoded within Ukraine's legislation. Our goal is to provide a new picture of Ukrainian governance that may be used by subject matter experts as a complement to existing forms of political data. To do so, we apply probabilistic topic modeling to compress over a decade of Ukrainian legislation into patterns of word usage. We then apply a recently developed calculation of novelty to measure how different each draft law is from the draft laws which precede it. We find an interesting pattern of legislative changes and identify some of the drivers of these changes. Finally, we discuss the relationship between our results and the broader context of Ukrainian political changes and suggest steps to explore this relationship further.

Keywords: Computational social science · Political science · Ukraine ·
Topic modeling · Legislation

1 Introduction

As a country, Ukraine comprises an ongoing series of socio-political changes. In particular, the parliament of Ukraine—the Verkhovna Rada—consists of an ever-changing array of political factions in which membership is fluid [1]. In this paper, we focus our analysis of Ukrainian politics between the years 2006 and 2018. Within this time period, multiple events with political salience occurred, including the 2014 ousting of Viktor Yanukovych as president. More broadly, this time period encompasses the fifth, sixth, seventh, and the majority of the eighth convocation, each representing the tenure of a newly-elected parliament.

In this study, we computationally analyze the draft legislation produced in each of the aforementioned convocations in order to examine how changing linguistic patterns within the legislation text might provide a complementary window into the country's political changes during this time. This analysis is made possible by the public availability of documents relating to registered bills on the Verkhovna Rada website[1].

---

[1] https://rada.gov.ua.

© Springer Nature Switzerland AG 2019
R. Thomson et al. (Eds.): SBP-BRiMS 2019, LNCS 11549, pp. 50–59, 2019.
https://doi.org/10.1007/978-3-030-21741-9_6

To conduct our analysis, we apply topic modeling [2] to the corpus of draft laws in order to identify patterns of word usage contained within. We then calculate how novel each draft law is in light of the laws that preceded it, utilizing the measurement of novelty put forward by [3]. We find that periods of elevated average novelty exist which correspond to salient periods of political change within the country. We also show that convocations VI, VII, and VIII are each characterized by distinct trends in average novelty. We show that a series of draft laws related to how elections are conducted account for one distinct period of elevated average novelty, and we identify which parliamentary committees are most responsible for introducing novel legislation. These findings serve to paint a quantitative picture of legislative evolution based on language patterns, which emerge from a massive collection of documents. This resulting portrait serves as a useful complement to traditional political science analysis, providing a view of legislative change that is inaccessible through the close reading of a smaller number of documents.

An examination of legislative novelty is important for many reasons. A bill that is highly novel may represent the introduction of new legislative discourse or a new combination of extant legislative discourses. Such bills may provide early signals of legislative shifts and are therefore likely to be salient for a variety of analyses. Additionally, periods of higher and lower legislative novelty may indicate periods of legislative exploration in which the seeds of new legislative goals are planted or periods of focus on a particular legislative path respectively.

The remaining sections of this paper are organized as follows. In Sect. 2, we provide the necessary background for understanding our methodology and a brief review of other works which analyze political text. In Sect. 3, we describe the methodology used and present our results in Sect. 4. In Sect. 5, we discuss possible interpretations of these results and suggest further steps. A brief conclusion follows in Sect. 6.

2   Related Work

2.1   Topic Models and Political Text

To carry out our analysis, we use the topic modeling algorithm, latent Dirichlet allocation (LDA) [2], to identify a fixed number of word-usage patterns (*i.e.*, topics) in our corpus and represent each draft law as a distribution of these topics. Importantly, LDA can be thought of as operationalizing certain sociological concepts including framing, polysemy, heteroglossia, and a relational approach to meaning [4].

A broad overview of the use of computational methods for analyzing political text is provided by [5], which references examples of topic model extensions used to analyze speeches made in the U.S. Senate [6] and press releases from U.S. senators [7]. As illustrated in these examples, the use of topic models in analyzing political text often takes the resulting topics as the primary outputs for interpretation. In this study, however, the use of topic modeling is primarily a means to transform documents into low-dimensional, semantically useful representations in order to carry out additional calculations.

2.2   Textual Novelty

The methodology we employ centers on the notion of novelty put forward by [3]. In that paper, the authors applied LDA to a corpus of speeches made during the first parliament of the French Revolution. With each speech represented as a distribution of topics, they define and calculate two related measures: a speech's novelty, N, and its transience, T. Both measures are based on the Kullback-Leibler divergence (KLD), which is an asymmetric measure of difference between two probability distributions also known as relative entropy. Novelty can be thought of as a quantity of how surprising a distribution is in light of the past, whereas transience relates to how surprising a distribution is in light of the future. In the present study, we only make use of novelty. The formulation of novelty given in [3] is itself based on a measure of textual novelty used within a cognitive framework describing how an information-seeking agent explores an environment of ideas [8]. A formal definition of novelty is provided in Sect. 3.

3   Methods

3.1   Data

As previously noted, a Verkhovna Rada website is maintained that enables users to view details about registered bills and download relevant documents, including the text of draft laws. For convocations V, VI, VII, and VIII, we download all available documents for each bill that corresponds to a draft law. In some cases, a single bill will have more than one available draft. In such cases, we download all available drafts. Curiously, there are a large number of draft laws from convocation V that do not have any draft documents available. For that reason, we restrict the bulk of our results to the convocations following it (*i.e.*, VI, VII, and VIII). However, the available draft laws from convocation V are still included in our analyses for two reasons: First, they provide useful training documents for constructing topic models, and second, they serve as a backdrop for calculating novelty for the early draft laws in convocation VI. The resulting corpus consists of 17,485 documents representing 17,164 draft law bills over a period of almost 12 years. A convocation-level breakdown of the corpus is given in Table 1.

Table 1. Description of data collected by convocation.

Convocation	Unique draft law bills	Total documents available	Earliest reg. date	Latest reg. date
V	514	533	May 25, 2006	Sep 4, 2007
VI	6,202	6,314	Nov 23, 2007	Dec 6, 2012
VII	3,888	4,015	Dec 12, 2012	Nov 26, 2014
VIII	6,560	6,623	Nov 27, 2014	May 17, 2018

In addition to these draft laws, we also make use of supplemental data generously provided to us by Dr. Tymofiy Mylovanov and his team at VoxUkraine[2]. These supplemental data include voting results (positive or negative) and the main committee from which the draft law originated. We were able to cross-reference our analysis with 2,974, 701, and 1,144 draft laws from the supplemental data for convocations VI, VII, and VIII respectively.

3.2 Topic Modeling

After collecting the text of each available draft law, we then use LDA (as implemented in [9]) to infer some number of topics in order to represent each draft law as a distribution of topics. To do so, we perform some minimal preprocessing of the text. We tokenize the text and remove all punctuation. We remove a limited number of stopwords (*i.e.* functional words that carry little semantic information). This is done primarily for convenience when manually inspecting topics, given that stopword-removal ultimately has a superficial effect on topics [10]. We do not perform stemming, which has also been shown to have little effect on topic models and may even have negative effects [11].

In some cases, it may make sense to evaluate the number of topics chosen, k, in order to find an optimal value, where what constitutes 'optimal' is likely contingent on what question is motivating the use of topic modeling. Here, however, we are not necessarily concerned with precise interpretations of each topic, but rather with topic modeling as a useful way of reducing the dimensionality of the documents. We therefore explore multiple choices of k in order to assess how sensitive our results are to each choice. Here, we train topic models with 20, 40, 60, 80, and 100 topics.

3.3 Novelty

Once topic modeling has been performed, the novelty of a given draft law can be calculated following the methods in [3]: for some number of preceding documents, the novelty of a document's topic distribution, d, is the average of the KLD from each preceding document's topic distribution to d. In the case of draft laws, a particular law's novelty represents how surprising that law's topics are, given an expectation of the preceding laws' topics. The number of preceding draft laws defines the scale at which novelty is computed and is denoted as the window width, w, a positive integer. The smallest possible scale, $w = 1$, is simply the KLD of draft law, $d^{(j)}$, relative to draft law $d^{(j-1)}$, which is given by

$$KLD\left(d^{(j)}|d^{(j-1)}\right) = \sum_{i=1}^{K} d_i^{(j)} \log_2 \left(\frac{d_i^{(j)}}{d_i^{(j-1)}}\right). \qquad (1)$$

---

[2] https://voxukraine.org.

For w greater than 1, the KLD is calculated between $d^{(j)}$ and each of the preceding w documents then averaged, so that the novelty of the jth draft law, represented by its topic distribution, is given by

$$\mathcal{N}_w(j) = \frac{1}{w}\sum_{s=1}^{w} KLD\left(d^{(j)}|d^{(j-s)}\right). \tag{2}$$

Just as we explore various choices of k to understand how sensitive our results are to the number of topics, we similarly explore different choices of w to see the effects of the scale at which novelty is computed. In this study, we calculate novelty using $w = 50, 100, 200, 400, 800, 1,000,$ and $2,000$.

We order draft laws based on their registration dates, representing a bill's formal birth into the legislative process. For days on which multiple bills were registered, we further order them based on their assigned bill number (however, this becomes more or less irrelevant for the scales at which we compute novelty).

4   Results

When comparing convocations VI, VII, and VIII in terms of novelty, we find that the mean novelty of convocation VI is greater than both VII and VIII. This finding is robust to changes in k and w (see Table 2).

Table 2. Comparison of novelty mean (std. deviation) for different values of k and w.

Convocation	$k = 20$		$k = 100$	
	$w = 100$	$w = 800$	$w = 100$	$w = 800$
VI	**5.541** (2.333)	**5.465** (2.322)	**8.327** (2.373)	**8.345** (2.357)
VII	4.916 (2.405)	4.872 (2.375)	7.525 (2.566)	7.433 (2.508)
VIII	4.997 (2.352)	4.999 (2.352)	7.613 (2.471)	7.628 (2.474)

In order to identify periods of especially high novelty over the ordering of bills, we plot each bill's novelty along with a moving average line (Fig. 1). For bills that have multiple drafts, we only include the draft which has the highest novelty. This is because, given a law that has a draft with high novelty, each other draft is likely to also have high novelty. Including both drafts in this case would artificially inflate the average novelty.

When examining the moving average of novelty within a window of 100 laws, we see several interesting features. The bills registered at the beginning of both convocation VI and VII have elevated average novelty. This is especially true for the beginning of convocation VII. However, convocation VIII does not display elevated average novelty within its first few bills, but rather at the very end of 2017. This basic pattern—elevated average novelty at the beginnings of convocations VI and VII and in late 2017 of convocation VIII—is robust to changes in k and w (see Fig. 2).

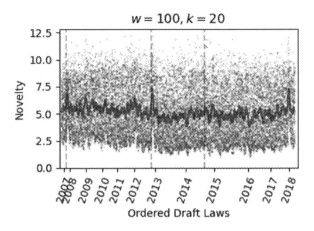

Fig. 1. A scatterplot of each draft law's novelty calculated at the scale of $w = 100$ and with $k = 20$ along with a moving average line of novelty values from 100 draft laws. The vertical dotted lines, from left to right, represent the first bill registered in convocation VI, VII, and VIII.

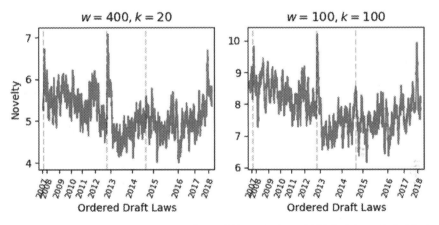

Fig. 2. Two examples of moving average lines within a window of 100 draft laws. Left: Moving average for novelty calculated at a scale of $w = 400$ and $k = 20$. Right: $w = 100$ and $k = 100$. While differences exist between the two, several features are stable across different values of w and k. Vertical dotted lines indicate the first bill of a new convocation.

In addition to these three peaks of average novelty, we also find that the average remains fairly elevated throughout much of convocation VI, which accords with our finding that convocation VI comprises higher novelty bills on average than convocations VII and VIII (see Table 2). What is notable when examining the moving average is that, after the peak in average novelty at the beginning of convocation VII, the average drops off significantly. After this drop, the average novelty oscillates between slight increases and more drops. Each drop in novelty signifies the presence of bills that exhibit similar topic distributions. This drop in average novelty is also robust to changes in k and w (Fig. 2).

While changes in legislative novelty are interesting in of themselves, it may also be of interest to examine precisely which bills and their corresponding topics most account for certain periods of interest. For example, we find that the period of elevated average novelty among bills registered in late 2017 is partly due to several draft laws on the subject of how elections are conducted (*e.g.,* bill 7366-1). Voters in Ukraine may cast their vote for a particular party without being provided a full list of candidates from the party, which has become the subject of a debate about whether such party lists should be made open[3]. These high-novelty draft laws are concerned precisely with this debate. Importantly, these laws were identified solely by their high novelty without prior knowledge of their political salience. While this may not necessarily be the case for every draft law with high-novelty, this example does suggest a link between legislative novelty and salience.

For the subset of bills for which voting and committee data was available, we make two comparisons. First, we compare the novelty of draft laws with positive voting results to those with negative voting results for both $k = 20$ and $k = 100$ with $w = 100$ in each case. We find that for both values of k, the mean novelty of draft laws with positive voting results is slightly less than the mean novelty of draft laws with negative results. However, this difference is much more pronounced for $k = 20$ than for $k = 100$ (Table 3).

Table 3. Comparison of novelty mean (std. deviation) for draft laws with voting results.

	Positive & negative results	Positive results	Negative results
$k = 20$	5.253 (2.447)	5.095 (2.407)	**5.616** (2.500)
$k = 100$	8.4015 (2.510)	8.4014 (2.529)	**8.4017** (2.468)

Second, we compare the novelty of draft laws based on which main committee produced the law. Here, we find that the choice of k is important. When comparing the ordering of each committee by its mean novelty for $k = 20$ and $k = 100$, we find fairly similar rankings of the committees except for two extreme cases: the two committees with the lowest mean novelty at $k = 20$ become the two most novel committees for $k = 100$. This is because when only 20 topics are available, the topics relevant for the types of bills produced by these two committees become subsumed by a broad and more generic topic. However, as the topic-granularity becomes finer with increasing values of k, topics more specific to these committees emerge as their own topics. Thus, we feel confident that the ordering based on $k = 100$ is much more informative than $k = 20$.

Excluding committees with fewer than ten bills available for analysis, the two committees with the highest average novelty are the Committee on European Integration (11.148 bits) and the Committee on Foreign Affairs (10.996 bits) (see Fig. 3). The Committee on Budget has the lowest mean novelty of 6.247 bits.

---

[3] We thank Dr. Tymofiy Mylovanov for pointing this out to us.

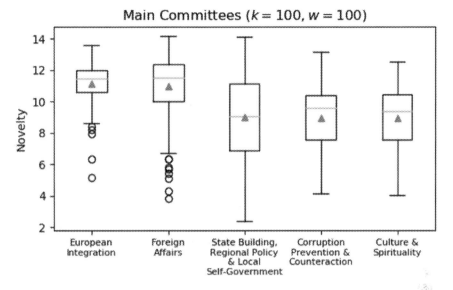

Fig. 3. A comparison of novelty distributions among the bills of the five main committees with the highest mean novelty values. Horizontal lines within the boxes denote the distribution median and triangles denote the mean. The committees are ordered from highest mean novelty on the left.

5 Discussion

Our findings suggest several interesting interpretations about the unfolding of the Ukrainian legislative process over the past decade. Each of the findings we have described illustrate how the analysis of legislative novelty may provide a complementary window into Ukrainian politics that is useful to political scientists. We therefore offer the following interpretations of our findings in the hope that they stimulate more in-depth analysis by political scientists and serve as an example of what types of interpretations become possible when analyzing legislative novelty.

First, we found that convocation VI comprises higher-novelty draft laws on average than convocations VII and VIII. This suggests that convocation VI dealt with a large number of newly encountered issues or dealt with extant issues in novel ways. Additionally, the elevated average novelty of convocation VI may signal the exploration of multiple legislative paths. The sharp increase in average novelty coinciding with the beginning of convocation VII may indicate a severe departure in legislative goals from convocation VI. This is supported by the fact that this peak in average novelty remains severe even at the scale of $w = 2,000$. The observed drop in average novelty directly after this peak indicates that, once the legislative direction changed at the outset of convocation VII, it stabilized. This is because low average novelty indicates repetition of the legislative topic distributions–the newly established legislative direction continues to be followed for a time. Interestingly, we see the average novelty begin a gradual ascent leading into the 2014 Ukrainian Revolution. Following this ascent, the oscillations in average novelty in convocation VIII indicate periods of fixed legislative themes (low novelty) interspersed with periods of legislative change, culminating in the previously described peak in late 2017.

Second, we found that, among a subset of draft laws for which voting data was made available, there is a slight bias towards passing draft laws with lower-than-average novelty. Such a bias indicates a possible reluctance to pass bills that constitute severe departures from established legislative norms. However, this bias appears to diminish for larger values of k, so further analysis of voting data is needed.

Third, we found that, of the main committees in which draft laws originated, the Committee on European Integration and the Committee of Foreign Affairs are responsible for the highest novelty bills on average (again, for a subset of bills). This is notable in light of the 2013 protests which would lead up to the 2014 revolution and eventual ouster of President Viktor Yanukovych. The protests were initially motivated by the decision to break association talks with the European Union, widely seen as a capitulation to Russian interests[4]. The high average novelty of these committees suggests that they have been drivers of legislative innovation and change across these convocations. In other words, the greatest legislative changes undergone by Ukraine largely deal with how the country has managed its relationships abroad. While this interpretation may strike some as obvious, it is important to note that these committees were identified purely through quantitative means.

Several weaknesses exist in this current study, which we intend to address in future work. First, we have only considered bills representing draft laws. While draft laws constitute a great deal of all Ukrainian bills, the inclusion of other bill types will provide an even bigger picture of Ukraine's political evolution. Additionally, we will continue to increase the number of extra-textual features incorporated from each bill (e.g., bill sponsorship, supporting committees, etc.). Finally, we intend to more closely collaborate with relevant subject matter experts in order to bolster the interpretations of our results.

6   Conclusion

As the textual artifacts of a complex political process, Ukrainian draft laws encode the paths explored through a political space on the part of the Verkhovna Rada. By condensing each draft law into a distribution of inferred topics, we can measure how surprising a given law is relative to some number of preceding laws using the notion of novelty from [3]. When we analyze approximately twelve years of draft laws in this way, an interesting picture of Ukrainian political evolution emerges: a period of high average novelty throughout convocation VI, a sharp increase in average novelty followed by low average novelty throughout convocation VII, and a gradual increase in average novelty throughout convocation VIII reaching a crescendo in late 2017. We also see that, of the parliamentary committees, those that are most responsible for driving legislative changes are those that deal with how Ukraine relates to other countries.

---

[4] For an example of how this was reported in American media, see https://www.npr.org/sections/thetwo-way/2013/11/25/247184300/ukraine-protests-continue-over-suspension-of-eu-talks.

Acknowledgements. The authors wish to thank Dr. Tymofiy Mylovanov for explaining the political context of certain bills and for the supplemental draft law data provided by him and his team at VoxUkraine. This research is funded in part by the U.S. Office of Naval Research (N00014-17-1-2675) and the Jerry L. Maulden-Entergy Endowment at University of Arkansas – Little Rock. Any opinions, findings, and conclusions or recommendations expressed in this material are those of the authors and do not necessarily reflect the views of the funding organizations. The authors gratefully acknowledge the support.

References

1. Herron, E.S.: Causes and consequences of fluid faction membership in Ukraine. Europe-Asia Stud. **54**(4), 625–639 (2002)
2. Blei, D.M., Ng, A.Y., Jordan, M.I.: Latent Dirichlet allocation. J. Mach. Learn. Res. **3**, 993–1022 (2003)
3. Barron, A.T.J., Huang, J., Spang, R.L., DeDeo, S.: Individuals, institutions, and innovation in the debates of the French Revolution. PNAS **115**(18), 4607–4612 (2018)
4. DiMaggio, P., Nag, M., Blei, D.: Exploiting affinities between topic modeling and the sociological perspective on culture: application to newspaper coverage of U.S. government arts funding. Poetics **41**, 570–606 (2013)
5. Grimmer, J., Stewart, B.M.: Text as data: the promise and pitfalls of automatic content analysis methods for political texts. Polit. Anal. **21**(3), 267–297 (2013)
6. Quinn, K.M., et al.: How to analyze political attention with minimal assumptions and costs. Am. J. Polit. Sci. **54**(1), 209–228 (2010)
7. Grimmer, J.: A Bayesian hierarchical topic model for political text: measuring expressed agendas in Senate press releases. Polit. Anal. **18**(1), 1–35 (2010)
8. Murdock, J., Allen, C., DeDeo, S.: Exploration and exploitation of Victorian science in Darwin's reading notebooks. Cognition **159**, 117–126 (2017)
9. Řehůřek, R., Sojka, P.: Software framework for topic modelling with large corpora. In: Proceedings of the LREC 2010 Workshop on New Challenges for NLP Frameworks, pp. 45–50 (2010)
10. Schofield, A., Magnusson, M., Mimno, D.: Pulling out the stops: rethinking stopword removal for topic models. In: Proceedings of the 15th Conference of the European Chapter of the Association for Computational Linguistics: vol. 2, Short Papers, pp. 432–436. ACL, Spain (2017)
11. Schofield, A., Mimno, D.: Comparing apples to apple: The effects of stemmers on topic models. Trans. Assoc. Comput. Linguist. **4**, 287–300 (2016)

Synthesizing Machine-Learning Datasets from Parameterizable Agents Using Constrained Combinatorial Search

Victor Hung[(✉)], Joshua Haley, Robert Bridgman, Norb Timpko, and Robert Wray

Soar Technology, Inc., Ann Arbor, MI 48105, USA
{victor.hung, joshua.haley, robert.bridgman,
bob.timpko, wray}@soartech.com

Abstract. The tedious, often hand-modeled, activity of designing and implementing simulation scenarios can benefit from modern-day data-driven methods, i.e., machine-learning (ML). We envision a toolchain that exploits information obtained during live operations, such as the observed maneuvers, techniques, and procedures of all interacting players in live operational settings, that serves as input into an ML-based scenario authoring process. We present a mechanism, called the Parameter Diversifier (PD), that takes a base scenario structure and synthesizes the comprehensive datasets needed for the supervised machine-learning of a scenario authoring model. The design of the PD explores and exploits low-level agent state search space as it relates to it high-level implications at the scenario level. This work demonstrates an explicit sampling of the scenario parameter search space to build an implicit model for use in simulation scenario generation.

Keywords: Scenario generation · Machine-learning · Behavior modeling

1 Introduction

Distributed training scenarios define the simulation environment settings needed for an exercise that reflects learning objectives. Designing these scenarios has customarily been a painstaking, detail-oriented task, as it is largely a manual process performed by a subject matter expert. [4] However, utilizing data-driven methods such as statistical machine-learning (ML) can provide a less labor-intense scenario building process that shifts many steps in scenario construction to automated generation.

We are working toward an automated simulation Scenario Generation (SGen) capability that can leverage data captured from operational settings and generalize and transfer these observations to the simulation realm. Specifically, we aim to create an ML infrastructure to intake real-world data and output a simulation scenario product. This approach would reduce the need for specialized expertise to create complex, realistic scenarios and the time needed to develop and test such scenarios.

In support of the longer-term goal, we are focusing our efforts on the challenge of generalization, which involves the existence of scenario element recognition models

R. Thomson et al. (Eds.): SBP-BRiMS 2019, LNCS 11549, pp. 60–69, 2019.
https://doi.org/10.1007/978-3-030-21741-9_7

that can automatically identify and capture high-level events and behaviors from empirically observed activity. Our SGen effort directly aligns with this vision, employing a supervised ML dataflow that processes a statistically large corpus of tagged scenario data samples to produce a model of behavior recognition. The tag for each these samples is a machine-interpretable encoding of a high-level entity behavior that a human trainee can identify in the simulation. Given enough examples of these tagged features, we can ultimately build a model of entity behavior recognition that can be employed for automated simulation in SGen.

To achieve the required volume of data needed to drive the ML modeling process, we developed the Parameter Diversifier (PD). Inspired by previously established generative techniques [9], the PD synthesizes a large tagged dataset to aid in the construction of a predictive model. It performs an exhaustive exploration of the universe of low-level entity state scenario parameters to provide a sufficiently significant number of data samples with a suitable distribution of high-level scenario feature type representation. Entity state parameters are used to instantiate an entity behavior representation, which are embodied as hierarchical task networks (HTN).

The PD outputs datasets that are packaged as parameterized initial conditions, which are executed in the simulation environment. This environment includes not only physical simulation but also behavioral simulation. It includes sophisticated Semi-Automated Force (SAF) behavior representations covering many entity types and exercise goals, increasingly created by personnel with expertise in these platforms and tactics. The SAFs represent thousands of hours of expert contribution embedded within the simulation environment's behavior capabilities. Thus, importantly, the PD seeks to leverage the existing behavioral capabilities of the simulation and that prior investment, and, in addition to the physical simulation, to produce new training scenarios.

2 Related Work

Scenario Generation has been a recent focus of research in the behavior representation community. From the work of our predecessors, we identify two classes of scenario authoring approaches towards data-driven SGen: manipulation and synthesis.

SGen methodologies that employ *manipulation* of scenario data directly change the values of characterizing scenario variables. By identifying these elemental data features, a scenario author can utilize data-driven methods to produce new, non-trivial training content. [11] A common approach to scenario data manipulation is the *transform* method. Scenario authors can generate new scenarios by directly altering values found in a base scenario. Tomizawa [12] automated this transform process by adding a storyboard-based user interface that processed an instructor's high-level mission objectives to generate the low-level details for a scenario, using a rule-based, slot-filling methodology. The takeaway from this work is that a base scenario can be transformed to generate novel variations. The process in which these transforms are planned or discovered is a challenge that we address with the PD.

Other manipulation approaches have focused on varying two particular scenario structures: *initial conditions* and *cognitive interactions*. Researchers have identified these scenario data elements as the more impactful parameters that define a scenario's novelty. [7] Initial conditions dictate the position of entities, obstacles, checkpoints, waypoints, and weapons within the simulation's environment. These data elements can easily and directly be modified via direct manipulation of existing entities. Cognitive interactions encompass the more abstract features of a scenario, such as the rules by which simulation entities behave. Cognitive interactions are especially relevant to the PD; ML requires superficially different simulation scenarios whilst maintaining a consistency in cognitive interaction goals. In this sense, we understand how these cognitive interaction data elements are directly related to the high-level behaviors that an entity exhibits in a simulation.

Encapsulating agent cognitive interaction rules into a data structure was further explored by Wallace [13] by way of hierarchical behavior representations (HBRs). Given the similarity of HTNs to HBRs, we can leverage Wallace's findings by harnessing his understanding of the relationship between entities' state data and behavioral representations. From this work, we can envision a function that that transforms agglomerations of low-level entity state data into compact high-level behavior representations.

Synthesis SGen methods attempt to discover effective scenario data elements. In contrast to manipulation methodologies, they require little knowledge about existing scenarios. Synthesis typically employs data-driven methods to generate scenario components for SGen purposes. Jennings-Teats [6] procedurally generated game element content in a 2-D platformer using a machine-learned model of level difficulty. Sorenson and Pasquier [10] researched dynamic 2-D game level design with a genetic algorithm (GA) to drive variation into the game element parameters. Zook et al. [16] presented a GA-based military simulation effort that applied combinatorial optimization as the means to select and order events in a training scenario. These synthesis efforts illustrate various computational approaches to automatic scenario development. A common thread is the need for a consequential amount of domain knowledge for generating useful results in a reasonable amount of time.

We recognize the value of both manipulation and synthesis SGen efforts, and we understand that manipulation methods can benefit from the automation aspects seen in synthesis. Synthesis methods can be improved by reducing their required amount of a priori domain knowledge, whereas manipulation systems have been implemented to simply require a base example scenario. The PD design embraces the melding of manipulation and synthesis methods, as it leverages a small amount of quantitatively defined expert knowledge to perform a computationally intense scenario data ML modeling activity.

3   A Representative Domain

We frame our SGen work within the domain of a distributed simulation environment and its associated exercise scenario underpinnings. A distributed simulation environment provides a significant foundation for achieving this goal for military training, and it typically provides a high-fidelity simulation of the operational characteristics of friendly force and oppositional force capabilities.

We specify *scenario* to mean the sequence of events that a trainee will experience when undertaking an authored/designed training experience. This use of *scenario* is distinct from a scenario definition data structure or *initial conditions* that define how a simulation should initialize itself for a particular training exercise. [3] The PD will produce scenario definition files, but the overall scenario authoring goal is to produce a sequence of events, the scenario, that is consistent with the captured operational experience while also allowing a trainee "free play" that allows variation from the events captured in the data.

Producing an appropriate sequence of events, rather than initial conditions only, is a challenge for most simulation systems, as they offer few "levers" for manipulating the sequence of events. In our approach to the PD, we assumed that the scenario file provides a way to specify events and actions that conditionally should be undertaken during execution. Such scenario adaptation capability [14] can be used to dynamically modify the execution of individual behaviors within a scenario. Scenario adaptation plays a critical, enabling role that allows scenario synthesis to specify different components of simulation entity behaviors to be used at different times during scenario execution.

As noted above, using existing SAF behavior representations could significantly reduce the computational requirements for synthesis. However, in order to use the HTNs (and especially sub-components within those networks), the SGen system needs to have some understanding of what those behavior representations do and how they are composed or how they could be decomposed. For example, imagine a monolithic "Automobile" platform model that did not encapsulate subsystems within the "Automobile" such as brakes, exhaust, acceleration, etc. If a SGen system wanted to use this monolithic "Automobile" model, it would either have to use it only as exactly specified, or it would need to perform some experimentation to determine how the monolithic model was put together and could be stimulated to produce various desired effects.

This analogy roughly summarizes the challenge of developing machine understanding of behaviors for the simulation environment in order to use them for SGen. In light of this, we have developed an ML pipeline designed to help explore and to understand the role of specific HTNs. Simulation behaviors along with SME inputs, are used to create a large collection of scenario files. The resulting scenario files are executed in the simulation engine. Key events and execution details are captured as the scenario executes, and they are saved in a database of logged scenario traces. As a group, these logs are then parsed and condensed into a form useful for ML. A deep learning classifier then builds a characterization of the HTN that can be employed in SGen [5].

4 Requirements Analysis

We are working within a simulation environment that provides a rich library of prebuilt behavior representations that roughly can be described as HTNs. [1] These representations enable entities in a simulation to perform in a realistic manner. An HTN consists of a network of primitive *tasks* (orders to perform an action) and *preconditions* (tests to determine if certain conditions have been met). Edges define links between tasks (collectively, the "nodes" in the network). By weaving sets of these behavioral primitives together, simulation entities produce contextually appropriate behavior in support of realistic and appropriate training.

Tasks and preconditions are parameterized through the use of HTN Parameters. These parameters empower simulation users with the ability to tailor certain aspects of a training scenario. Simulation HTNs thus provide the "raw materials" for use in synthesizing the behaviors that would be used in a new scenario. However, non-trivial human analysis would be needed to transform real-world data into its simulation scenario counterpart, under the constraints put forth by these HTNs and the simulation environment itself. This often happens when some events that could occur in any observed situation might differ in training-specific ways from the available HTNs. For example, an observed entity might exhibit movements in specific conditions that are not encapsulated within a single HTN. A human analyst would need to manually extract and repurpose HTNs (and fragments of HTNs) to achieve a scenario that reproduced the observed event. Because this mapping is costly and slow, it is relatively rare for observed situations that could be relevant to current training to make it back into training scenarios. We want to automate this mapping, a task that requires machine understanding of the role individual HTN Parameters (and various values of those parameters) play in an HTN and, thus, across various scenarios in which that HTN is used.

HTNs are constructed with predefined values for their Parameters. Users can modify these values to see slightly different behavior execution. However, making specific HTN Parameter changes is largely a manual effort. An alternative approach to exploring this mapping challenge is to develop an ML pipeline that can read all of the tasks, preconditions, and associated HTN Parameters for any entity, along with other supplemental information about a scenario, and observe the simulation output created by relationships. The resulting ML-based model will effectively map the sequences of low-level entity states to the particular high-level behavior that produced the data track.

The *behavior space* produced by a single HTN is the union of all behavioral event sequences that can be produced by that graph for any valid set of Parameters and values for that graph. To model this entity-state-sequence to behavior-characteristics mapping, we must fully explore the space exposed by the Parameters associated with the HTN. The PD, which takes in a base scenario and its accompanying behavior set to perform a combinatorial exploration across all of the HTN Parameters associated with those behaviors, is the software utility we developed to meet that need.

5   Parameter Diversifier

To generate the HTN mappings outlined in the previous section via ML, we need samples that reflect how HTNs "behave" with various settings and under various conditions. The PD generates HTN Parameter assignments within HTN initial conditions across a scenario. The output of the PD is a set of scenarios. Executing these scenarios will result in a wide or "diverse" set of behavioral traces. These traces will typically only provide a sample of all possible scenarios that could be generated from a given collection of HTNs in the scenario. The SGen model will use these "samples" of the behavior traces to learn a characterization of the HTN.

A scenario configuration and launch component executes each of the scenario files generated by the PD. Within the simulation environment, the HTN executes the decision-making on behalf of individual simulation entities (e.g., an automobile) within an executed scenario as defined by the PD. Entity state (e.g., location, speed, and heading) and HTN state (task/precondition executed) are automatically captured in a generic messaging infrastructure within the simulation environment; a "sniffer" captures records of these events as they occur and saves them to a database. This trace of the events (after some transformation) is used as input to the ML system to learn a characterization or summary of the HTN. Thus, the PD is critical for producing sufficient variation to support effective learning, but it also needs to generate representative examples that attempt to minimize the number of scenario executions needed to create the individual sample traces used by learning.

5.1   Illustrative Example

In order to describe the approach for machine-learned behavior characterization, we present an example behavior that contains a set of specific actions (specific relationships of tasks and preconditions) with exposed HTN Parameters. These parameters serve as "levers" that the PD algorithm will manipulate in specific variations of the general behavior or maneuver.

We identify two example actions for an automobile: *Pass* and *Follow*. *Pass* has the HTN Parameter *DistanceThreshold*, which represents a safe following distance. When this threshold is crossed, a lane change occurs based on the value of *WheelTurnAngle*. Next the automobile will drive straight in its new lane, with a relative passing speed that falls between the value of *PassLowThresh* and *PassUpperThresh*, at which point it will move in front of the other automobile to complete the passing. The simpler *Follow* maneuver has the HTN Parameter *ManeuverStart* that tells the automobile to follow and hold its position at a safe relative distance (*FollowDistance*).

Consider an example HTN that can generate several related maneuvers: *Pass*, *SlowPass*, *DoubleLanePass*, *Follow*, and *SlowFollow*. These maneuvers are a subset of an HTN that can generate these maneuvers. Tasks and preconditions control the execution of each maneuver. By design, the same types of tasks and preconditions are used across different maneuvers; the difference in behavior between the maneuvers lies in the values of the HTN Parameters and the node control flow designed for each maneuver.

The goal of the PD is to generate scenario files that specify various values for HTN Parameters used in these maneuvers such as *ManeuverStart, PassLowThresh, FollowDistance,* etc. For any individual scenario generated by the PD, some of these HTN Parameters may not be relevant (e.g., *PassUpperThresh* is only relevant to a scenario in which the *Pass* maneuver will be used). The PD uses subject-matter input to specify what values are relevant to what (classes of) scenarios as well as ranges for the parameters. A range for one value can depend on another value. For example, the *PassUpperThresh* range is dependent on the *PassLowThresh* range.

```
HTN Parameter Configuration Knowledge:
1. Param1 has values {true, false} when Param2 equals true
2. Param2 has values {true, false}
3. Param3 has values {0, 2, 4, 6}
4. Param4 has value {'A', 'B'} when Param3 equals 2 and Param2
equals false
```

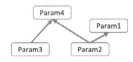

Fig. 1. HTN parameter configuration (left) and Dependency graph (right)

5.2   HTN Parameter Diversification

For each scenario in the simulation environment, the scenario file holds all of the information necessary to reproduce the simulation, including the entities and their associated behaviors, and the (initial) value of all of the Parameters used by the HTNs. This specification thus provides the information required for training an ML algorithm. The algorithm just needs to be able to vary the values of each Parameter and observe the resulting data from runs of the simulation against those scenario files.

Starting with a base scenario, the PD will exhaustively assign values from a pre-specified operationally relevant range for each HTN Parameter. These ranges are defined in a configuration file, which maintains these values in an XML format. Continuous variables are discretized by having the SME specify step size as well as minimum and maximum bounds. Figure 1 (left) gives a notional example of knowledge that can be derived from a configuration file, given a scenario with four HTN Parameters.

Although the PD is equipped to generate a separate scenario that covers every single combination of possible HTN Parameter values, there are some instances where a Parameter value does not need to be evaluated. For example, the HTN Parameter *Param1* is not relevant whenever *Param2* is set to a "false" value. In this case, we can discard generating any scenario variations over the range of *Param1* whenever *Param2* equals "false". Figure 1 (right) gives an example of a dependency graph that reflects the conditional infrastructure dictated by the Fig. 1 (left) HTN Parameters.

Table 1. HTN Parameter diversification

Diversification of P
function **Diversify** (L, V, T)
pop *head* of L into I *(variable of interest)*
set Satisfied to false
for each dependency d in D$_I$
if d is satisfied, set Satisfied to true
if Satisfied is true **then**
for each value v in V$_I$
set V[I] to v
if L is empty **then**
write scenario for variation
else
call **Diversify** (L, V, P)
add I to *head* of L
else
set V[I] to NR
if L is empty **then**
write scenario for V
else
call **Diversify** (L, V, P)

To improve diversification and to limit complexity, the PD algorithm utilizes *Not Relevant* (NR) values whenever a failure in the conditional dependencies required for an HTN Parameter's need for diversification is detected. Using Kahn's algorithm on the HTN Parameter dependency graph (Fig. 2, right), a partial ordering of the Parameters is produced to properly assign NR values in a systematic manner. For the example, in Fig. 2, we compute a partial ordering of *Param3, Param2, Param1, Param4*.

After determining the topological sort of the HTN Parameters to handle NR assignments, the PD is now ready to effectively generate a pruned version of the Cartesian product of its Parameters' values (over the ranges specified by the configuration file). The PD's diversification algorithm is dictated by conditional rules that relate HTN Parameters with one another, as per their dependency relationships, which allows us to utilize NR value assignments to prune the search space. The pseudocode listed in Table 1 depicts the final step of HTN Parameter assignment in the diversification algorithm, given a partially ordered HTN Parameter list (L), the list of HTN Parameters (P), and the HTN Parameter set value assignments (V).

5.3   Machine Learning Artificial Biasing

We identified two optimizations in the ML workflow that dealt with artificial biasing during the supervised training phase: (1) random distribution of NR values, and (2) prevention of imbalanced training sets. With the inclusion of Parameter dependencies, we discovered that NR have a bearing on our ML processes as they have no effect on the empirical outcome of the HTN. Initially, we handled NRs by assigning them the same default value. This practice, however, ended up artificially biasing the ML model toward these default values. By uniformly distributing (i.e., Gaussian methods) the actual values of these NRs in the PD's generated scenarios, we were able to prevent this artificial value biasing effect.

We also noted that an imbalance of training data examples may occur when utilizing the PD's exhaustive search process. The number of behavior type samples that will be generated by the PD is directly related to the number of variables, and the range of values for each variable, that are associated with the behavior. Past research has asserted that a sampling imbalance can lead to artificial biasing into the ML workflow. [2] To prevent this effect, we incorporate upsampling to bring balance to the training sets, thereby creating a more uniformly distributed representation of behavior state examples produced by the PD.

5.4   Diversified Scenario Execution

After performing the diversification process on a base scenario file, the PD transforms each HTN Parameter assignment into a unique scenario file. The diversification can continue to iteratively repeat until training of the ML algorithm converges, which typically lasts for thousands of runs. Upon the generation of the diversified and uniformly represented collection of scenarios, each scenario is played out in the distributed simulation environment. The data artifacts from the completion of these scenario executions produced gigabytes of entity track logs, to be used for the eventual ML SGen modeling process, as detailed in Haley et al. [5].

6   Results and Lessons Learned

The PD was an integral component for the ML modeling procedure. The engineering involved in the PD development and its ensuing results brought to light lessons pertaining to ML data requirements and knowledge engineering. First, by utilizing domain knowledge, the PD's constrained combinatorial approach pruned the search space by orders of magnitude, as compared to a naïve approach. Table 2 summarizes this pruning effect. In this table, we consider an HTN consisting of three binary HTN Parameters, each one associated with a continuous parameter. The HTN's Configuration Knowledge dictated that all of the parameters are relevant in a mutually exclusive manner, and it also gave valid ranges and step sizes for the continuous parameters (enumerated A, B, and C). A naïve method would yield 64 million permutations, many of which are non-sensical for the domain.

Utilizing a knowledge-driven approach without the concept of parameter dependencies, we reduce this search space down to 32,000 permutations. This added some measure of valid ranges for the domain and reasonable step sizes to prevent selection of too many trivially unique values. While meaningful, this method still yields a large amount of redundancy.

When factoring the full Configuration Knowledge, including the dependency constraints, we further reduced the number of permutations to fifty. To illustrate this effect, consider an HTN Parameter A that depends upon the value of another Parameter or the values of a set of Parameters. It does not make sense to continue varying Parameter A when its dependency conditions have not been met, making Parameter A's value an irrelevant factor in the grand scheme of the ML modeling process. Given that there is both a time and computation cost associated with each individual permutation, such a reduction is a force multiplier in optimizing the exploration of only meaningful parts of the scenario parameter space.

Table 2. Permutation analysis of HTN Parameters

	Naïve method	Knowledge-based dependency-free method	Full knowledge-based method
Value ranges			
Parameter A	(0, 1000)	(0, 1)	(0, 1)
Parameter B	(0, 1000)	(0, 10000)	(0, 10000)
Parameter C	(0, 1000)	(0, 200)	(0, 200)
Step size {A, B, C}	{5, 5, 5}	{0.1, 500, 10}	{0.1, 500, 10}
Dependencies	None	None	A XOR B XOR C
# Permutations	64×10^6	32×10^3	50

Scenario diversification provides a first step to building an ML model that realizes automated Scenario Generation by mapping the low-level, entity-state event traces that occur in a specific scenario with high-level behaviors that can reproduce those events. The Parameter Diversifier performs this variation process, utilizing the structured nature of scenarios and a small amount of expert knowledge.

Acknowledgements. The authors thank Dr. Heather Priest and Mr. Samuel Parmenter for their contributions to our approach. The opinions expressed here are not necessarily those of the Department of Defense or the sponsor of this effort: Naval Air Warfare Center Training Systems Division. This work was funded under contracts N68335-17-C-0574.

References

1. Erol, K., Hendler, J.N., Nau, D.S.: HTN planning: complexity and expressivity. In: 12th National Conference on Artificial Intelligence (1994)
2. Fernandez, A., Garcia, S., Hernandez, F., Chawla, N.V.: SMOTE for learning from imbalanced data: progress and challenges, marking the 15-year anniversary. J. Artif. Intell. Res. **61**, 863–905 (2018)
3. Folsom-Kovarik, J.T., Woods, A., Wray, R.E.: Designing an authorable scenario representation for instructor control over computationally tailored narrative in training. In: Proceedings of the 29th International FLAIRS Conference. AAAI Press, Key Largo (2016)
4. Graffeo, C., Benoit, T., Wray, R.E., Folsom-Kovarik, J.T.: Creating a scenario design workflow for dynamically tailored training in socio-cultural perception. In: Proceedings of the 2015 Cross-Cultural Decision Making Conference. Springer, Las Vegas (2015)
5. Haley, J., Hung, V., Bridgman, R., Timpko, N., Wray, R.E.: Low level entity state sequence mapping to high level behavior via a deep LSTM model. In: 20th International Conference on Artificial Intelligence, Las Vegas (2018)
6. Jennings-Teats, M., Smith, G., Wardrip-Fruin, N.: Polymorth: a model for dynamic level generation. In: AAAI Conference of Artificial Intelligence and Interactive Digital Entertainment (2010)
7. Juul, J.: Variation over time: the transformation of space in single-screen action games. In: von Borries, F., Walz, S.P., Brinkmann, U., Böttger, M. (eds.) Space Time Play. Birkhäuser, Basel (2007)
8. Mayer, N., et al.: What makes good synthetic training data for learning disparity and optical flow estimation? Int. J. Comput. Vis. **126**(9), 942–960 (2018)
9. Shrivastava, A., Pfister, T., Tuzel, O., Susskind, J., Wang, W., Webb, R.: Learning from simulated and unsupervised images through adversarial training. In: IEEE Conference on Computer Vision and Pattern Recognition (2017)
10. Sorenson, N., Pasquier, P.: Towards a generic framework for automated video game level creation. In: Di Chio, C., et al. (eds.) EvoApplications 2010. LNCS, vol. 6024, pp. 131–140. Springer, Heidelberg (2010). https://doi.org/10.1007/978-3-642-12239-2_14
11. Summerville, A., et al.: Procedural content generation via machine learning (PCGML). IEEE Trans. Games **10**(3), 257–270 (2018)
12. Tomizawa, H.: Automated SGen In A Simulation. Master's thesis, University of Central Florida (2006)
13. Wallace, S.: Behavior bounding: an efficient method for high-level behavior comparison. J. Artif. Intell. Res. **34**, 165–208 (2009)
14. Wray, R.E., Bachelor, B., Jones, R.M., Newton, C.: Bracketing human performance to support automation for workload reduction: a case study. In: Schmorrow, D.D., Fidopiastis, C.M. (eds.) AC 2015. LNCS (LNAI), vol. 9183, pp. 153–163. Springer, Cham (2015). https://doi.org/10.1007/978-3-319-20816-9_16
15. Wray, R.E., Priest, H., Walwanis, M.A., Kaste, K.: Requirements for future SAFs: beyond tactical realism. In: Interservice/Industry Training, Simulation, and Education Conference, Orlando (2015)
16. Zook, A., Lee-Urban, S., Riedl, M.O., Holden, H.K., Sottilare, R.A., Brawner, K.W.: Automated SGen: toward tailored and optimized military training in virtual environments. In: International Conference on the Foundations of Digital Games, pp. 164–171 (2012)

Exploiting Emojis for Sarcasm Detection

Jayashree Subramanian$^{(\boxtimes)}$, Varun Sridharan$^{(\boxtimes)}$, Kai Shu, and Huan Liu

Computer Science and Engineering, Arizona State University, Tempe, AZ, USA
{jsubram5,vsridh19,kai.shu,huan.liu}@asu.edu

Abstract. Modern social media platforms largely rely on text. However, the written text lacks the emotional cues of spoken and face-to-face dialogue, ambiguities are common, which is exacerbated in the short, informal nature of many social media posts. Sarcasm represents the nuanced form of language that individuals state the opposite of what is implied. Sarcasm detection on social media is important for users to understand the underlying messages. The majority of existing sarcasm detection algorithms focus on text information; while emotion information expressed such as emojis are ignored. In real scenarios, emojis are widely used as emotion signals, which have great potentials to advance sarcasm detection. Therefore, in this paper, we study the novel problem of exploiting emojis for sarcasm detection on social media. We propose a new framework ESD, which simultaneously captures various signals from text and emojis for sarcasm detection. Experimental results on real-world datasets demonstrate the effectiveness of the proposed framework.

1 Introduction

Social media plays a major role in everyday communication. While images and videos are common in social media sites such as Facebook[1] and Twitter[2], the text is still dominating the communication. Communication through text may lack non-verbal cues, and *emojis* can provide richer expression to mitigate this issue. Emojis are a set of reserved characters that are rendered as small pictograms that depict a facial expression [1,11]. In social media, sarcasm represents the nuanced form of language that individuals state the opposite of what is implied.

Sarcasm detection is an important task to improve the quality of online communication. First, it helps us to understand the real intention of the user's feedback. For example, user reviews can contain examples such as "Wow this product is great", "It is very fast", "Totally worth it", etc. These comments, however, are being said in a sarcastic tone. Second, sarcastic posts may influence people's emotions and reactions to the political campaign [10].

The majority of existing sarcasm detection algorithms focuses on text information [8].

---

[1] https://www.facebook.com/.
[2] https://twitter.com/?lang=en.

© Springer Nature Switzerland AG 2019
R. Thomson et al. (Eds.): SBP-BRiMS 2019, LNCS 11549, pp. 70–80, 2019.
https://doi.org/10.1007/978-3-030-21741-9_8

These include identifying the traits of the user from their past activities, responses texts, etc. Most of them have tried to train deep neural network models using the text to analyze sarcasm. To overcome the challenges faced by all of these methods and for better performance, the Emoji can be considered to detect sarcasm. Emojis help us to find the tone of speech, the mood of the user and identify sarcasm in a better way. For example, comments such as, "Wow!! This is beautiful 😊 😊", "You can do this, I trust you 😊 😊", "It's big proud 😊" are examples of sarcastic comments. The above comments without the emoji convey us a different meaning and are taken in the positive sense since it has the keywords "beautiful", "proud", "trust", "wow". However, with emoji, they strongly help us to identify the sarcasm in the comments. Human thought process and emotions are best conveyed through Emojis and these emotional signals are much stronger than the text. These emotional signals will help the model to learn more accurately about the intention, thought process of the user than by merely looking at the text.

In this paper, we address the problem of identifying sarcasm in social media data by exploiting Emojis. In essence, we investigate: (1) how to learn the representation of text and emojis separately; (2) how to take advantage of the emoji signals to improve sarcasm detection performance. In an attempt to solve these two challenges, we propose a novel Emoji-based Sarcasm Detection framework ESD, which captures text and emoji signals simultaneously for sarcasm detection.

Our contributions are summarized as follows:

- We provide a principled way to model emoji signals for social media post;
- We propose a new framework ESD which integrates text and emoji signals into a coherent model for sarcasm detection; and
- We conduct experiments on real-world datasets to demonstrate the effectiveness of the proposed framework ESD.

2   Related Work

2.1   Sarcasm Detection

Automatic sarcasm detection is the task of predicting sarcasm in text. It is an important step in sentiment analysis, considering the prevalence and challenges of sarcasm in a sentiment-bearing text [8]. [4] identifies sarcasm using bi-directional recurrent neural network by extracting all the contextual features from the history of tweets. [3] The sarcasm is handled by creating word embeddings for the tweets and is fed to the DNN, CNN, RNN, LSTM models. The performance of these models are compared with each other and it is observed that the combination of CNN + LSTM + DNN gives the highest F1 score. However, the built model fails to classify comments like 'Thank God it is Monday!' as sarcastic comments. [6] This paper investigates how sentiment, emotional, personality features can be combined to detect sarcasm using deep Convolutional neural networks. The sarcasm is identified using the user's past activities, behavioral and psychological features. The SCUBA model performs with an accuracy of 82% with all features when compared to other baseline models [14].

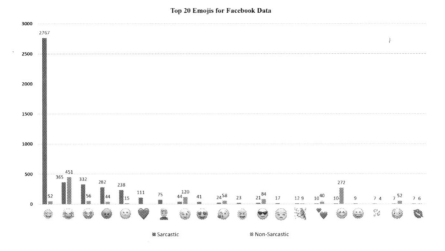

Fig. 1. Comparison of top 20 emojis for Facebook data.

2.2   Emoji Analysis of Social Media

Emojis have become an important tool that helps people to communicate and express their emotions. The study of emojis, as they pertain to sentiment classification and text understanding, attracts attention [1, 9, 12]. Hu *et al.* [7] proposes an unsupervised framework for sentiment classification by incorporating emoticon signals. Hallsmar *et al.* [5] investigates the feasibility of an emoji training heuristic for multi-class sentiment analysis on Twitter with a Multinomial Naive Bayes Classifier. Eisner *et al.* [2] learns emoji representation by running skip gram on descriptions of emojis provided in the Unicode standard. The resulting emoji representation along with the word embedding from Google News are used to perform sentiment analysis and the results show that emoji representation can improve sentiment analysis.

Kelly *et al.* [9] shows that emojis can be used as appropriations which can help facilitate communications using the interview data.

3   Preliminary Analysis of Emoji Usage

Emojis serve as a medium for us to express certain opinions that can't be expressed by our voice or body language. Emojis are the major contributing factor to the improvement in accuracy of our model because the neural network learns the connection between text and emojis. This analysis is performed to research in depth about the types of emojis used across the comments in the Twitter and Facebook data set. This gives us a clear picture of the most frequently used emojis in both sarcastic as well as non-sarcastic comments which in turn helps us to rank emojis based on their count of occurrences in the comments.

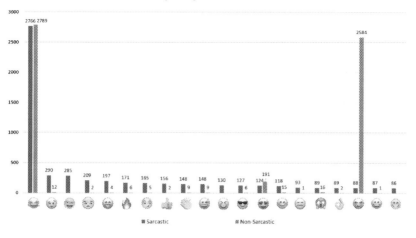

Fig. 2. Comparison of top 20 emojis for Twitter data.

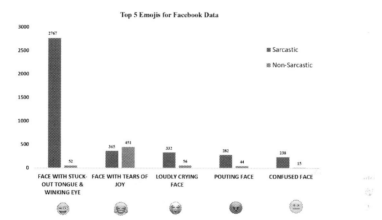

Fig. 3. Comparison of top 5 emojis for Facebook data.

The top 20 and top 5 emojis used in our Twitter/Facebook data are visualized through the graphs. The following insights are obtained from the graphs (Figs. 1, 2, 3 and 4).

– On comparison of emojis used across entire Facebook and Twitter data, the usage of Face with tongue out emoji is the highest (2.7K) among the sarcastic comments. The Face with tears of Joy, Loud crying face (2.6K), Grinning and Pouting face are the three specific emojis that are most frequently used with non-sarcastic comments.

– The number of other emojis used in sarcastic comments like winking face, the smirking face is found to be uniformly distributed across the Twitter data whereas emojis such as the loud crying face, pouting face, the confused face is observed to be uniformly distributed for the Facebook data.

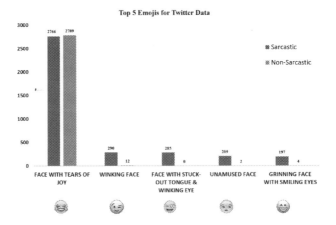

Fig. 4. Comparison of top 5 emojis for Twitter data.

- The usage of Face with stuck out tongue emoji is the first highest for Facebook data and third highest for Twitter data. However, the face with tears of joy emoji is being increasingly used in both sarcastic and non-sarcastic comments across the platforms.
- It is also clearly observed that the amount of Face with tongue out emoji in sarcastic comments is very high which is nearly 54 times its usage in non-sarcastic comments for Facebook data. For twitter non-sarcastic comments, the count of this emoji is in-fact zero. This proves the fact that most of the comments having this emoji are clearly being sarcastic in nature.

4   Text and Emoji Embedding for Sarcasm Detection

In this section, we introduce the details of the proposed framework ESD for sarcasm detection on social media. It mainly consists of three components (see Fig. 5): a text encoder, an emoji encoder, and a sarcasm prediction component. In general, the text encoder describes the mapping of words to latent representations; the emoji encoder illustrates the extraction of emoji latent representations, and the sarcasm prediction component learns a classification function to predict sarcasm in social media posts.

4.1   Text Encoder

The entire dataset, which is basically a list of sentence vectors, is divided into a list of tokens which contains both English words as well as Emojis. In order to get the embedding for the English words alone, we first filter the data with the help of python libraries like Enchant and the NLTK work tokenizer.

The embeddings are obtained from GloVe [13] and are stored in an embedding matrix W. Essentially, an embedding is a mapping from a word to a vector.

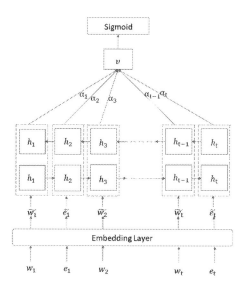

Fig. 5. The proposed framework ESD for sarcasm detection takes a list of words $[w_1, w_2...w_t]$ and emojis $[e_1, e_2, ..e_t]$ as input and converts them into word $[\tilde{w}_1, \tilde{w}_2, ...\tilde{w}_t]$ and emoji $[\tilde{e}_1, \tilde{e}_2, ...\tilde{e}_t]$ embeddings. $[h_1, h_2, ...h_t]$ denotes the list of concatenated vectors which are passed through the bi-directional GRU. The attention weights $[\alpha_1, \alpha_2, ...\alpha_t]$ are then multiplied and summed with the vector representations to give the context vector, v. This vector v is finally passed to the sigmoid function for classification.

Therefore, it is important to store the embeddings in the same order that they occur in the sentence. These embeddings represent the words in a transformed space. This helps our model to capture relationships between words which are not possible otherwise. Given a list of sentence vectors S_i, where each sentence contains T_i words, there exists a word embedding $\tilde{w}_i$ for every word w_i.

$$w_i \rightarrow \text{GloVe}(w_i) \rightarrow \tilde{w}_i \tag{1}$$

4.2   Emoji Encoder

We first extract the emojis from sentences and pass them through certain filters so as to remove redundant characters and retain only the emojis. This is achieved with the help of a python library called Enchant. The embeddings for the filtered list of emojis are retrieved using emoji2vec. These embeddings are then stored in the embedding matrix W along with the word embeddings. The emoji embeddings play a major role in determining sarcasm. These embeddings help us in understanding complex emotions which cannot be derived from words alone. Given a list of sentence vectors S_i, for every emoji e_i, there exists an emoji embedding $\tilde{e}_i$:

$$e_i \rightarrow \text{emoji2vec}(e_i) \rightarrow \tilde{e}_i \tag{2}$$

Table 1. The statistics of datasets

Datasets	Twitter	Facebook
No. of Sarcasm	6, 592	2, 668
No. of Non-sarcasm	10, 267	2, 803

4.3   Sarcasm Detection

Once the embedding matrix W is complete, we pass these embeddings as an input to a bi-directional GRU. The embedding $\tilde{s}_i$ could either be a word embedding $\tilde{w}_i$ or an emoji embedding $\tilde{e}_i$. We utilize the bi-directional GRU to obtain a forward hidden state $\overrightarrow{h_{it}}$ and a backward hidden state $\overleftarrow{h_{it}}$. The concatenated vector $h_i = [\overrightarrow{h_{it}}, \overleftarrow{h_{it}}]$ represents the information of the whole sentence centered around w_i.

$$\overrightarrow{h_i} = \overrightarrow{GRU}(\tilde{s}_i), i \in [1, T] \tag{3}$$

$$\overleftarrow{h_i} = \overleftarrow{GRU}(\tilde{s}_i), i \in [T, 1] \tag{4}$$

Our model uses an embedding layer of 300 dimensions for mapping both words and emojis to vector space. We use the bidirectional GRU with dropout, which takes the vector representation of each word and emoji in the dataset as an input. Finally, an attention layer takes the previous layers as input and weighs each word according to its importance or relevance in the text. The Activation function, Sigmoid and Adam optimizer are used to build the model.

$$u_i = \tanh(W h_i + b_w) \tag{5}$$

$$\alpha_i = \frac{\exp(u_i^T u_w)}{(\sum_i \exp(u_i^T u_w)} \tag{6}$$

$$v = \sum_i \alpha_i h_i \tag{7}$$

Here, u_i is the score obtained by applying a hyperbolic tangent function over the product of h_i, the vector representation of the word, and the weight matrix W, and a bias b_w is added to the product. This score is generated for each state in the bi-directional GRU. v is the final representation vector of the text.

v is also referred to as the context vector. v is the weighted summation of the product of attention weight α_i and word representation h_i. These context vectors are computed for every word and are given as an input to the final sigmoid layer for classification.

5   Experiments

In this section, we present the experiments to evaluate the effectiveness of the proposed ESD framework. Specifically, we aim to answer the following evaluation questions:

- Is ESD able to improve the sarcasm detection performance by modeling text and emoji information simultaneously?
- How effective are the text and emoji features, respectively, in improving the sarcasm detection performance of ESD?

5.1   Datasets

Data was collected from Twitter for a 5-month period and from Facebook for two years from 2015 to 2017 using web scraping in Python[3]. The sarcastic pages such as 'sarcasmLOL', 'sarcasmBro' from Facebook and tweets with hashtags, 'sarcasm', 'sarcastic' were taken from Twitter. The data was scraped, preprocessed and the data containing only text plus emoji were extracted. The data preprocessing involved removal of hyperlinks, special characters, hashtags, retweets, etc. The statistics of the datasets are given in Table 1.

5.2   Comparison of Sarcasm Detection Methods

The representative state-of-the-art sarcasm detection methods that are compared with ESD, are listed as follows:

- FSNN [3]: FSNN stands for Fracking Sarcasm using a Neural Networks, which uses a Convolutional Neural Network (CNN) followed by an LSTM and a Deep Neural Network (DNN) to detect sarcasm in a sentence.
- CASCADE [6]: CASCADE stands for Contextual Sarcasm Detection in Online Discussion Forums. CASCADE uses CNNs to capture the user's personality features to boost the performance of classification.
- RCCSD [4]: RCCSD stands for The Role of Conversation Context for Sarcasm Detection, which uses conditional LSTM networks with sentence-level attention on conversational context and response.

Table 2. Best performance comparison for Sarcasm detection

Datasets	Metric	FSNN	CASCADE	RCCSD	ESD
Twitter	Accuracy	0.891	0.753	0.763	0.991
	Precision	0.910	0.798	0.768	0.998
	Recall	0.904	0.802	0.791	0.976
	F1	0.899	0.867	0.820	0.987
Facebook	Accuracy	0.878	0.745	0.768	0.971
	Precision	0.901	0.771	0.733	0.975
	Recall	0.889	0.789	0.745	0.979
	F1	0.893	0.842	0.772	0.969

---

[3] https://github.com/jsubram/Sarcasm-Detection-Using-Emoji.

5.3   Performance Comparison

We compare ESD with state-of-the-art sarcasm detection methods. The metrics for evaluation are Precision, Recall, F1 and Accuracy. The dataset is split into training, testing, and validation in the ratio 6:2:2. All models are trained for 50 epochs with early stopping and their results are shown in Table 2.

- In general, ESD outperforms other baselines. We use word embeddings, to learn the representations of each word, and emoji embeddings to learn complex sentiments in the sentence that are not easily learned by word embeddings alone. We also add an attention layer to focus on the part of the sentence which has sarcasm, thereby enhancing our model's performance.
- FSNN has the highest Accuracy, Precision, Recall and F1 score amongst all the three baseline models. This is due to the fact that the model architecture of FSNN is much deeper and more complex than the other two models. The good results show that the depth of the neural network helps in the better learning of word representations.
- CASCADE performs slightly better than RCCS because it utilizes CNN to capture complex stylometric and personality features of the user. The results demonstrate the importance of extracting features from sentences in detecting sarcasm.
- RCCSD uses an Attention-based LSTM to model both context and response. The key feature of RCCSD is its attention layer. The attention is used to identify the sarcastic part in the response. This indicates that having an attention layer on top of an LSTM can help our model to focus on the part of the sentence which contains sarcasm.

5.4   Assessing Text and Emoji Components

The Text and Emoji Components are obtained as embeddings using the Word2Vec and Emoji2Vec methods. We test our baseline features on 9 different widely used machine learning algorithms such as Support Vector Machines (SVM), Decision Tree Classifier, etc. From Table 3, we observe the following:

- Only word embeddings: When we train our model with only word embeddings, we observe that the model struggles to learn sarcastic features in the data since it is difficult to infer sarcasm using only words.
- Only emoji embeddings: When we train our model with only emoji embeddings, we observe that the model performs better than it performed with word embeddings. This is because emojis are able to convey complex emotions that are essential to detect sarcasm.
- Both word and emoji embeddings concatenated: The word embeddings and emoji embeddings are concatenated horizontally and are given as input to the model. We observe that the model is able to perform considerably better than it did with only word and only emoji embeddings because the model is able to relate complex emotions with the contextual meaning. This enables the model to detect sarcasm more accurately.

Table 3. The results of average F1 scores

Datasets	Classification algorithm	Text	Emoji	Text+Emoji
Twitter	SVM	0.7674	0.8288	0.8465
	Decision tree classifier	0.7430	0.8463	0.8803
	Random forest	0.7773	0.8522	0.8729
	Adaboost classifier	0.7690	0.8466	0.8964
	Gradient boosting classifier	0.7821	0.8549	0.8876
	K neighbors classifier	0.7699	0.7376	0.8934
	Stochastic gradient descent	0.7804	0.8635	0.8810
	Bayesian classifier	0.7557	0.8454	0.8943
	ExtraTreesClassifier	0.7755	0.8518	0.8936
Facebook	SVM	0.6496	0.9283	0.9302
	Decision tree classifier	0.6460	0.9523	0.9432
	Random forest	0.6809	0.9596	0.9615
	Adaboost classifier	0.7023	0.9578	0.9578
	Gradient boosting classifier	0.7223	0.9597	0.9670
	K neighbors classifier	0.5429	0.8093	0.8751
	Stochastic gradient Descent	0.6361	0.9561	0.9506
	Bayesian classifier	0.6497	0.9189	0.9287
	ExtraTreesClassifier	0.6865	0.9523	0.9597

6   Conclusion

Emojis provide a new dimension to social media communication. We study the role of emojis for sarcasm detection on social media. We propose a new deep learning model by introducing an attention layer which helps to model the text and emojis simultaneously for sarcasm detection. The empirical results on real-world datasets demonstrate the effectiveness of the proposed framework.

Acknowledgements. This material is based upon work supported by, or in part by, the ONR grant N00014-17-1-2605 and N000141812108.

References

1. Barbieri, F., Kruszewski, G., Ronzano, F., Saggion, H.: How cosmopolitan are emojis?: exploring emojis usage and meaning over different languages with distributional semantics. In: Proceedings of the 2016 ACM on Multimedia Conference, pp. 531–535. ACM (2016)
2. Eisner, B., Rocktäschel, T., Augenstein, I., Bošnjak, M., Riedel, S.: emoji2vec: learning emoji representations from their description. arXiv preprint arXiv:1609.08359 (2016)

3. Ghosh, A., Veale, T.: Fracking sarcasm using neural network. In: Proceedings of the 7th Workshop on Computational Approaches to Subjectivity, Sentiment and Social Media Analysis, pp. 161–169 (2016)

4. Ghosh, D., Fabbri, A.R., Muresan, S.: The role of conversation context for sarcasm detection in online interactions. arXiv preprint arXiv:1707.06226 (2017)

5. Hallsmar, F., Palm, J.: Multi-class sentiment classification on twitter using an emoji training heuristic. Master's thesis, KTH Royal Institute of Technology School of Computer Science and Communication, May 2016

6. Hazarika, D., Poria, S., Gorantla, S., Cambria, E., Zimmermann, R., Mihalcea, R.: Cascade: contextual sarcasm detection in online discussion forums. arXiv preprint arXiv:1805.06413 (2018)

7. Hu, X., Tang, J., Gao, H., Liu, H.: Unsupervised sentiment analysis with emotional signals. In: Proceedings of the 22nd International Conference on World Wide Web, pp. 607–618. ACM (2013)

8. Joshi, A., Bhattacharyya, P., Carman, M.J.: Automatic sarcasm detection: a survey. ACM Comput. Surv. (CSUR) **50**(5), 73 (2017)

9. Kelly, R., Watts, L.: Characterising the inventive appropriation of emoji as relationally meaningful in mediated close personal relationships (2015)

10. Lee, H., Kwak, N.: The affect effect of political satire: sarcastic humor, negative emotions, and political participation. Mass Commun. Soc. **17**(3), 307–328 (2014)

11. Morstatter, F., Shu, K., Wang, S., Liu, H.: Cross-platform emoji interpretation: analysis, a solution, and applications. arXiv preprint arXiv:1709.04969 (2017)

12. Novak, P.K., Smailović, J., Sluban, B., Mozetič, I.: Sentiment of emojis. PLoS ONE **10**(12), e0144296 (2015)

13. Pennington, J., Socher, R., Manning, C.D.: Glove: global vectors for word representation. In: EMNLP, vol. 14, pp. 1532–43 (2014)

14. Rajadesingan, A., Zafarani, R., Liu, H.: Sarcasm detection on Twitter: a behavioral modeling approach. In: Proceedings of the Eighth ACM International Conference on Web Search and Data Mining, pp. 97–106. ACM (2015)

Condorcet Optimal Clustering with Delaunay Triangulation: Climate Zones and World Happiness Insights

Max Bassett, Blake Newton, Joseph Schlessinger, Jacob Schmidt,
Scott Lynch[✉], Patrick Kuiper, Ryan Miller[✉], Steven Morse, James Pleuss,
Travis Russell, and William Pulleyblank

United States Military Academy, West Point, NY 10996, USA
{scott.lynch,ryan.miller2}@westpoint.edu

Abstract. Condorcet clustering methods have the attractive features
of producing clusterings which place similar points in the same cluster
and dissimilar points in different clusters as well as not requiring *a priori*
specification of the number of clusters. They have the disadvantages of
being combinatorially hard and the method produces only convex clusters. We propose a novel modification to this method, which improves it
significantly on both accounts and works particularly well when applied
to social network type data sets. Specifically, we reduce the domain of
the clustering to be over a Delaunay triangulation, whose size scales as
$O(n^{\lfloor m/2 \rfloor})$ where n is the number of records and m is the number of
attributes used for the clustering. The triangulation also limits focus to
local structure, which allows for non-convex clusterings. We demonstrate
its use in comparison to other well-known heuristic methods using several
constructed datasets, then use it to cluster real-world datasets.

Keywords: Clustering · Condorcet metric · Delaunay triangulation

1 Introduction

The problem of effectively determining sensible clustering of data has received
significant attention in the fields of pattern recognition and machine learning,
both in theory and application. Clustering often forms a first step in data analytics and its value to the application depends on the quality of the results of
subsequent steps which are influenced by the clustering. Because the terms "effective" and "sensible" often depend on the application, dozens of viable approaches
exist, each with strengths and limitations, and many of them share theoretical
connections. We propose a methodology based on a quality measure for a clustering that provides a flexible platform for problem-specific structure, and for which

This research was completed in partial fulfillment of the United States Military
Academy's Network Science Minor program and sponsored by the West Point Network Science Center.

R. Thomson et al. (Eds.): SBP-BRiMS 2019, LNCS 11549, pp. 81–91, 2019.
https://doi.org/10.1007/978-3-030-21741-9_9

we can compute an *optimal* clustering. We then use this method to examine two real-world case studies involving climate zones and world happiness data.

1.1 Background

Informally, clustering is the process of partitioning a dataset into clusters such that records in the same cluster are similar to each other, and records in different clusters are dissimilar. However, specifying "similarity" of data can have many different interpretations, and this has led to a diverse and active field of research. We begin by briefly outlining several known techniques that are relevant to the methodology presented in this paper [7,11].

Given a dataset $\mathcal{D}$ and distance metric d, centroid-based clustering methods such as k-means partition $\mathcal{D}$ into k clusters for which the sum of the within-cluster distances is small. Hierarchical methods such as agglomerative clustering circumvent some of these limitations by organizing data points into a hierarchical structure of similarity based on a distance metric d. The number of clusters can then be selected *post-hoc*, and convex limitations may be avoided by careful selection of the linkage method. In spectral clustering, a kernel map is derived from the spectral properties of the distance matrix associated to the pair $(\mathcal{D}, d)$. A clustering is obtained by applying this kernel followed by some other clustering algorithm such as k-means [16,19]. While non-convex clusters can be produced with these methods, other limitations may be imposed by the choice of post-kernel clustering algorithm. Density-based methods, such as DBSCAN [5] and HDBSCAN [14], cluster points which are connected by sufficiently dense paths within $\mathcal{D}$, suitably defined. The clusters formed need not be convex, and the number of clusters need not be specified in advance.

Two techniques play a crucial role in our own contributions: Condorcet clustering and Delaunay triangulation. Condorcet clustering is a clustering method inspired by the Condorcet criterion for ranking, wherein candidates are selected based on the number of head-to-head runoffs they would win judging by voters' rank orderings [13]. When applied to the clustering problem, the goal is to minimize a penalty function over the space of all possible clusterings of $\mathcal{D}$ where roughly, a penalty is applied any time two points in the same cluster are far apart and any time two points in different clusters are close together. Condorcet clustering is performed by finding a clustering which attempts to minimize the sum of the penalties, often through exhaustive combinatorial search [3].

One common modification to many clustering algorithms, in particular the family of agglomerative and spectral techniques outlined above, is to incorporate prior knowledge of structure within the data [6]. The goal here is to improve both the sensibility of the clustering, by emphasizing certain structure, and the speed or complexity of the algorithm, by reducing the search space. We describe a modification in this vein to Condorcet clustering based on notion of the Delaunay triangulation of a set of points, a well known topic in computational geometry which we describe further in the Methods section.

1.2   Contributions

We first offer a mixed-integer programming (MIP) formulation of the Condorcet criterion as applied to clustering, which produces optimal clusterings without prior specification of the number of clusters which is similar to those in [3,13,18], which are generalizations of the clique partitioning problem. Secondly, we show that using a subset of the decision variables of the optimization problem both reduces the complexity and improves the sensibility of the resulting "optimal" clustering. Specifically, we propose using a Delaunay triangulation of the data, which reduces the number of decision variables and constraints to be $O(n^{\lfloor m/2 \rfloor})$ in the size (n) and dimension (m) of the data. We show experimentally that Condorcet Clustering with Delaunay Triangulation (CCDT) can lead to more "sensible" results, and perform qualitatively as well or better than other popular clustering algorithms. We perform the comparison on several datasets with $m = 2$ and $m = 3$, both contrived and real-world. We also offer an effective heuristic for choosing the parameters.

2   Methods

2.1   Condorcet Clustering

We are given a data set $\mathcal{D}$ of n points or records, and a distance metric $d : \mathcal{D} \times \mathcal{D} \to [0, \infty)$ where d_{uv} represents the distance between the distinct datapoints u and v. We assume $d_{uv} = d_{vu}$, $d_{vv} = 0$, and that d obeys the triangle inequality, $d_{uv} + d_{vw} \geq d_{uw}$. Let E_n be the set of all unordered pairs of points, $|E_n| = n(n-1)/2$.

A *clustering* is a partition of $\mathcal{D}$ into some number of sets, that is, an equivalence relation defined on $\mathcal{D}$. We can represent the clustering by a map assigning each $v \in \mathcal{D}$ to one of the partitions (its "cluster"). Alternatively, we can represent this by defining a boolean variable x_{uv} for each pair $\{u, v\}$ of points in E_n where $x_{uv} = 1$ if points u and v are in the same set, and $x_{uv} = 0$ if they are not. Transitivity is enforced by requiring $x_{uw} = 1$ for pairs of points satisfying $x_{uv} = 1$ and $x_{vw} = 1$ for each triangle, i.e. triple of pairs, $\{u, v\}, \{u, w\}, \{v, w\}$ in E_n.

We can restrict our consideration to a subset $S \subseteq E_n$ of all ordered pairs, in order to take advantage of known structure in the data and thereby sparsify the connectivity matrix X. We often interpret each point as a vertex in an undirected graph $G = (V, E)$, with an edge $uv \in E$ wherever $x_{uv} = 1$. Now $X = \{x_{uv}\}$ corresponds to the adjacency matrix of G, and clusters correspond to the vertex sets of the connected components of G. Under this graph theoretic interpretation, $E \subseteq S \subseteq E_n$, and when $S = E_n$, the connected components will necessarily be *cliques*, by the transitivity requirement.

A *Condorcet metric* penalizes points in the same set if they are too far apart, and penalizes points in different sets if they are too close together. It is motivated by the Condorcet method of voting, which elects the candidate who would win in the most number of head-to-head elections against the other candidates based on the voters' rank ordering [1,13].

We represent the Condorcet metric $\mathcal{C}(X)$ of a clustering X over the subset $S \subseteq E_n$ as follows:

$$\mathcal{C}(X) = \sum \Big(\big(d_{uv} - R \; : \; uv \in S, \; x_{uv} = 1, \; d_{uv} > R\big) + \\ \big(r - d_{uv} \; : \; uv \in S, \; x_{uv} = 0, \; d_{uv} \le r\big)\Big), \tag{1}$$

where R and r are real parameters such that $0 \le r \le R$. For pairs u, v of points, if they are in the same set, but $d_{uv} > R$, we incur a penalty $d_{uv} - R$; if they are in different sets, but $d_{uv} < r$, we incur a penalty $r - d_{uv}$. The Condorcet metric as defined in Eq. (1) is the sum of these penalties over all pairs of points. This "metric" is thus a function $\mathcal{C} : X \to \mathbb{R}_+$, not to be confused with the inter point distance metric d.

In comparison with other clustering metrics, such as the silhouette score [22] or the Calinsky-Harabasz score [4], the Condorcet metric also favors convex clusters (at least, if there is no imposed connectivity structure, which we will address later), but differs in that it allows for the specification of parameters R, r quantifying "farness" and "closeness".

We now minimize $\mathcal{C}(X)$ as an integer linear programming problem. The formulation is a variation of the version given in [1],

IP Formulation.

$$(P) \quad : \quad \min \sum \big(d_{uv} x_{uv} \; : \; uv \in S\big)$$
$$- R \sum \big(x_{uv} \; : \; uv \in S, d_{uv} > R\big)$$
$$- r \sum \big(x_{uv} \; : \; uv \in S, d_{uv} < r\big)$$
$$\text{s.t.} \quad x_{uv} + x_{vw} - x_{uw} \le 1 \text{ for all } u, v, w \in S$$
$$x_{uv} \in \{0, 1\}$$

When $R = r$, the objective function of reduces to

$$\sum \big((d_{uv} - R)\, x_{uv} \; : \; uv \in S\big).$$

This shows that the Condorcet clustering problem is a variant of the *clique partitioning problem* (CPP): partition a graph $G = (V, E)$, $E \subseteq S$, into cliques, which minimize the sum of the weights of the edges in the cliques, where now the weights are $d_{uv} - R$. The constraint polytope of the CPP has been studied, see [8,9,15,21]. For its applications to clustering see most recently [17]. Also, several efficient algorithms have been developed although we do not implement them directly here [2,23].

Our formulation requires a post-processing step of labeling clusters based on the connected components of the undirected graph given by adjacency matrix corresponding to X.

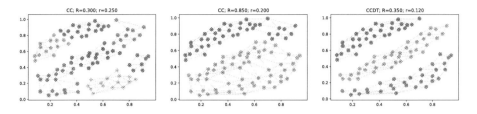

Fig. 1. Condorcet clustering with $R \approx r$ produces convex clusters similar to k-means (left), and with $R \gg r$ approximates an effect of focusing on local structure by ignoring middle-distance edges (middle). With Delaunay triangulation (CCDT, right) focus is on local structure and gives the most sensible clustering. Lines depict edges with $x_{uv} = 1$, $uv \in \mathrm{DT}(\mathcal{D})$.

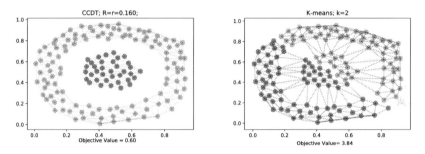

Fig. 2. Concentric circles using CCDT and k-means. Lines depict edges with $x_{uv} = 1$, $uv \in \mathrm{DT}(\mathcal{D})$.

2.2   Sparsification

We may consider the full dataset $\mathcal{D}$ but select a subset $S \subset E_n$ of pairwise connections which preserves some skeletal structure, which we will refer to as *sparsification*. In general this has the potential this benefit of reducing the computational time and improving the sensibility of the clustering. An example of a popular S is the k-nearest-neighbors subgraph; this is a common modification to, for example, hierarchical clustering methods, see [6].

We propose using a well-known method of triangulation of the data, called the Delaunay triangulation. Given a data set $\mathcal{D}$, its Delaunay triangulation is a graph G with vertex set $\mathcal{D}$ and edge set chosen such that the smallest circle inscribing any three points of $\mathcal{D}$ corresponding to a 3-cycle of G contains no other points of $\mathcal{D}$. We will denote this edge set $\mathrm{DT}(\mathcal{D})$ [12].

A Delaunay triangulation has several desirable properties. First, there are algorithms for constructing the triangulation in $O(n \log n)$. Second, the triangulation scales efficiently in the number of edges and simplifies computation to $O(n^{\lfloor m/2 \rfloor})$. Third, the Delaunay contains several important structures, the minimum spanning tree, the convex hull, and the 1-nearest-neighbors graph.

An optimal clustering which minimizes the Condorcet metric above, but over the subset $S = \mathrm{DT}(\mathcal{D})$ consisting of the Delaunay triangulation, is denoted by

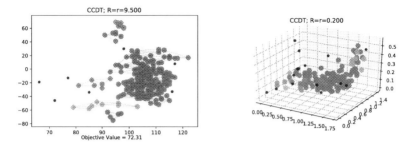

Fig. 3. Climate zone data [left] [20] using CCDT and World Happiness Data [right] using CCDT. CCDT detects climate outliers (black dots, e.g. Mt. Washington, NH) and smaller climate groups (tropical, arctic). In the happiness dataset, CCDT detects clusters of countries based in mutual expectation in the dimensions.

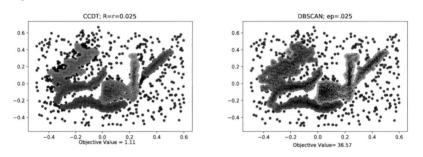

Fig. 4. Dataset in [14] using CCDT and DBSCAN. Both methods produce similar results with DBSCAN preferring slightly larger clusters.

Condorcet Clustering with Delaunay Triangulation, or CCDT, and is the focus of the remainder of the paper.

Selection of r and R. The sets produced in the clustering depend on the values of r and R. An inappropriate selection for these parameters can lead to less desirable clusterings. Here we an offer a standardized approach for parameter selection, which can be tailored to other data sets.

First, some observations. Note the objective function of P_1 or P_2 effectively ignores any pair of points with distance $r \leq d_{uv} \leq R$. Setting r, R closer to the minimum edge length favors a clustering that keeps pairs of points whose distance is large in different clusters, while setting r, R closer to the maximum edge length favors a clustering that keeps pairs of points for which the distance is small in the same cluster.

With this in mind, we found it heuristically effective to select $r \leq R$ within one standard deviation of the mean of d_{uv}, $uv \in S$. Although this works in practice for all of the data sets we examined, further exploration of this topic is needed.

Table 1. Condorcet metric score comparison. Score based on $R = r$ values used in CCDT.

Method	Elliptical (n = 100)	Concentric circles (n = 150)	McInnes dataset (n = 2309)
Condorcet with Delaunay	0.54	0.57	1.11
Condorcet full graph	32.24	75.3	-
Agglomerative clustering	1.29	1.54	36.60
Spectral clustering	2.14	0.57	40.40
DBSCAN	0.82	0.57	36.70
k-means	2.87	3.70	41.90

3   Results

In this section, we will first demonstrate experimentally the effect of the Delaunay triangulation and parameters R, r on the discovered clusters. We then present a comparison between CCDT and other methods, in $\mathbb{R}^2$, using three contrived data sets to show the effectiveness of CCDT. We finally then analyize two real-world data sets using CCDT.

3.1   Effect of Triangulation and Parameter Selection

To demonstrate the effect of the Delaunay sparsification and the parameters r, R on the optimal clustering, consider the elliptical bands data set in Fig. 1.

The left two subfigures depict the results of a "full" Condorcet clustering, i.e. $S = E_n$, under different selections for r, R. We note it has the unsurprising tendency to select convex clusters, especially with r close to R, since we are essentially minimizing within-cluster distance. However, as we increase R (that is, the distance at which we begin imposing penalties for within-cluster far-ness), the method reluctantly begins identifying the bands and even selects some non-convex structure (middle subfigure).

Contrast this with the rightmost subfigure of Fig. 1, which depicts the optimal clustering over a Delaunay triangulation, with r and R set nearly equal. Now the method easily discovers the clear structure of parallel bands. In a sense, the triangulation of the data has the effect of increasing R in a data-specific way, and thus eases the selection of the r, R parameters.

Due to this observation, and to simplify the presentation, we will use CCDT with $r = R$ for the majority of the remainder of the results.

3.2   Comparison of CCDT to Heuristic Methods

We now compare CCDT to several well-known and popular heuristic methods, specifically, k-means, agglomerative clustering, and DBSCAN.

Figure 2 shows the result of clustering a data set with two concentric rings of data, using CCDT (left) and k-means (right). The triangulation allows CCDT to ignore long edges and discover the non-convex patterns, whereas unsurprisingly, k-means (with $k = 2$) divides the data into two convex shapes. We emphasize again here that we did not need to specify or constrain CCDT to find two clusters. We also note that we selected $R = r = 0.170$ following the rule-of-thumb outlined in the Methods section of this paper (Fig. 3).

Next, in Fig. 4, we use the contrived data set presented in [14] to compare CCDT and the popular density-based algorithm DBSCAN. This data set is slightly larger than the previous two ($n = 2,309$), and contains several unusual patterns which make it challenging for most clustering methods: there are non-convex patterns, overlapping patterns, and a uniform scattering of "noisy" points. Density-based methods like DBSCAN perform well on these types of data, as evidenced by the clustering in the right subfigure. However, CCDT also has no trouble: the triangulation assists it in focusing on local structure. Further, we were able to select $R = r = 0.025$ using the rule given in Methods, whereas the selection of eps $= 0.025$ for DBSCAN required some trial-and-error (and is still combining two groups which appear more sensibly separated).

Lastly, in Table 1, we summarize the performance of these algorithms on all data sets, from the point-of-view of the Condorcet metric. Naturally, CCDT has the best value under the Condorcet metric for all data sets since it is directly optimizing it. Of interest is comparing performance of other methods under this metric: spectral and DBSCAN, for example, find the same clustering as CCDT in the "concentric circles" data set. We also wish to point out that we do not have results for Condorcet clustering *without* triangulation on two of the data sets due to computational limitations.

3.3 Clustering Social and Environmental Data Using CCDT

We used this clustering method on two real world data sets. The first data set had 263 records whose attributes included the record maximum of and record minimum temperature of US cities [20]. The CCDT created one primary cluster which represented the majority of the stations in the continental US There were also a number of smaller clusters that represented Hawaii, Alaska, and Pacific island territories. Although all of the Continental US was clustered together, CCDT made finer distinctions between the climate of Hawaii and the Pacific Islands such as Guam because of a difference in their record low temperatures. CCDT also selects several singleton clusters, which we may interpret as "outliers" analogous to methods like DBSCAN. For example, one outlier station is Mt. Washington, New Hampshire, the highest mountain in the northeastern U.S. (Fig. 3).

Next, in Fig. 3, we examine the World Happiness Report 2015 [10]. This data set has multiple dimensions, but for the purposes of visualization only three dimensions were used. GDP per capita, Social Support, and Corruption were the chosen dimensions. See [10] for details on the dimension descriptions.

Interestingly, when CCDT was used to cluster the data set, it extracted more than just a rank ordering of the countries. After clustering with $r = R = .2$ one large cluster was found consisting of 57% of the records. This large cluster included most the happier countries, but also included some of the less happy countries. All of the countries in this cluster have similar ranks for all three dimensions, regardless of their overall happiness. For example, Angola, one of the unhappiest country in this cluster is consistently unhappy in all three categories ranking 118th in Social Support, 110th in Corruption and 99th in GDP per capita. A number of other smaller clusters were also identified, most of these clusters included similar outliers. For example there is a cluster of 7 countries with much higher GPD scores relative to their other two dimensions. This cluster includes Luxembourg, Hong Kong, and many of the OPEC countries. Another large cluster includes 29 countries which are mostly unhappy and have economies that are even lower than expected based on the other two dimensions.

For a more rigorous analysis we examined the clusters as they related to a best fit line through the data. In context, the best fit line is a representation of the idea that we would expect the various dimensions of happiness to be correlated. If a country is happy in most of the dimensions of happiness, it is expected that they would be happy in all dimensions. The primary cluster was most in line with this expectation. Let $\vec{r}(t) = \langle x, y, z \rangle t$ be the vector equation of the best fit line of the dataset and let d_{ri} be the minimum distance from the ith data point to the best fit line. The primary cluster has the common trait of having relatively small d_{ri}. In fact if $\bar{d}_{rj}$ is the average distance from the best fit line to each point in cluster j then the primary cluster had the minimal $\bar{d}_{rj}$. Not only can the clusters be examined by the magnitude of their distance from $\vec{r}(t)$, but also their direction. For those clusters that are far from $\vec{r}(t)$ we would expect that they are a similar direction from $\vec{r}(t)$. Let $\overrightarrow{PQ}$ be the displacement from any point P in the dataset to a point Q, which is the closest point on the line $\vec{r}(t)$ to P. We consider this the error vector for the point P. Then let Υ_{ij} represent the average of all the pairwise dot products of error vectors in cluster i, with error vectors in cluster j, If Υ_{jj} is the average pairwise dot product of all error vectors in cluster j, then a large value for Υ_{jj} means cluster j has large error vectors all oriented in the same direction. We find that a number of the smaller seemingly random clusters can be explained by examining Υ_{jj}. One of the seemingly insignificant clusters includes only Mongolia and Paraguay. The countries are ranked far apart (100 and 53 respectively). But they share a trait in that they have a much higher level of family trust relative to their other two metrics. The $\bar{d}_{rj}$ is large compared to the primary cluster, and Υ_{jj} is in the top quintile when compared to all v_{ij}, Similarly, the previously discussed cluster with many OPEC countries also had a large $\bar{d}_{rj}$ with an Υ_{jj} again in the upper quintile.

The CCDT method of clustering provides meaningful insights into both datasets, Importantly, CCDT can group clusters not based on surface level information but based on the underlying structure of dataset. This is an improvement on existing methods which looked for convex shapes in three dimensions. This method can should be considered for other social and societal datasets.

4 Conclusion and Future Work

We have presented a variant of Condorcet clustering which optimizes the Condorcet metric using integer programming, but over a reduced domain corresponding to a Delaunay triangulation of the data. We presented an IP formulation of this method and demonstrated through experimental evidence on various contrived and real-world datasets that this reduced search space results in improved computational complexity and more sensible clusters. We also showed that the method is effective in two and three dimensions. Computationally our method scales for higher dimensions, however, most practical applications can be reduced to lower dimensions through practices such as principal component analysis and are visualized in two or three dimensions.

We wish to further explore the selection of the parameters r, R beyond the experimental evidence presented in this paper; specifically, the implications concerning intuitive results produced by values near the mean edge length. Other sparsification methods to replace (or augment) our use of the Delaunay triangulation should also be explored. For example, Gomory-Hu cut trees, maximum spanning trees, or simple thresholding. An approximation of the Delaunay triangulation could also be used to decrease the complexity of this preprocessing step without affecting the quality of the triangulation. For planar graphs, the Delaunay triangulation algorithm performs well, but as the dimension increases, finding the exact Delaunay becomes computationally expensive.

References

1. Ah-Pine, J., Marcotorchino, J.F.: Overview of the relational analysis approach in data-mining and multi-criteria decision making. In: Web Intelligence and Intelligent Agents. InTech (2010)
2. Atamtürk, A., Nemhauser, G.L., Savelsbergh, M.W.: A combined lagrangian, linear programming, and implication heuristic for large-scale set partitioning problems. J. Heuristics **1**(2), 247–259 (1996)
3. Bertsimas, D., Allison, K., Pulleyblank, W.R.: The Analytics Edge. Dynamic Ideas LLC (2016)
4. Caliński, T., Harabasz, J.: A dendrite method for cluster analysis. Commun. Stat.-Theory Methods **3**(1), 1–27 (1974)
5. Ester, M., Kriegel, H.P., Sander, J., Xu, X., et al.: A density-based algorithm for discovering clusters in large spatial databases with noise. In: KDD, vol. 96, pp. 226–231 (1996)
6. Franti, P., Virmajoki, O., Hautamaki, V.: Fast agglomerative clustering using a k-nearest neighbor graph. IEEE Trans. Pattern Anal. Mach. Intell. **28**(11), 1875–1881 (2006)
7. Friedman, J., Hastie, T., Tibshirani, R.: The Elements of Statistical Learning. Springer Series in Statistics, vol. 1. Springer, New York (2001). https://doi.org/10.1007/978-0-387-21606-5

8. Grötschel, M., Wakabayashi, Y.: A cutting plane algorithm for a clustering problem. Math. Program. **45**(1–3), 59–96 (1989)
9. Grötschel, M., Wakabayashi, Y.: Facets of the clique partitioning polytope. Math. Program. **47**(1–3), 367–387 (1990)
10. Helliwell, J.F., L.R., Sachs, J.: World happiness report 2015 (2015)
11. Jain, A.K.: Data clustering: 50 years beyond k-means. Pattern Recogn. Lett. **31**(8), 651–666 (2010)
12. Lee, D., Schachter, B.J.: Two algorithms for constructing a Delaunay triangulation. Int. J. Comput. Inf. Sci. **9**(3), 219–242 (1980)
13. Marcotorchino, F., Michaud, P.: Agregation de similarites en classification automatique. Rev. Stat. Appl. **30**(2), 21–44 (1982)
14. McInnes, L., Healy, J., Astels, S.: HDBSCAN: hierarchical density based clustering. J. Open Source Softw. **2**(11), 205 (2017)
15. Mehrotra, A., Trick, M.A.: Cliques and clustering: a combinatorial approach. Oper. Res. Lett. **22**(1), 1–12 (1998)
16. Meila, M., Shi, J.: Learning segmentation by random walks. In: Advances in Neural Information Processing Systems, pp. 873–879 (2001)
17. Miyauchi, A., Sonobe, T., Sukegawa, N.: Exact clustering via integer programming and maximum satisfiability. In: AAAI Conference on Artificial Intelligence (2018)
18. Miyauchi, A., Sukegawa, N.: Redundant constraints in the standard formulation for the clique partitioning problem. Optim. Lett. **9**(1), 199–207 (2015)
19. Ng, A.Y., Jordan, M.I., Weiss, Y.: On spectral clustering: analysis and an algorithm. In: Advances in Neural Information Processing Systems, pp. 849–856 (2002)
20. NOAA: Comparative climatic data. National Centers for Environmental Information (2015). https://www.ncdc.noaa.gov/ghcn/comparative-climatic-data
21. Oosten, M., Rutten, J.H., Spieksma, F.C.: The clique partitioning problem: facets and patching facets. Netw. Int. J. **38**(4), 209–226 (2001)
22. Rousseeuw, P.J.: Silhouettes: a graphical aid to the interpretation and validation of cluster analysis. J. Comput. Appl. Math. **20**, 53–65 (1987)
23. Sukegawa, N., Yamamoto, Y., Zhang, L.: Lagrangian relaxation and pegging test for the clique partitioning problem. Adv. Data Anal. Classif. **7**(4), 363–391 (2013)

Using Common Enemy Graphs to Identify Communities of Coordinated Social Media Activity

Lucas A. Overbey[✉], Bryan Ek, Kevin Pinzhoffer, and Bryan Williams

Naval Information Warfare Center (NIWC) Atlantic,
P.O. Box 190022, North Charleston, SC 29419-9022, USA
lucas.overbey@navy.mil

Abstract. Increased use of and reliance on social media has led to a responsive rise in the creation of automated accounts on such platforms. Recent approaches to identification of individual automated accounts has relied on machine learning methods utilizing features drawn predominantly from text content and profile metadata. In this work we explore a novel use of graph theoretic measures, specifically common enemy graphs, to identify and characterize groups of accounts exhibiting shared behavior in online social media, particularly those exhibiting characteristics of automation and/or potential coordination. In addition, we develop edge weight variants of fuzzy competition graphs to further characterize common group behavior of automated accounts within subnetworks of social media ecosystems.

Keywords: Graph theory · Social network analysis · Social bots ·
Competition graphs · Community detection

1 Introduction

The recent proliferation of social media has led to a corresponding rise in the use of automated (i.e., bot) accounts. Bots can be created for many reasons including entertainment or information sharing purposes, enhancing perceived levels of influence of human users, or to act with nefarious intent, for example, to spread misinformation, manipulate crowds, propagate social hysteria, and cause group polarization [6,8,13]. Aggregations of bots often act as amplifiers, for flooding channels to crowd out other sources, gaming social media platform algorithms to promote content ahead of credible sources, identify target audiences, or attack targets with automated messaging based on the content of user's messages. In such cases, groups of bots, or *botnets*, may be created en masse with common and/or shared programming to enact similar behaviors.

Funded by the Office of Naval Research.

Much of the recent research related to social media bots has focused on individual identification. These methods frequently use common heuristic features (e.g. ratio of retweets and posting regularities) and supervised machine learning techniques [7,9,15,16]. Existing work has largely utilized temporal and content-based features with only recent consideration of network measures [2,3]. And once a bot is identified, little work has focused on the organizational structures of these networks. Al-Khateeb and Agarwal [1] investigated a Twitter dataset and found interactions between bots and a "broker" node, revealing discernible group-level coordination. They showed while botnets may or may not share direct follower/following relationships, they exhibit common shared behaviors to enact amplification, but in such a way as to not be blatant enough to cause accounts to be removed by platform owners. However, there is still a need to develop mechanisms to identify shared behavior or coordination within botnets.

Social media structure can be represented by a directed graph with vertices representing actors (accounts) and edges reflecting direct communications (replies/retweets/mentions). Edge weights can be expressed by the quantity of such interactions [10]. Using this framework, we develop graph theoretic approaches to identify groups of accounts exhibiting shared behaviors indicative of coordinated behaviors. Specifically, we rely on common enemy graphs, a variant of competition graphs that involves a transformation of an original directed graph to identify and weight indirect relationships between vertices. Competition and common enemy graphs are concepts originally developed for modeling ecological systems [5]. Our goal is not necessarily to identify bots, but to develop a means to assess degree of common behavior potentially indicative of possible coordination. We develop novel edge weighting schemes on common enemy graphs and derive filtering mechanisms to assess extent and proportion of common communication behaviors. We evaluate our results incorporating an existing bot identification approach [7]. We find that our approach does identify botnets and other groups of accounts exhibiting abnormally large levels of in-common communication behaviors.

2   Common Enemy Graphs

Robert Paine initially defined a food web as a model of an ecosystem with a directed graph modeling predator-prey relationships [11]. Later, Cohen introduced the notion of a competition graph as a way to analyze relationships between predators within an ecosystem [5]. A competition graph of an ecosystem is a mapping of a food web onto an undirected graph, such that vertices represent predators and edges represent resource competition between them. Conversely, a common enemy graph represents prey that share common predators. Figure 1 shows example common enemy and competition graphs for a small ecosystem.

Let a digraph D be given as an ordered pair (V, A) such that V is the set of vertices and A is the set of ordered pairs of elements representing the arcs in D.

Definition 1 (Competition Graph). *The* competition graph *(of D) is an undirected graph $C(D) = (V, E)$ on the same vertices as D where E is a set of*

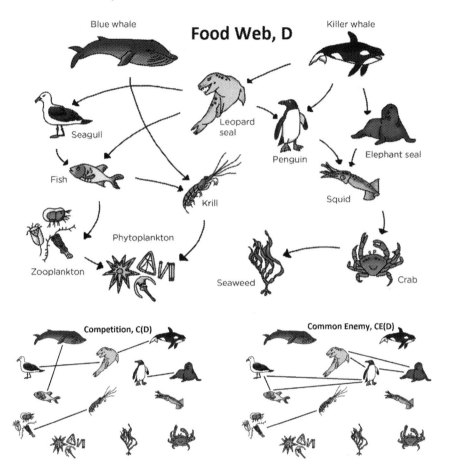

Fig. 1. A basic food web, here with directed edges from predator to prey. The competition graph identifies connections between species that compete for resources. The common enemy graph identifies when species share a threat source.

unordered pairs of elements representing edges in $C(D)$. $(x,y) \in E$ if and only if for $x, y \in V$, $x \neq y$, and $\overrightarrow{xv}, \overrightarrow{yv} \in A$ for some $v \in V$.

Definition 2 (Common Enemy Graph). *The* common enemy graph *(of D) is an undirected graph $CE(D) = (V, E)$ on the same vertices as D where E is a set of unordered pairs of elements representing edges in $CE(D)$. $(x,y) \in E$ if and only if for $x, y \in V$, $x \neq y$, and $\overrightarrow{vx}, \overrightarrow{vy} \in A$ for some $v \in V$.*

3   Application to Social Media

In the context of the information environment, competition is over attention and is spread via information or message passing. For example, in Twitter, as mentioned previously, we can construct a graph made up of user accounts (vertices) and direct communications (directed edges). We specifically focus on a directed retweet network, in which a directed edge is drawn from the originator of a message to retweeters (additional sharers) of that message. While a retweet does not indicate agreement, it reveals that the retweeter has at least seen the message and responded in a specific way that is indicative of interest in the content [14].

Common enemy graphs are equivalent to competition graphs through a reversal in direction of arcs in the directed graph D. But when considered in an applied setting (with fixed arc direction), the derived graphs have different meanings. While competition graphs do provide useful information, they indicate windows of prominence rather than communities of amplification. As such, we focus solely on common enemy graphs and their interpretations in this paper.

Edges in the common enemy graph of a retweet network identify users that have retweeted one or more messages from the same account(s). Botnets are frequently employed as information spreaders and amplifiers [1]. Therefore, as both the proportion and quantity of common retweet behavior represented in the common enemy graph grows, the likelihood that a group of accounts can be discounted as coincidentally exhibiting coordinated behaviors diminishes, and the likelihood that the coordination is either automated or intentional or both increases.

3.1 Weighting Schemes

Samanta et al. [12] introduced the notion of fuzzy competition graphs to provide a weighting on the edges and vertices based on fuzzy set theory. This form allowed a certain level of measurement for *how much* two vertices had in common outside of their neighborhood size. We make use of this general idea of measuring the magnitude of the connection but without measuring degree of belonging. Let $\sigma(v, x)$ denote the weight of the arc from v to x in D. Rather than a proportion or fraction of belonging, we denote $\sigma(v, x)$ as the number of times that account v was retweeted by account x. We do not normalize because the number of retweets between two accounts is not restricted and strongly dependent on activity and time window (bounds) for data collection. Given this initial arc weight, we define two novel weighting schemes to account for both the quantity (magnitude of shared commonality) and proportion (degree of shared commonality) of in-common behavior.

Quantity. Given that account x retweets account v $\sigma(v, x)$ times, and that account y retweets the same account v $\sigma(v, y)$ times, we can define the edge weight in the common enemy graph $\omega_1(x, y)$ as

$$\omega_1(x, y) = \max\{\min\{\sigma(v, x), \sigma(v, y)\} : v \in V\}, \tag{1}$$

where V is the set of all vertices in $CE(D)$. Consequently, prolific retweeters of the same account(s) will have extremely high edge weights to other prolific retweeters, whereas sporadic or isolated retweeters of the same account will be connected at a much lower strength.

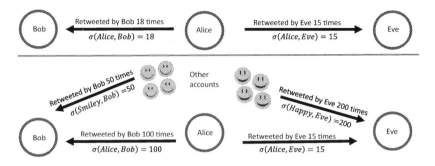

Fig. 2. Two very different social activity settings that result in the same weighted common enemy connection between Bob and Eve: $\omega_1(\text{Bob, Eve}) = 15$. Neither the large increase in retweets by Bob nor the extraneous retweets by either party were reflected in ω_1. Assuming other retweets are of distinct sources, $\omega_2(\text{Bob, Eve}) = \frac{2 \cdot 15}{15 + 18} \approx 0.909$ for the top setting and $\omega_2(\text{Bob, Eve}) = \frac{2 \cdot 15}{150 + 215} \approx 0.082$ for the bottom setting.

Proportion. Because we want to account for high amplification rates, and because information or message passing is a nonconservative process (the amount of message passing is not a fixed quantity within the system) [10], we do not have an upper bound on ω_1. Thus, the proportion of similarity is not reflected in ω_1. For example, two accounts may be prolific retweeters of one similar account *and* several different accounts. Hence, they share a lot in common from a relative amplification sense to one other account but much less in common in terms of overall retweet behavior (see Fig. 2).

To account for proportion, we introduce a second edge weighting scheme,

$$\omega_2(x, y) = 2 \frac{\sum_{v \in V} \min\{\sigma(v, x), \sigma(v, y)\}}{\sum_{v \in V} [\sigma(v, x) + \sigma(v, y)]}. \tag{2}$$

This fraction counts the sum of common in-edges (common retweets of a source) that x and y have in common. The commonality is then divided by the total weighted indegree (total number of retweets) by x and y. $\omega_2(x, y) = 0$ indicates x and y retweet no one in common, while $\omega_2(x, y) = 1$ indicates that x and y retweet the same sources at the same rates.

For our analysis, we use ω_1 and ω_2 in conjunction. A high ω_1 could result from prolific activity that is not necessarily coordinated. A high ω_2 could be the result of a few select tweets in common and nothing else to differentiate the accounts. Having high values for both ω_1 and ω_2 is more indicative of cooperative activity for a common or intentionally conceived goal.

4   Methodology

4.1   Data Collection

We obtained data using the Twitter streaming API from October 3rd, 2017 to November 27th, 2017 (8 weeks) using the search term "nato," or "#nato."

Table 1. Gathered Twitter data by week and the unique totals after merging all weeks.

Week	Tweets	Retweets	Accounts	$|E(D)|$	$|V(D)|$
10/03/2017–10/09/2017	126,629	72,271	82,216	61,908	54,109
10/10/2017–10/16/2017	116,320	71,185	85,137	62,090	62,004
10/17/2017–10/23/2017	145,422	79,481	96,304	68,504	66,019
10/24/2017–10/30/2017	160,094	73,255	92,629	64,778	65,125
10/31/2017–11/06/2017	151,621	68,430	93,532	61,196	58,544
11/07/2017–11/13/2017	173,247	89,165	109,722	76,880	72,517
11/14/2017–11/20/2017	353,693	255,882	201,482	225,835	157,713
11/21/2017–11/27/2017	270,604	195,149	171,947	168,113	136,077
Total	1,497,630	904,818	476,899	**778,680**	**476,043**

Table 1 lists the total data collected by week. While the original intention of this search was to obtain data surrounding individuals discussing the North Atlantic Treaty Organization (NATO), the search also returned extraneous data, including individuals with the string "nato" in their user name and messages associated with the Cebuano language in the Philippines ("nato" translates to "us" in Cebuano). While these messages are superfluous to NATO, they can still contain botnets or groups of coordinating accounts, so we did not remove them. Additionally, they provided a larger pool of topical content to draw from for analysis.

4.2 Common Enemy Graph Formation

We filtered messages to only include retweets and constructed the digraph D, exploring various time intervals. In the interest of scope, we only describe one-week time intervals below. Heuristically, one week of data in our set resulted in a balance between enough data to see dynamics of coordinated behaviors while also not being so long that these behaviors shifted. We then form the digraph D with arcs from users with the original message to the user retweeting that message, weighted by the number of times in the set that they were retweeted, $\sigma(v, x)$. We also removed any isolated pairs, as these accounts are irrelevant to an assessment of coordinated retweet behaviors.

We construct $CE(D)$ from Definition 2 with edge weight values according to Eq. (1). The resulting graph typically consists of dense subnetworks which represent promoted accounts. Groups of paid-for or created bots to boost an accounts apparent influence [6] would show up here as a very dense component.

One drawback of using data collected in this manner is that we can miss activity, outside of "nato", between actors that could change their connection strengths. The work-around would be to draw all tweets by a particular account and include those in the network. This would produce a more accurate ω_2 measuring the proportion of *total* activity in common. However, this measure can still be used as a proxy for proportional similarity given the data/network bounds.

4.3   Filtering and Community Detection

We introduce a cutoff value Ω to help remove noise and increase the likelihood that remaining nodes represent bot-influenced communities (e.g., with atypical levels of in-common retweet behavior). If an edge in CE has weight $\omega_1 < \Omega$, then that edge is removed. If a vertex becomes isolated, it is removed as well. The Ω cutoff is a generally straightforward removal of accounts not of interest for assessing common behaviors or coordination.

If an edge passes the filtering, it is given a new weight corresponding to how similar its incident vertices behave, i.e. $\omega_2(x, y)$. Using these edge weights, we perform Louvain modularity clustering [4] on each component individually.

5   Results

Figure 3 shows resulting CEs from two of the eight weeks evaluated and after applying the filter Ω. Similar structures and types of communities appear in the other weeks. We explored multiple values of Ω and here applied an $\Omega = 14$. This value corresponds to a high degree of in-common behavior (an average minimum of two shared retweets per day). This Ω will not exclude extremely active non-automated accounts but will likely capture many high-activity, amplification-driven bots as well. Each CE consists of several disjoint components indicative of highly shared retweet behavior. Node colors show Louvain communities indicating level of similarity between retweet behaviors. We then compared the communities discovered through this approach to those found using a machine learning bot identification approach (botometer) [7]; see Table 2. We describe characteristics of some of these individual components below.

5.1   Communities

F (Puppet Master). These are bots set to retweet one master account which was captured in the cluster as well. This group existed solely for the purpose of amplification of the source. This community appeared week after week with little variation in membership or connectivity; see Fig. 4. Between the week ending October 16 and the week ending October 23, the previously more loosely connected account (hypothesized master account) was suspended from Twitter. The botnet temporarily fell apart, and then one of the puppet accounts was (likely) re-coded to be the master, leading to a looser connection to the others. A few other puppet accounts were created and/or suspended by Twitter during the time span as well, reflecting the other small changes in the botnet's structure. 100% of these accounts were identified as bots or deleted accounts by botometer.

B (Poetry). This large group appeared only in the first week of data. They are almost entirely bots retweeting messages related to Filipino poetry. Clustering separated two groups of accounts that behaved similarly but not identically. We cannot identify whether these accounts are a part of a single botnet or two related

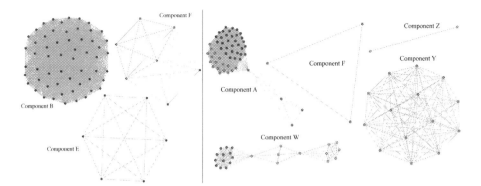

Fig. 3. The components in existence in weeks ending 10/09/2017 and 11/27/2017 respectively. Each component has had Louvain clustering performed. (Color figure online)

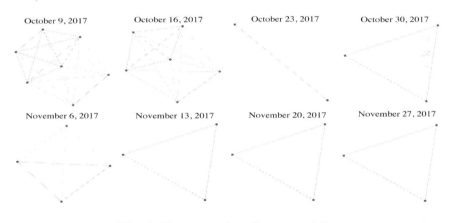

Fig. 4. The progression of component F.

botnets. 81% of these accounts were identified as bots or deleted by botometer. When they are evaluated as two separate communities, one community had 89% bots/deleted and the other had 74% bots/deleted via botometer.

W (Turkish Political). A group of accounts that appear to be Turkish based and political in nature. Using Louvain clustering on this component split it into three groups, one of which (righthand group of six) contained a much higher proportion of bots according to botometer (67% vs less than 31%).

Y (Suspended). We have no better description for this group other than that it consists almost entirely of suspended accounts at the time we completed this analysis. Botometer says that the lone two not-suspended accounts are bots.

Table 2. Accounts in each group labeled as a bot (botometer score ≥ 2.5 out of 5) or deleted. Botometer (formerly BotOrNot) was produced by Indiana University [7].

Community	F	B	W	Y	A	E
Size	9	57	26	14	62	16
Bot (%)	4 (44)	13 (23)	2 (8)	2 (14)	10 (16)	1 (6)
Deleted (%)	5 (55)	33 (58)	8 (31)	12 (86)	16 (26)	4 (25)

A (Political). The largest group in our evaluation, this group consisted of highly partisan political messages. This group contained a high number of very active but non-automated accounts along with several automated accounts set up to amplify specific messages. Using clustering, we identified three separate groups (consistently across several weeks), of which one was completely identified by botometer as not bots and one which contained a much higher proportion of positively identified bots (59% vs 25% vs 0% for the three clusters).

E (Pop Star). This group grew in the first weeks before disappearing. It appears to be a group of dedicated fans of a famous pop star. The one account identified as a bot could be an extremely active and singularly engaged real person.

To show that CE produces further insight than the original digraph D, we also conducted Louvain modularity clustering on D (by considering it as an undirected graph and removing edges of weight $< \Omega = 14$). Most groups did appear with this simpler method. However, they were represented almost exclusively by star graphs with their sources at the center. Common enemy graphs show the coordination directly as well as providing a measure, ω_2, for how strong the connection is relative to each individual's activity.

Bots are often embedded within natural communities [15]. Thus, finding a single component still leaves the task of identifying suspicious accounts. Performing clustering on the star graph components of the simplified method just yields the entire component as a single cluster. Louvain clustering on common enemy components yields separation between subgroups of accounts coordinating at different levels; some levels contain much higher proportions of bots.

6   Summary/Discussion

As the presence of bots grows in social networks, the battle of identification versus obfuscation grows. Bots continue to evolve in sophistication to avoid detection by the current acme of account-based identification methods. However, to retain influence over a community, certain network connections must be made. Our technique is a step toward building a network-based system of identification.

We also compared our approach to Louvain clustering on the original digraph D using a somewhat similar initial filtering approach. This approach was able to identify some of the communities we also found using common enemy graphs, but reveals them as star structures rather than direct connections between common behaviors, and it cannot further separate them into potentially more or less automation-driven behavioral groups. Given that bots are often embedded within natural communities [15], the coarse separation by the simplistic filtering provides less insight than our approach.

A singular bot is not common in effective influence campaigns [1]; they are typically part of a larger botnet to achieve amplification and spread messages more effectively. By applying filtering to common enemy graphs, we can identify a larger group of bots. Our technique does not explicitly say that an account is a bot or not. However, further research could be conducted to combine characteristics of a bot in a common enemy graph into bot identification. Note that botometer (like all bot detection algorithms), for which we compared our results, is not infallible. Some of the accounts we identified may have been bots that were not classified as such using botometer or vice versa. In fact, our approach could potentially be used as one of several features as inputs into a machine learning approach to identify individual bots or botnets. Alternatively, given an identified bot, common enemy graphs can be used to identify other accounts exhibiting very similar behaviors, possibly in coordination.

Our techniques identify groups that are working, not necessarily together, but alongside each other. Group tactics may be teased out and characterized; for example, Components A and W above exhibit very similar compositions, behavior, and CE network properties. Additionally, the common enemy graph paradigm can be explored for other types of constructions of social media networks, for example, replies or shared links. These techniques could also be combined with an approach that simultaneously evaluates content or looks for changes of coordination behaviors over time (e.g. dynamics such as Fig. 4). We will explore these additional considerations in future research.

References

1. Al-Khateeb, S., Agarwal, N.: Understanding strategic information manoeuvres in network media to advance cyber operations: a case study analysing pro-russian separatists' cyber information operations in crimean water crisis. J. Baltic Secur. **2**(1), 6–27 (2016)
2. Beskow, D.M., Carley, K.M.: Bot conversations are different: leveraging network metrics for bot detection in Twitter. In: 2018 IEEE/ACM ASONAM, pp. 825–832. IEEE (2018)
3. Beskow, D.M., Carley, K.M.: Bot-hunter: A Tiered Approach to Detecting & Characterizing Automated Activity on Twitter. SBP-BRiMS (2018)
4. Blondel, V., Guillaume, J.L., Lambiotte, R., Lefebvre, E.: Fast unfolding of communities in large networks. J. Stat. Mech.: Theory Exp. **10**, 155–168 (2008)
5. Cohen, J.E.: Interval graphs and food webs: a finding and a problem. RAND Corporation Document 17696 (1968)
6. Confessore, N., Dance, G.J.X., Harris, R., Hansen, M.: The Follower Factory. The New York Times, January 27 2018

7. Davis, C.A., Varol, O., Ferrara, E., Flammini, A., Menczer, F.: Botornot: a system to evaluate social bots. In: Proceedings of the 25th International Conference Companion on World Wide Web, pp. 273–274. International World Wide Web Conferences Steering Committee (2016)
8. Ferrara, E., Varol, O., Davis, C.A., Menczer, F., Flammini, A.: The rise of social bots. Commun. ACM **59**(7), 96–104 (2016)
9. Morstatter, F., Wu, L., Nazer, T.H., Carley, K.M., Liu, H.: A new approach to bot detection: striking the balance between precision and recall. In: IEEE/ACM International Conference on Advances in Social Network Analysis and Mining (ASONAM) (2016)
10. Overbey, L.A., Greco, B., Paribello, C., Jackson, T.: Structure and prominence in Twitter social networks centered on contentious politics. Soc. Netw. Anal. Mining **3**(4), 1351–1378 (2013)
11. Paine, R.T.: Food web complexity and species diversity. Am. Nat. **100**(910), 65–75 (1966)
12. Samanta, S., Pal, M.: Fuzzy k-competition graphs and p-competition fuzzy graphs. Fuzzy Inf. Eng. **5**(2), 191–204 (2013)
13. Starbird, K.: Examining the alternative media ecosystem through the production of alternative narratives of mass shooting events on Twitter. In: AAAI International Conference on Web and Social Media (ICWSM) (2017)
14. Starbird, K., Palen, L.: (How) will the revolution be retweeted? Information diffusion and the 2011 Egyptian uprising. In: CSCW (2012)
15. Varol, O., Ferrara, E., Davis, C.A., Menczer, F., Flammini, A.: Online human-bot interactions: detection, estimation, and characterization. In: AAAI International Conference on Web and Social Media (ICWSM) (2017)
16. Wu, L., Hu, X., Morstatter, F., Liu, H.: Detecting camouflaged content polluters. In: AAAI International Conference on Web and Social Media (ICWSM) (2017)

Chronological Semantics Modeling: A Topic Evolution Approach in Online User-Generated Medical Data

Cheng-Yu Chung$^{(\boxtimes)}$ and I-Han Hsiao

Arizona State University, 699 S. Mill Avenue, Tempe, AZ 85281, USA
{Cheng.Yu.Chung,Sharon.Hsiao}@asu.edu

Abstract. Online medical discussion forums/question answering sites have become one of the major resources for people to look for healthcare information. These sites typically contain tremendous user-generated content (UGC) that possesses complex domain-specific information in layman's terms, which is the opposite of formal medical records kept in hospitals (i.e. Electronic Health Record). The goal of this project is to dissect semantics and extract valuable information systematically from UGC composed in unstructured and unconstrained format. We propose an automatic medical content analyzer that takes into account language semantics as well as progression (evolution) of medical events. The preliminary evaluation on the WebMD dataset shows that evolution-based recommendation uncovers broader domain semantic which might be ignored when using word-level or concept-based features.

Keywords: Topic evolution · Text processing · User-generated content

1 Introduction

With the increasing volume of medical data, the potential of innovative and effective approaches for knowledge discovery has become one prevalent topic in the medical field [16]. The Electronic Health Records (EHRs) carry a patient's medical history in electronic form, which enables abundant opportunities for knowledge discovery by modern computer technologies. This feature is not only appealing for medical professionals to conduct diagnosis with systematic EHR support, but also paves the way for researchers to determine unprecedented medical evidence at a reasonable cost. Various paradigms for EHRs data analysis have been proposed and implemented in recent years [6,7,15,16]. In contrast to the private and formal medical records, user-generated content (UGC) produces a massive amount of public and consumer-centric content. For example, Stack Overflow is one of the biggest Q&A websites for programmers; Yahoo Answers and Quora cater to users coming from a wide range of domains; WebMD Answers is one of the most popular online health and wellness Q&A sites, etc. Users share ideas, exchange professional knowledge, or just seek solutions to problems, from general to specific through these online services. Essentially, UGC-based platforms

© Springer Nature Switzerland AG 2019
R. Thomson et al. (Eds.): SBP-BRiMS 2019, LNCS 11549, pp. 103–112, 2019.
https://doi.org/10.1007/978-3-030-21741-9_11

provide flexible and faster content production and delivery. With these features, we have seen many companies building their business on such healthcare-related UGC.

We argue that online medical content would become more important in the following years as more organizations contribute to online healthcare. However, to our knowledge, there is limited work focusing on applying medical data analysis to consumer-centric medical content. Thus, our goal is to investigate technological solutions to facilitate analyzing large amount of textual UGC in healthcare information space. Moreover, people nowadays are more accustomed to seeking health or medical information online than in the past [11,12]. Some surveys even show young adults may tend to use Internet as a trustworthy source of medical information [4,5]. Information such as general facts and personal experiences are also used to support healthcare decision-making [4]. Nonetheless, in spite of many advantages, such social medical data is usually messy, unstructured, and much more difficult to be analyzed than well-formatted EHRs [5,9,11]. The quality of online healthcare information has always been a critical concern to many users. Although the reliability of online medical information is contradictory, however, the nature of fast, low cost and unlimited access to UGC still dominates a popular trend in social medical content. The evidence not only encourages more research efforts in studying consumer-centric medical content but also stimulates new and creative methods to systematically analyze large corpora of unstructured medical-related data.

One of the common approaches to wrangle unstructured texts is to apply natural language processing. But reducing the ambiguity and vagueness in textual content has always been challenging. Several techniques like ensemble classification [8], probabilistic topic model [1] and semantic network [3] were proposed and have been utilized in various applications. Instead of order-free composition, some researchers noticed the importance of temporal structure in texts [10,11]. The topic evolution within one subject or between multiple subjects reveals how an event evolves over time. It discloses the progression information, which is usually ignored if we apply order-free processing. To the best of our knowledge, there are rare studies focusing applying evolutional analysis on online medical content. Therefore, the objective of this work is to apply evolutional analysis into recommendation of online medical UGCs. We believe the chronological information generated by topic evolution discovery can bring benefits to data analysis on unstructured and messy UGCs. To work toward this end, we aim to answer the following research questions:

1. How to determine temporal segmentation in online user-generated medical content?
2. Does topic evolution algorithm improve the recommendation for users to find similar experiences?

The rest of paper is organized in the following order: In Sect. 2, we review several works related to our study; Sect. 3 proposes our methodology designed to answer our research questions; Sect. 4 evaluates the feasibility of the proposed semantic evolution algorithm for user-generated medical content; and finally, we summarize the work in Sect. 5.

2   Literature Review

2.1   Medical Text Analysis

There are many approaches proposed and designed for medical content analysis, such as visualization, visual analysis, and natural language processing. West and colleagues [15] argued that to discover underlying knowledge from such unusually large amount of data, "innovation in visualization techniques beyond graphs and charts" is required. Cao et al. [2] presented the visual analytical tool, FacetAtlas, compassing multifaceted medical corpora by elaborating global and local patterns of different concepts in a sophisticated visualization. Their evaluation and expert feedbacks showed that the assistive tool had great potential in medical education and advantages in the diagnostic process of physicians. Wiesner and Pfeifer [16] proposed a practical framework of recommender for personal health record systems (PHRS) and indicated basics and challenges when coping with sensitive information. Hripcsak et al. [6] employed precedent feature and concludes its positive impact on revealing clinical associations. Perer and colleagues [13] developed Care Pathway Explorer and demonstrated how their system integrated data visualization and sequence mining with EHRs and discovered published and unknown knowledge behind the data. All of research above have shown a promising avenue to uncover the potential in the enormous medical data.

2.2   Topic Evolution

A sequence of events typically indicates different phenomenon from different perspectives. In clinical field, an unusual frequent occurrence of a disease within a period may reveal an emergency condition (i.e. an epidemic disease) in an area or a population. In biology, trends are crucial to understand the mechanism behind an ecosystem. In the field of data mining, Temporal Text Mining (TTM) is coined to indicate the analysis of "discovering temporal patterns in text information." [10] Being different from grouping data by timestamps and conducting analyses, TTM usually requires a more sophisticated method due to the fact that textual data are usually polymorphic and ambiguous. In previous works, Mei and Zhai [10] proposed a probabilistic method to extract evolutionary theme patterns (ETP) automatically. The authors divided the problem into smaller progressive aspects and constructed an overall framework by providing solutions to each one of these based on Hidden Markov Model and topic probabilistic model. By testing on news articles and scientific research papers, their method showed the ability to summarize textual data through meaningful temporal structures.

Unlike probabilistic model, Introneand and Drescher [7] proposed Topic Evolution Analysis (TEvA) to track topic evolution in discussion groups. They used semantic networks to represent topic in text and built evaluation based on an algorithm of community analysis in social networks. Their work was tested in dialogs of discussion and showed how topic dynamics were related to decision making and knowledge construction. In this work, we adapt TEvA for evolutionary analysis because the properties of their data are comparable to our dataset, which is a kind of discussion.

3   Methodology

This study focuses on analyzing unstructured user-generated medical content. We choose the WebMD Answers as our subject because it is an open-access, consumer-centric, medical discussion forum targeting general users. Any WebMD Answers's user can freely ask questions about their conditions. Experts or other users can answer the questions according to their experience or knowledge. We crawled all the posts generated before Dec 10, 2014. There are 25319 posts in total. A post on WebMD Answers contains the question title, question content, topic(s) given by WebMD and answer(s). For example, a post may ask "What is the best diet schedule?" The content that asker provides is "Can you provide a full guideline for diet schedule of a diabetes patient?" Its topics may be "diet" and "diabetes". In this work, we only utilize questions' content because our goal is to analyze unstructured content by topic evolution and aim to provide a patient similar experiences from others.

To integrate topic evolution into recommendation routine, we choose TEvA as the basis and develop sequence-based criteria for the recommender. The whole process comprises the following phases: semantic labeling, temporal entity identification, semantic network creation, topic evolution recognition, and recommendation generation. The detail of TEvA algorithm is described in [7]. Here we only describe the domain-specific semantic labeling and recommendation generation.

3.1   Semantic Labeling

TEvA identities topics and their evolutionary connections in text by an unsupervised approach, which seems to be a preferred feature for some NLP application. However, such topics are usually difficult to interpret and process without experienced experts because they are merely collections of plain words. It also requires excessive manpower to transform unsupervised topics into uniform concepts following a unified standard. To solve this issue, we used cTAKES [14] to automatically label words in a topic. Only five tags of cTAKES are included in the current design, which are Anatomical, SignSymptom, Procedure, DiseaseDisorder, and Medication. After all words are labeled, the majority of tags becomes the representative of the topic. This approach eliminates the need of manual processing and improves our interpretation to the unsupervised topics.

3.2   Recommendation Generation

Our work focuses on generating recommendation for medical-related UGC by utilizing topic evolution. After obtaining potential topics and their evolutions in an article, we need proper recommendation criteria to capture the evolutional property and relevant content. Because topic evolutions are represented as nodes and links, we assume that the most relevant content to one article will be those with shared evolution sequences. Given a subject article A, if there is another article B contains most evolution sequences of A, B will be considered the most relevant content to A.

Instead of costly sequence matching algorithm, we simplified the problem to set the comparison. First, we break evolution sequences into many two-item tuples. Each tuple then represents some parts of evolution patterns. For example, a post p may have topic evolution like

$$E_p = (e_1, e_2, e_3, \ldots, e_n) \tag{1}$$

where e_i for $1 \leq i \leq n$ is a topic. This topic evolution will be broken into a set

$$S_p = (e_1, e_2), (e_2, e_3), \ldots, (e_{(n-2)}, e_{(n-1)}), (e_{(n-1)}, e_n) \tag{2}$$

For any two texts i and j we can have two sets of topic evolutions Si and S_j. Then, we calculate the Jaccard index between S_i and S_j to get the evolutional similarity between these two texts. By computing Jaccard index against all other texts, we can rank the similarity values and find the most relevant posts for a given text.

4   Evaluation

The evaluation consisted of two parts. First, we examined whether our temporal segmentation, which divided a text into multiple snippets by time entities, was effective and made separated segments contain the same amount of information. We believed this is one imperative prerequisite of this study because the topic evolution depended largely on chronological sequence of events. We hoped our segmentation could make all divided snippets hold similar amount of information but not just trivial collections of words. Different segmentation approaches were evaluated by clustering and Silhouette scores. In the second part, we compared the performance of the proposed evolution-based recommender with those generated by word-based or label-based recommenders. Considering the cost, we chose an offline approach which is commonly used in recommender evaluation.

4.1   Comparing Different Segmentation Approaches

To build topic evolution, appropriate ordering events in a text were required. Because it was difficult to process the order of events in a text, we assumed that all events were in chronological order and we could separate them by temporal entities. However, we wondered whether this approach was appropriate. Was it possible that our segmentation mishandled the situation and divided segments that were nothing but nonsense? To answer these questions, we conducted an experiment to compare the amount of information hold in divided segments (which are the snippets generated by separating one text with temporal entities).

We hypothesized that if the segmentation is meaningful, it would balance information contained in each generated segment out of a given text. Specifically, if all segment contains equal or similar amount of information, the Silhouette score should be low. On the other hand, if one dataset could be clustered effectively, we might infer that a portion of data points (segments in our case) shared more or less information than those of other portions. Thus, the segmentation made segments not contain the same amount of information. This is not a good segmentation under our assumption.

To evaluate our proposed temporal segmentation method, two other generic segmentation methods were used as primary and secondary baselines, including equal-length segmentation and sentence segmentation. The *equal-length segmentation* divides a post into equal-length passages in order. We set 27 as the predefined length because we observed the average length of segments made by temporal segmentation was 27. The second approach, *sentence segmentation* was literally to separate a post sentence by sentence. Finally, the third method was the proposed *temporal segmentation*. For the outcome of each method, we ran the hierarchical clustering and calculated the Silhouette scores by using from 1 cluster to 800 clusters. The result is showed in Fig. 1.

The result adhered to our expectation. The temporal segmentation outperformed its siblings because it got the lowest Silhouette score than the others (lower was better). From this result we could infer that the temporal segmentation might be meaningful, at lease when compared to equal-length and sentence segments.

4.2   Evaluating Recommendation Performances

To evaluate how topic evolution affected content-based recommendation, we compared the proposed recommender with two baselines. The first baseline was to retrieve content based on Latent Semantic Index (LSI) word embedding (labeled as *R1*). A secondary baseline was based on word embedding generated by only semantic labels (labeled as *R2*). The last recommendation method utilized the topic evolution we built in this work (labeled *R3*).

LSI recommender is trained by LSI word embedding. Another recommender was trained by semantic labels, which were aforementioned high-level cTAKES tags. We used cTAKES tags for not only analyze the evolution but also treated them as standalone training data for recommender. The last recommender was based on evolution and the detail is described in the previous section. In addition, it should be noted that because of some empty lists made by evolution-based recommendation, only 540 posts without any empty recommendation list were considered in this test.

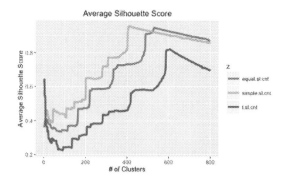

Fig. 1. Averaged silhouette scores of clustering by different segmentation. The red line (equal.sl.cnt) stands for equal-length segmentation, the green line (simple.sl.cnt) is for sentence segmentation, and the blue line (t.sl.cnt) is the temporal segmentation. The x-axis shows numbers of clusters used to calculate Silhouette score. The y-axis shows averaged Silhouette scores. This figure shows temporal segmentation might be a better approach comparing to sentence and equal-length segmentations. (Color figure online)

We took topic tags given by WebMD as comparable metric across different recommendation. Several examples are reported in the Fig. 3. Our assumption was that users find posts relevant if they had the same WebMD topic. For each input post, we calculated the precision value of top-20 similar items to show how accurate one recommender was. The result is shown in Fig. 2 (left). The recommendation generated by semantic labels (R2) outperformed others, which means using semantic labels as recommender criterion might be helpful to find more relevant content than using other ways. The content-based recommendation (R1) performed the worst, and the precision from evolution-based recommendation (R3) did not seem to be appealing at first glance.

However, when we checked the retrieved labels in Fig. 3, we could see that the posts recommended by evolution might actually be relevant to input post content but were assigned to different WebMD topics. Since the precision score did not consider synonyms, it was possible the recommendation was relevant but appeared in different forms. To further test this derived assumption, we first measured the relationship between different WebMD topics and then calculated the topic relevance to the target tag (the WebMD tag of an input text). The result is reported in Fig. 2 (right).

From the Fig. 2 (right), we could find the recommendation made by evolution-based recommendation (R3.Sim) was relevant to the input text even though it got the lower precision than the score of semantic-label recommendation. R3 shows comparable relevance to R2. We did not find significance in statistical test between R2.Sim and R3.Sim, which inferred that the outcomes of semantic-label recommender and the evolution-based recommender were equally relevant to the input posts. We concluded that evolution-based recommendation could suggest posts which were relevant but might have different assigned topics. A single tag might be too ambiguous for user-generated medical content and could not fully reveal the underlying information. However, the proposed evolution-based recommendation considered the granularity of sentences and produce broader but relevant results for a given text. We believed such property was helpful when it came to medical diagnosis that analyzed compound evidences.

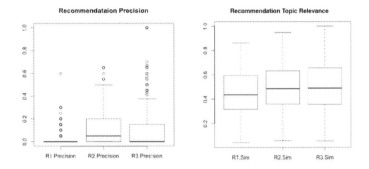

Fig. 2. Recommendation precision (Left) and Relevance (Right) of the three recommenders. R1, R2, and R3 stands for the results generated by LSI, semantic labels, and topic evolution, respectively. The y axes show the values of recommendation criteria (precision or relevance).

5 Summary

In this work, we proposed a topic-evolution based recommender for user generated medical content. Topic evolution algorithm learned topic changing traces in texts. To make recommender apprehend such information, we integrated a topic-evolution algorithm, TEvA, used the outcomes to train a recommender, and compared its performance with conventional recommenders trained by high-level labels and low-level word embedding. The result showed the evolution-based recommender could produce broader but relevant content to the input text in the offline evaluation.

Topic evolution might be essential when it comes to medical content analysis. Traditional approaches like word-level or concept-level features ignore the granularity of sentences and the progress of events described in a text snippet. In our experiment, we showed that topic-based recommender can reach comparative results to concept-based recommender. However, the performance in real application still requires further studies.

5.1 Limitations and the Future Works

There are several limitations in this study. The first is that our dataset may not be sufficient to generalize the results. Our dataset consists of 25319 posts made by general users. Nevertheless, due to the nature of user-generated content, some

Content	WebMD.Topic	By.Plain.Content	By.Semantic.Label	By.Evolution
I have been on the same birth control for nearly 3 years now. Never had period problems. Recently I will have a period 2 weeks before I am suppose to for 2 days and also have a normal period at the right time for about 5 days. I have bleeding throughout the month. I also have lower abdominal pain randomly. I get sever cramps and bloating with my normal period. I can loose 5 LBS just from having a period. Sometimes I have a dark bloody discharge throughout the month. HELP! Should I see a MD?	abdominal-pain	period confusion bloating condom period electrogastrogram condom breast ovulation irregularity	birth-control abdominal-pain abdominal-pain abdominal-pain abdominal-pain back-pain cramps birth-control abdominal-pain abdominal-pain	abdominal-pain birth-control cramps ejaculation birth-control birth-control period ovulation cramps
I have lost 20 lbs through weight watchers. I lost 10lbs then joined weight watchers for a total loss of 30 lbs. I have high blood pressure for a number of years controlled by medication. Diovan 80 mg once a day. Just recently I have noticed my blood pressure going up ex. 177/112 normally 130/80. My Dr increased the Diovan to two tabs a day one am and one pm. This Rx is helping somewhat. I am just wondering if significant weight loss can add to the high blood pressure? Thank You, Sandra Deustachio	blood-pressure	chest burn nausea ligament burn brain breast back-pain back-pain	antibiotic coldness blood-pressure blood-pressure blood-pressure blood-pressure back-pain diet chest	blood-sugar breastfeed dizziness enema drinking back-pain period pregcy ovulation burn
My father is a 15 year pancreatic cancer survivor. In 1999 he had a successful whipple, radiation and chemo. Last year he was diagnosed with stomach cancer. He had surgery to have most of his stomach removed and another round of chemo. Now they have found the stomach cancer is still there and there are growths ON the pancreas. Is this pancreatic cancer again? What are viable treatment options - he can't have anymore radiation there and no one is talking about surgery, just more chemo. Survival?	pancreatic-cancer	back-pain burn ligament arm movement back-pain nausea labor drinking appendicitis	cancer hair cancer bladder breast-cancer cervical-cancer prostate-cancer cramps breast dog	family bowel-movement cancer diarrhea nausea chest bladder irregularity corpulence back-pain

Fig. 3. Examples of original content and recommended topics from different recommenders.

of them do not meet the quality requirement as analysis subjects. For example, one issue is that the content is not long enough. In the field of natural language processing, short text has been considered a difficult issue when building linguistic models. When we applied our recommender to such texts, we could not even generate any result.

Second, our evaluation metric is limited. There are two major approaches to evaluate a recommender: online and offline evaluations. For an online evaluation, a prototypical recommender is deployed in the real system and researchers might conduct a controlled test to compare multiple recommenders. In an offline evaluation, people use the existing evidence like user rating to do cross validation and select the right model. However, in this study we cannot conduct online evaluation because we do not own the website to check whether improvement exists. We are not fully qualified for offline evaluation because there is no user rating like data existing for our dataset (or it is too scarce to use). Thus, we resorted to a semi-optimal solution which used WebMD topic tag to do precision evaluation.

References

1. Blei, D.M., Ng, A.Y., Jordan, M.I.: Latent dirichlet allocation. J. Mach. Learn. Res. **3**(Jan), 993–1022 (2003)
2. Cao, N., Sun, J., Lin, Y.R., Gotz, D., Liu, S., Qu, H.: Facetatlas: multifaceted visualization for rich text corpora. IEEE Trans. Vis. Comput. Graph. **16**(6), 1172–1181 (2010)
3. Danowski, J.A.: Wordij version 3.0: semantic network analysis software. University of Illinois at Chicago, Chicago (2013)
4. Entwistle, V.A., et al.: How information about other people's personal experiences can help with healthcare decision-making: a qualitative study. Patient Educ. Couns. **85**(3), e291–e298 (2011)
5. Fahy, E., Hardikar, R., Fox, A., Mackay, S.: Quality of patient health information on the internet: reviewing a complex and evolving landscape. Australas. Med. J. **7**(1), 24 (2014)
6. Hripcsak, G., Albers, D.J., Perotte, A.: Exploiting time in electronic health record correlations. J. Am. Med. Inf. Assoc. **18**(Suppl._1), i109–i115 (2011)
7. Introne, J.E., Drescher, M.: Analyzing the flow of knowledge in computer mediated teams. In: Proceedings of the 2013 Conference on Computer Supported Cooperative Work, pp. 341–356. ACM (2013)
8. Lin, Y.L., Chung, C.Y., Kuo, C.W., Chang, T.M.: Modeling health care Q&A questions with ensemble classification approaches. In: Twenty-Second Americas Conference on Information Systems (2016)
9. Lukyanenko, R., Parsons, J., Wiersma, Y.F.: The IQ of the crowd: understanding and improving information quality in structured user-generated content. Inf. Syst. Res. **25**(4), 669–689 (2014)
10. Mei, Q., Zhai, C.: Discovering evolutionary theme patterns from text: an exploration of temporal text mining. In: Proceedings of the Eleventh ACM SIGKDD International Conference on Knowledge Discovery in Data Mining, pp. 198–207. ACM (2005)
11. Nowrouzi, B., Gohar, B., Nowrouzi-Kia, B., Garbaczewska, M., Brewster, K.: An examination of scope, completeness, credibility, and readability of health, medical, and nutritional information on the internet: a comparative study of wikipedia, webmd, and the mayo clinic websites. Can. J. Diab. **39**, S71 (2015)
12. O'Neill, B., Ziebland, S., Valderas, J., Lupiáñez-Villanueva, F.: User-generated online health content: a survey of internet users in the United Kingdom. J. Med. Internet Res. **16**(4), e118 (2014)
13. Perer, A., Wang, F., Hu, J.: Mining and exploring care pathways from electronic medical records with visual analytics. J. Biomed. Inf. **56**, 369–378 (2015)
14. Savova, G.K., et al.: Mayo clinical text analysis and knowledge extraction system (ctakes): architecture, component evaluation and applications. J. Am. Med. Inf. Assoc. **17**(5), 507–513 (2010)
15. West, V.L., Borland, D., Hammond, W.E.: Innovative information visualization of electronic health record data: a systematic review. J. Am. Med. Inf. Assoc. **22**(2), 330–339 (2014)
16. Wiesner, M., Pfeifer, D.: Health recommender systems: concepts, requirements, technical basics and challenges. Int. J. Environ. Res. Public Health **11**(3), 2580–2607 (2014)

Massive-Scale Models of Urban Infrastructure and Populations

Daniel Baeder, Eric Christensen, Anhvinh Doanvo, Andrew Han,
Ben F. M. Intoy[✉][iD], Steven Hardy, Zachary Humayun, Melissa Kain,
Kevin Liberman, Adrian Myers, Meera Patel, William J. Porter III,
Lenny Ramos, Michelle Shen, Lance Sparks, Allan Toriel, and Benjamin Wu

Deloitte Consulting LLP, Arlington, Virginia, USA
bintoy@deloitte.com

Abstract. As the world becomes more dense, connected, and complex,
it is increasingly difficult to answer "what-if" questions about our cities
and populations. Most modeling and simulation tools struggle with scale
and connectivity. We present a new method for creating digital twin simulations of city infrastructure and populations from open source and com-
mercial data. We transform cellular location data into activity patterns
for synthetic agents and use geospatial data to create the infrastructure
and world in which these agents interact. We then leverage technologies
and techniques intended for massive online gaming to create 1:1 scale
simulations to answer these "what-if" questions about the future.

Keywords: Simulation · Urban mobility · Pattern mining

1 Introduction

There is tremendous value to studying complex systems, but these systems cannot be understood simply by analyzing their individual components [2]. If a complex system cannot be studied analytically, an alternative option is to explore the
system experimentally. However, these experiments may be difficult to perform
due to cost or infeasibility. An easier, more rapid, and more scalable solution
would be to instead model and then simulate the system.

As used in this document, "Deloitte" means Deloitte Consulting LLP, a subsidiary of
Deloitte LLP. Please see www.deloitte.com/us/about for a detailed description of our
legal structure. Certain services may not be available to attest clients under the rules
and regulations of public accounting. This publication contains general information
only and Deloitte is not, by means of this publication, rendering accounting, business,
financial, investment, legal, tax, or other professional advice or services. This publication is not a substitute for such professional advice or services, nor should it be used
as a basis for any decision or action that may affect your business. Before making
any decision or taking any action that may affect your business, you should consult a
qualified professional advisor. Deloitte shall not be responsible for any loss sustained
by any person who relies on this publication.

© Springer Nature Switzerland AG 2019
R. Thomson et al. (Eds.): SBP-BRiMS 2019, LNCS 11549, pp. 113–122, 2019.
https://doi.org/10.1007/978-3-030-21741-9_12

Modeling and simulation are commonly used techniques for gathering data from, and making predictions inside of, complex systems. This is especially important if the data are sparse, private, or difficult to collect.

There are many cases where we can use simulation to model scenarios and their possible outcomes [1,4,11]. However, these simulations use static or random data to drive entity behavior. With data on the movements of individuals becoming more readily available, implementing realistic and anonymized human travel behavior into models is now a possibility [5].

Currently, it is neither efficient nor feasible to know the position of every single person in a large population. However, by using a subset of the population, it is possible to generate general travel schedules [8,15]. This is advantageous because it allows a user to simulate large populations without using personally identifiable information. It also gives a user the ability to generate different, but statistically similar, data sets in order to test the robustness of the system.

A city is more than its population. Cities also include infrastructure such as power, roads, telecommunications, water, etc. that are interconnected and are controlled by people or autonomous systems, that, in turn, are influenced by infrastructure. This high degree of connectivity combined with uncorrelated dynamics means that changes cascade through and across the layers in the system. We are able to simulate these interactions and visualize all of these different layers in a format that allows for user interactivity. To make the simulation tractable at scale, we distribute the micro-simulations in a cloud environment [3].

2   From Geolocations to User Travel Schedules

In order to simulate realistic human behavior, we aim to construct synthetic traffic patterns for individuals in a metropolitan area using geolocation data sourced from cellular phones. These patterns take the form of generated individual schedules of stationary locations with associated start and end time values. These schedules are then used in a cloud-based simulation environment to model traffic behavior. Our work builds off of a process that uses cellular call detail records to train an input-output hidden Markov model (IOHMM) to generate models of individual activity [8,15]. Following the methodology in [8], we do not make any native assumptions that user behavior will adhere to a standard pattern (e.g., home-work-leisure-home). Instead, we intentionally capture irregularities in travel patterns across individuals to model aggregate behavior more completely. However, our approach diverges from earlier work in two important ways. First, we use individual position data derived from a mobile location service that provides more frequent and precise information. Second, we import our generated individual travel schedules to a simulated urban environment through which we can capture effects on behavior stemming from environmental changes (e.g., adding or removing roads on which users can travel). Figure 1 illustrates the process of transforming raw data into synthetic individual user schedules.

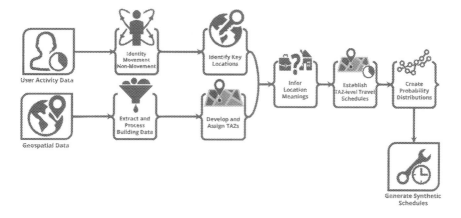

Fig. 1. Process of constructing synthetic user schedules

2.1   Identifying Stationary Periods

The raw geolocation data cover approximately 100,000 unique users per day. In their raw form, these spatio-temporal data points do not capture whether an individual is moving or stationary at any given point, nor do they record activity over time. We transform the data into time-ordered lists of start and end positions and times from which we can extract distances traveled and average velocities achieved.

To build user schedules from point-to-point data, we identify discrete periods of movement and non-movement. The simulation environment uses pathfinding algorithms that require defined start and end points, so we focus on isolating periods of non-movement, known as stay locations. A user's schedule of activities contains these stay locations and the amount of time spent at each. This process has the added benefit of significantly reducing the data size, as many data points can exist during a single movement or non-movement period.

Each user data point has an associated latitude and longitude position, but the positioning methodology used to create the raw data introduces some error. To account for this, we set a minimum threshold for detecting movement between data points at 0.05 miles. Travel distance less than that threshold over a period of time is classified as stationary behavior. We also control for several contingencies, including (1) instances where data points are so frequent that travel distance fails to exceed our threshold, and (2) short breaks in travel between periods of clear movement (i.e., an individual stopped at a traffic light), by classifying these periods as movement. There were 2,695,624 entries in our raw sample, which we reduced by 89% to 293,759 entries after completing this process. Even so, this initial data reduction is intended to be conservative in order to retain as much user activity data as possible.

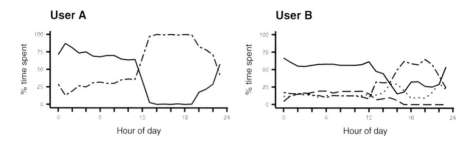

Fig. 2. Behavioral profiles of weekday stay locations for two users. User A has a clear two-location profile, with possible home and work locations. User B has a complex profile, with possible home and three work/leisure locations.

2.2   Identifying Key User Locations

We use a density-based clustering algorithm (DBSCAN) to cluster multiple data points that reflect a single user location but appear as multiple points due to geolocation error. The DBSCAN algorithm is well-suited for this problem because it operates independently of any preconceived notion of a system's number of clusters and the shapes of those clusters [12]. We parameterize the algorithm by specifying the maximum allowable radius (r) from each seed point (s) for points (p) to be considered part of the same cluster (C) defined in Eq. 1. We also specify the minimum percentage of the user's time (t), defined in Eq. 2, that is captured in the data for a group to be considered a cluster. Thus, points which meet the following criteria are combined into clusters:

$$C_r(s) : \{p \mid d(s,p) \leq r\}, \tag{1}$$

$$t_C \geq 0.1 * t \quad \forall_C, \tag{2}$$

where t_C is the time spent in cluster C and $d(s,p)$ is the great-circle distance between s and p.

The clustering process provides us with two classes of stationary behavior: clustered locations where user data points appear repeatedly and irregular travel locations where users spent time, but not with enough frequency to be captured as a cluster. From this output, we are able to link key user locations to time periods and generate behavioral profiles of users based on the percentage of each hour of the day that a user spends in each of their stationary locations. We are able to generate these behavioral profiles for any user in the dataset, as exemplified in Fig. 2, and, at this point, can derive user schedules without location labels for use in the simulation environment.

2.3   Inferring Cluster Meaning

User stay location clusters are coded numerically to indicate their importance to the user in terms of total time spent in each cluster, but the meaning of each cluster might differ across users due to high variability in the dataset. We specify a Gaussian mixture model (GMM) to group clusters across all users by temporal characteristics. A GMM is an appropriate choice in situations where there is a strong assumption of underlying groupings in the data, but those groups are unlabeled and unobserved. The algorithm iterates over an expectation-maximization process to assign unlabeled observations to clusters based on the probability of each point belonging to a cluster given its variance-adjusted distance from that cluster centroid. We conservatively estimate to expect two clusters, home and not-home, opting to handle irregular travel locations separately.

We link each stay location cluster to an individual building to provide further context to user activities. We use open source data on tax parcels to extract centroid positions for each one [13]. We then estimate building type as being "residential" or "non-residential" based on the percentage of total residential square footage indicated for each parcel. Lastly, we link each user stay location, both clustered and irregular travel points, to a building using a nearest neighbor approach, with distances between points calculated based on the World Geodetic System Ellipsoid. Thus, if a user's primary stay location shares temporal characteristics with locations we label as home, and is also linked to a residential building, we can more confidently say that location is the user's home. Accurate labeling of primary user locations is critical to properly train the IOHMM. A rules-based approach to home and work location labeling would make sense in regions where tax parcel data is unavailable

2.4 Assigning User Stay Locations to Traffic Analysis Zones

To simultaneously abstract from specific user locations and reduce our risk of overfitting, we establish traffic analysis zones (TAZ), opting to use Census geographies in order to simplify our overall methodology [14]. The selection criteria for appropriate TAZs is an objective function defined by a minimal overall number of potential user locations and a maximal retention of user transition information (i.e., avoiding recursive transitions whenever possible). We find that Census blocks are the optimal choice given these criteria (see Table 1). The result is a reduction in the location feature space by 95% with a small loss in user transition detail, as measured by the percentage of time captured in each user's top ten stay locations or zones. Census blocks are also highly homogeneous in terms of building type included within each block, which protects against user transition information loss, since individuals often travel between residential and non-residential locations. Lastly, the use of Census blocks could allow for a more in-depth investigation of the relationship between the travel patterns and socioeconomic characteristics of users since Census demographic and housing data is easily attached to block polygons.

We link each user stay location to a TAZ based on the position of the stay location in the case of irregular travel, or based on the position of the cluster centroid in the case of clustered stay locations returned from the DBSCAN algorithm. This results in the aggregation of close by irregular travel locations for each user to a single TAZ, capturing broader areas of frequent activity. We

Table 1. Traffic analysis zones. The third column is the percentage of user time captured in the top ten locations or zones. The fourth column is the percentage of zones with high building type concentration (95% or greater of one type).

Geography type	Feature size	User time captured	High building concentration
Building layer	1,373,831	0.847	1
Census blocks	63,305	0.925	0.734
Census block groups	3,238	0.934	0.485
Census-defined TAZs	1,114	0.935	0.376

then condense the data to a list of distinct locations with start and end times for each user, along with labels of estimated location meaning. From this matrix, we can calculate the distribution of times during which users transition between stay locations, both at the individual and sample population levels. We also calculate the geographic distribution of users' primary cluster locations. We will extend this methodology to a generative approach utilizing an IOHMM architecture as in [15].

3   Simulation Architecture

In order to achieve the scale, density, and interconnectivity present in modern cities in the simulation, we must distribute it over multiple compute nodes. For each simulation layer, we assign several computers to process and implement updates to the entity states at each time step. The world is partitioned into disjoint regions that are assigned to a computer and are identified for exclusive *write* access. A computer can change the state of all the entities, which may or may not interact with each other, within their *write* partition.

Areas on and near the boundary between *write* partitions contain information that is potentially relevant to state changes of entities that may not be within the same *write* partition. We identify these areas as areas with *read* access, where computers send information about state changes to other computers that have *read* access over those entities. When interactions and updates only depend on local information, a cloud distribution of this sort will scale easily to larger geographical areas.

To understand how our simulated population effects, and is affected by, the rest of its environment, we create a digital twin and model cars, traffic signals, telecommunications, and power at a microscopic level. Each car is a simulation agent that follows a set of logic either dictated by rules or by artificial intelligence (AI). By using the travel schedules generated in the previous section, the cars travel from destination to destination, routing between activity points via the road network.

We use OpenStreetMap (OSM) [10], an open- and crowd-sourced database, to generate the network of roads, buildings, and traffic lights in which the entities exist in the simulation. Each road in the network is assigned a weight that

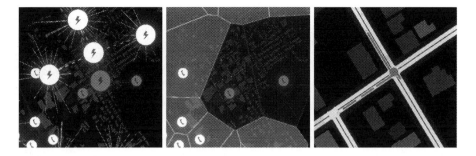

Fig. 3. Example images of power failure cascade. The leftmost image contains power and telecommunication nodes, with cartoon thunderbolt symbols representing power nodes, the phone symbols representing telecommunication nodes, and red denoting a node that is not working. The center image contains telecommunication nodes, with Voronoi areas with service in a lighter color and areas without in a darker color. The rightmost image is an intersection with cars (red and green rectangles along the roads) backing up due to a powered down traffic signal and loss of navigation. (Color figure online)

can take the form of the length or the expected time to traverse. A hierarchical bi-directional A-Star algorithm [6] provides the optimal path subject to either metric. As these cars traverse the road network, they stop at traffic lights and stop signs which are parsed from OSM. They are also restricted by a desired follow distance to the car in front of them and follow empirically derived kinematics [7]. We take the location of telecommunication towers from the Open-CellID database [9], and infer the power network structure from the building distribution given by OSM.

We can add dynamic, human-in-the-loop interaction with power and telecommunication entities, and changes to the road network. When users disable power network entities, the dependent traffic lights and telecommunication antennas are disabled (see example images in Fig. 3). In our simulation, cars connected to disabled telecommunication antennas change their routing behavior to prioritize high throughput roads and the disabled traffic lights toggle to behave as inefficient stop signs.

4 Using Data to Drive Entity Behaviors

We can use the empirically derived data to drive the entity behaviors. We achieve this by importing user schedules generated from geolocation data (detailed in Sect. 2) to dictate the agents within the simulation (detailed in Sect. 3). By using these generated data, the number of agents within the simulation can exhibit realistic behavior at any scale. This can lead to non-trivial, complex, and emergent behavior, such as rush-hour traffic, that may not be seen in simpler models.

The user can then perturb the system. An obvious example is to alter the road network and observe the differences of the traffic flows and patterns before

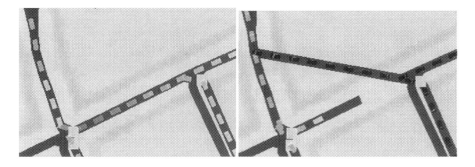

Fig. 4. A before and after example of rerouting by adding and removing a road.

and after the change. Figure 4 shows how traffic is rerouted with the addition and removal of roads. Changes to the road network can be adding or removing roads, altering the speed limit of a road, and automated adjusting of stop light timing. We can schedule changes to happen at a predetermined time during the simulation, allowing the study of shock events.

Just as we can perturb the environment, we can perturb the agents themselves. A simple example is an event where a subset of the agents converge on a location, which would approximate a sporting event, concert, political rally, etc. Another example would be to change the AIs of the cars. This could include a mixture of self-driving and human-driven cars and examining how their combined presence would impact traffic patterns. An advanced perturbation would be infrastructural in nature. An example would be a loss of the electrical power layer, which then causes failures in telecommunication and loss of navigation abilities for car agents. This causes the cars to either take less efficient paths or prefer commonly-known roads to reach their destinations.

Data are collected from the simulation following these perturbations, such as vehicle travel times, speeds, and distances. These data are gathered on a system-wide scale (such as an average speed) or on a local scale (such as the traffic volume of a single road). We can also gather data on an individual level (i.e. a single car). We can then read the database to produce real time level of service metrics. Examples of this include car density and road congestion heat maps, and power and telecommunication network loads.

5   Summary and Discussion

Because of its ability to provide analysis and insight both with, and in the absence of, previously existing data, we can use the simulation to answer questions about the future. These "what-if" questions are varied, from natural disaster planning ("What happens if we face a major storm and are forced to evacuate millions of people?") to military preparedness ("What are the mission impacts if we lose the ability to operate from certain roads or ports?") to business modeling

Table 2. Potential use cases

Sector/Industry	Use case
Defense	Evacuating an area in the event of a natural disaster or man-made disruption
Supply chain	Moving people, equipment, and supplies around the world in a dynamic supply, demand, and transportation network
Urban planning	Managing the impacts of massive area changes (ex. flooding, migration)
Urban planning	Managing the impacts of new construction projects (housing, commerce, industry)
Transportation	Improving mobility in a metropolitan area through changes to infrastructure, public transportation, and public-private partnerships
Transportation	Investigating consequences of a road closure during a large event (ex. the Olympics, a presidential visit, the building of a stadium or theme park)
Telecom	Optimizing the deployment of 5G infrastructure to serve cellular customers today, and self-driving vehicles and IoT devices tomorrow

("What are the implications if our company implements self-driven ride sharing or drone delivery?").

We build a digital twin of a city using information about building locations, cell towers, and the electrical grid. We drive entity behavior with user travel schedules generated from geolocation data. These data, captured from cellular phones, are initially analyzed to identify periods of movement and non-movement. We apply density-based clustering to group nearby points of non-movement into single locations if certain criteria are met. Next, we infer location labels (home or not-home) by temporal characteristics along with the inferred use type of the nearest building. Next, we link all stay locations to Census blocks to abstract away from exact coordinate locations, and calculate transition time and geographic distributions across the dataset.

With the help of cloud computing, we can run the simulation and model many infrastructural layers of a city, including traffic, telecom, and the electrical grid. Lastly, we can perturb the system by introducing some new future scenario to see how the system responds. The data and insights gleaned from this process can allow policymakers and corporations to take proactive measures to exploit opportunities and minimize vulnerabilities in the future. See Table 2 for potential use cases.

In the past, reactive and suboptimal decisions were often made because a problem was not anticipated, or because it was too difficult to prepare for. Even when ample time is available to make a careful decision, the interconnectedness of our world makes it difficult to anticipate the full impact of each option, or to identify the best one. It would not be safe or feasible to shut off the electrical grid over a large portion of Manhattan to observe how the telecom layer reacts, or to build new roads in Detroit for a traffic simulation and then remove them a few minutes later. However, the simulation platform allows these types of scenarios to be reproduced and tested safely and quickly. With the help of the capabilities described in this paper, these types of simulations are no longer limited by scale, complexity, or ability to consider collateral and cascading effects. The simulation platform allows decision makers to take action in a manner that is proactive rather than reactive. By combining widely available data, cloud computing, agent-based simulation methods, and data science, we demonstrate a new approach for proactive planning and decision making where future scenarios can be evaluated in a high-fidelity virtual environment at low risk and cost.

References

1. Adiga, A., Marathe, M., Mortveit, H., Wu, S., Swarup, S.: Modeling urban transportation in the aftermath of a nuclear disaster: the role of human behavioral responses. In: The Conference on Agent-Based Modeling in Transportation Planning and Operations. Citeseer (2013)
2. Anderson, P.W., et al.: More is different. Science **177**(4047), 393–396 (1972)
3. Attiya, H., Welch, J.: Distributed Computing: Fundamentals, Simulations, and Advanced Topics, vol. 19. Wiley, Hoboken (2004)
4. Barrett, C., et al.: Cascading failures in multiple infrastructures: from transportation to communication network. In: 2010 5th International Conference on Critical Infrastructure (CRIS), pp. 1–8. IEEE (2010)
5. Barrett, C.L., Eubank, S., Marathe, A., Marathe, M.V., Pan, Z., Swarup, S.: Information integration to support model-based policy informatics. Innov. J.: Public Sector Innov. J. **16**(1) (2011). https://www.innovation.cc/volumes-issues/vol16-no1.htm
6. Goldberg, A.V., Kaplan, H., Werneck, R.F.: Reach for a*: efficient point-to-point shortest path algorithms. In: 2006 Proceedings of the Eighth Workshop on Algorithm Engineering and Experiments (ALENEX), pp. 129–143. SIAM (2006)
7. Kesting, A., Treiber, M., Helbing, D.: Enhanced intelligent driver model to access the impact of driving strategies on traffic capacity. Philos. Trans. R. Soc. Lond. A: Math. Phys. Eng. Sci. **368**(1928), 4585–4605 (2010)
8. Lin, Z., Yin, M., Feygin, S., Sheehan, M., Paiement, J.F., Pozdnoukhov, A.: Deep generative models of urban mobility. IEEE Trans. Intell. Transp. Syst. (2017)
9. OpenCell ID. Data. http://opencellid.org/downloads. http://opencellid.org
10. OpenStreetMap contributors: Planet dump (2017). https://planet.osm.org. https://www.openstreetmap.org
11. Ren, Y., Ercsey-Ravasz, M., Wang, P., González, M.C., Toroczkai, Z.: Predicting commuter flows in spatial networks using a radiation model based on temporal ranges. Nat. Commun. **5**, 5347 (2014)
12. Simoudis, E., Han, J., Fayyad, U.M. (eds.): A density-based algorithm for discovering clusters in large spatial databases with noise. Proceedings of the Second International Conference on Knowledge Discovery and Data Mining. AAAI Press (1996)
13. Southeast Michigan Council of Governments: Data portal. http://maps-semcog. opendata.arcgis.com/datasets
14. US Census Bureau: Tiger/line shapefiles. https://www.census.gov/cgi-bin/geo/ shapefiles/index.php
15. Yin, M., Sheehan, M., Feygin, S., Paiement, J.F., Pozdnoukhov, A.: A generative model of urban activities from cellular data. IEEE Trans. Intell. Transp. Syst. **19**(6), 1682–1696 (2017). https://doi.org/10.1109/TITS.2017.2695438

Dynamic Resource Allocation During Natural Disasters Using Multi-agent Environment

Alina Vereshchaka[(✉)] and Wen Dong

Department of Computer Science and Engineering,
State University of New York at Buffalo, Buffalo, USA
{avereshc,wendong}@buffalo.edu

Abstract. Natural disasters are devastating for a country and effective allocation of critical resources can mitigate the impact. While traditional approaches usually have difficulties in making optimal critical resource allocation, in this paper we introduce a novel hierarchical multi-agent reinforcement learning framework to model optimal resource allocation for natural disasters in real-time. On the lower level a set of agents navigate with the continuous time environment using deep reinforcement algorithms. On the higher level, a lead agent takes care of the global decision-making. Our framework achieves more efficient resource allocation in response to dynamic events and is applicable to problems where disaster evolves alongside the response efforts, where delays in response can lead to increased disaster severity and thus a greater need for resources.

Keywords: Autonomous agents · Reinforcement learning ·
Resource allocation · Emergency management · Multi-agent system ·
Decision making · Natural disasters · Disaster management

1 Introduction

Natural disasters are important and devastating events for a country, which makes it difficult to make informed decisions in terms of allocating limited resources for mitigation efforts. According to the United Nations Office for Disaster Risk Reduction [1] natural disasters are happening more frequently, which has caused a rise of 151% over the last twenty years in economic losses from climate-related disasters. During 1998–2017, disaster-hit countries reported direct economic losses from natural disasters of US$2,245 billion and these hazards are responsible for 77% of the total losses in these countries. The USA has the greatest economic losses among other countries of US$ 944.8 billion. Over the last twenty years, disasters claimed more than 1.3 million lives, and more than 4 billion people were injured, rendered homeless or in need of assistance.

Scientific and technological advances provide various approaches for responding to the urgent need to mitigate the impacts of natural disasters. When reacting to real-time scenarios, traditional resource allocation approaches based on

© Springer Nature Switzerland AG 2019
R. Thomson et al. (Eds.): SBP-BRiMS 2019, LNCS 11549, pp. 123–132, 2019.
https://doi.org/10.1007/978-3-030-21741-9_13

human decision making systems usually have difficulties sorting out the most efficient resource assignment. One of the main reasons for this difficulty is the large number of data they have to go though. This inefficiency in turn can exacerbate the societal and economic costs of natural disasters.

In this paper we introduce the novel approach of reinforcement learning framework to model optimal resource allocation for natural disasters. Our hybrid model is a combination of social behavioural modeling and deep reinforcement learning. We propose a multi-agent real-time resource allocation framework to respond to the disaster scenarios. The model can be used by experts, decision makers, disaster managers and emergency personnel to assess the critical event and respond appropriately in order to mitigate the disaster effect in real-time. This framework can be applicable to various scenarios in response to the natural hazards, like allocating firetrucks in case of wildfire, allocating snow plows during the winter storms or allocating rescue crews during flooding.

We are concerned with the optimal recourse allocation problem in an online stochastic environment in the domain of disaster events. The goal is to design a system of multi-agents, which aim to maximize the cumulative reward, with a lead agent that will act as a decision maker based on the data provided by each individual agent. We are also incorporating a level of confidence for each of the sub-agents, along with sub-region importance that evolves over time, and the current and projected levels of hazards.

In our framework a multi-agent system models a lead agent's behaviour as a finite automaton. This model can be scalable, depending on the number of sub-regions. If the number of sub-regions is large enough, we can cluster sub-agents related to one domain and add a main-agent to each sub-region, thus increasing the number of levels in hierarchy. The main agent's ability to solve optimization tasks with a number of agents in the system demonstrates the flexibility of the model. Furthermore, no prior knowledge is needed for either the sub-agents or the lead agent. Because of these qualities in the framework, it can compute optimal strategies for decision making scenarios that arise during disaster response.

The main contribution of this paper is to propose a hierarchical multi-agent framework for solving resource allocation problems for natural disasters. On the lower level there is a set of multi-agents that navigates with the continuous time environment by using deep reinforcement algorithms. On the higher level, there is a lead agent that is responsible for decision making. We have also adjusted a policy gradient method to navigate within the time-series environment. Our framework can work on problems that incorporate nature and human life, e.g. wildfires, disease epidemics, and snowstorms, when the disaster evolves on the same timescale as the response effort, where delays in response can lead to increased disaster severity and thus greater demand for resources.

The rest of the paper is structured as follows. We begin with discussing the related works, followed by a preliminary material in Sect. 2. In Sect. 3, we present and analyze our approach and algorithms on incorporating autonomous agents on resource allocation problem. In Sect. 4, we give the results of our experiments. We conclude with a discussion on open problems in Sect. 5.

2   Background

2.1   Modeling Social Behaviour for Mitigation Efforts

There is abundant research that utilizes social behavior for disaster management. Recent studies show the benefit of extracting social media posts about injuries and help requests during disasters, such as floods, epidemic outbreaks [2,3], and post-earthquake conditions [4], and utilizing these posts to create a pool of reports for modeling a disaster relief response [5]. The effects of sending emergency broadcasts after disaster events has also been studied [6]. Petrovich et al. [7] introduced a framework that dynamically computes optimal strategies for decision making during wildfires. Furthermore, researchers have demonstrating that disaster management platforms can capture and manage the flow of information between rescue groups and the public [8].

2.2   Autonomous Multi-agents for Optimal Resource Allocation

The development of autonomous multi-agents that can effectively interact with other agents has recently been receiving more attention, as the number of domains where this type of architecture can be applied is increasing [9]. There have been recent advances in modeling multi-agent complex systems within the frameworks of Markov decision process and reinforcement learning, like discrete-event decision process [10–12], multi-players games [13], and multi-robot control [14]. Deep reinforcement learning has been used to solve a decentralized resource allocation mechanism for vehicle-to-vehicle communications [15]. Policy gradient methods have been widely applied to on-line environments, for example when solving micro-tolling assignment problems [16].

There has been a recent success [17] in solving ad hoc coordination problems, where an autonomous agent is able to find optimal flexibility and efficiency in a multiagent system when this agent is given limited access to the observed data. This proposed concept of Harsanyi-Bellman Ad Hoc Coordination [17] incorporates the benefits of both the Bayesian Nash equilibrium, along with the Bellman optimal control.

2.3   Decision Making Problem

First, we will introduce the notation we will use for the standard optimal control or reinforcement learning formulation. We consider innite-horizon Markov decision process (MDP), defined by the tuple $\langle S, A, O, P, r, \rho_0, \gamma \rangle$, where $s \in S$ denotes states, describing the possible configurations of all agents; $a \in A$ denotes actions, which can be discrete or continuous; $P : S \times A \times S \to \mathbb{R}$ is the state transition probability distribution, where states evolve according to the stochastic dynamics $p(s_{t+1}|s_t, a_t)$, which are in general unknown; O is a set of observations for each agents; $r : S \to \mathbb{R}$ is the reward function; $\rho_0 : S \to [0, 1]$ is the distribution of the initial state s_0; $\gamma \in [0, 1]$ is a discount factor. For our work we make assumption about the stochastic action choices, in which actions may be chosen with any probabilities to facilitate more robust prediction and planning. To choose an action, each agent uses a stochastic policy $\pi_\theta : O \times A \to [0, 1]$, which produces the next state according to the state transition probability. Each agent obtains rewards as a function of the state and agent's action $r : S \times A \to \mathbb{R}$, and receives a private observation correlated with the state $\mathbf{o} : S \to O$.

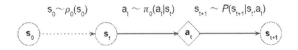

Fig. 1. Initial state s_0 is sampled from prior state visitation distribution $\rho_0(s_o)$; action a_t is sampled from the policy π_0 and the next state s_{t+1} is chosen based on the transition probability distribution $P(s_{t+1}|s_t, a_t)$

Solving a MDP means finding a policy π^* that maximizes the expected long-term reward $R = \sum_{t=0}^{T} \gamma^t r^t$, where T is the time horizon.

2.4   Policy Gradient Algorithms

Policy-gradient method is one of the most effective reinforcement learning techniques for complex real-world control problems with continuous, high-dimensional, and partially-observable properties, such as robotic control systems [18].

Policy gradient is directly adjusting the parameters θ of the policy (π) in order to maximize the objective $J(\theta) = E_{s \sim \rho_\pi, a \sim \pi_\theta}[R]$ by taking steps in the direction of $\nabla_\theta J(\theta)$:

$$\hat{g} = \hat{\mathbb{E}}_t \left[\nabla_\theta \log \pi_\theta(a_t|s_t) \hat{A}_t \right], \tag{1}$$

where π_θ is a stochastic policy and $\hat{A}_t$ is an estimator of the advantage function at timestep t. Here, the expectation $\hat{\mathbb{E}}_t[...]$ indicates the empirical average over a finite batch of samples. The estimator $\hat{g}$ is obtained by differentiating the objective:

$$L^{PG}(\theta) = \hat{\mathbb{E}}_t \left[\log \pi_\theta(a_t|s_t) \hat{A}_t \right] \tag{2}$$

3   Methodology

In this section, we first report on how we formulate the social behaviour problem of critical resource allocation as a reinforcement learning framework, and then we detail our algorithm for solving the problem.

We developed a two-level framework that involves a set of autonomous agents controlled by the lead agent, which makes a final decision on critical resource

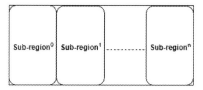

Fig. 2. The region that is affected by a disaster. It is divided into n sub-regions, thus every agent is responsible for one sub-region.

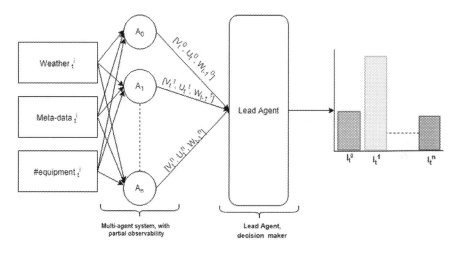

Fig. 3. An overview of our framework. Left: a view of the module with a bundle of independent agents. Note that the input data and parameters are unique for each agent. Right: a lead agent that makes an allocation decision based on the data broadcasted by each of the subagents.

allocation. To initialize our framework, we define the region that is affected by the disaster where the limited resources have to be allocated. Further, we divide the affected region into sub-regions to narrow down the areas (Fig. 2).

As shown on Fig. 3 our framework consists of two parts. On the lower level (left side) there is a set of multi-agents $(A_0, A_1, ..., A_n)$ that corresponds to the number of sub-regions (Fig. 2). Each agent predicts the volume of hazards (V_t^n) and the importance (U_t^n), taking as input an observation from the time-series environment. Each agent is treated as an autonomous agent in a multi-agent framework with partial observability. The total number of agents is bounded by the number of sub-regions. On the higher level (right side) there is a lead agent, which is trained to make resource allocation decisions for assigning volume distribution of involved resources. To make the decision the lead agent takes the volume of hazards (V_t^n) and the importance (U_t^n) as input from each of the lower level sub-agents.

Algorithm 1. PPO for multiple agents

Input   : initial policy parameter θ_0, clipped threshold ϵ
Output: projected hazard volume V_t^i, projected importance level U_t^i
for $k = 0,\ 1,\ 2,\ ...$ **do**
  Collect set of partial trajectories D_k on policy $\pi_k = \pi(\theta_k)$;
  Estimate advantage A_t using any advantage estimation algorithm;
  Compute policy update by taking K steps of minibatch SGD,

  $$\theta_{k+1} = \arg\max_\theta L^{CLIP}(\theta),\ \text{where}$$

  $$L^{CLIP}(\theta) = \mathbb{E}_t[min(r_t(\theta)A_t, clip(r_t(\theta), 1 - \epsilon, 1 + \epsilon)A_t].$$

end

3.1   Multi-agent System: Lower Level

A bundle of agents at the lower level perform predictions about the time-series environment. The processes are done independently for each sub-agent, thus making the framework applicable for real-time scenarios. We are using a modified version of a proximal policy optimization algorithm (PPO) [19]. Observations are unique for each agent and the parameters are not shared within them.

3.2   Resource Allocation: Higher Level

On the higher level of our architecture a lead agent takes predictions made by each of the sub-agents at the lower level and based on them makes a decision about the optimal resource allocation.

Now we will define the notations. Sub-agents $A_1, A_2, A_3, ..., A_n$, where n is the number of sub-agents, equivalent to the total number of sub-regions. For each time step t, agent i broadcasts (V_t^i, U_t^i) to the lead agent, where V_t^i is a projected volume of work that needs to be done at sub-region^i for the next time step, e.g. the volume of snow that has to be removed. U_t^i is a level of importance of sub-region^i, e.g. the projected traffic.

The lead agent gets as an observation W_t^i, a real volume of work that has been completed during the previous time step. The level of importance for each of the sub-regions is defined as a normalized product of the projected volume of work and the projected volume of traffic (Eq. 3), and the cumulative error to be minimized by the worker agents is the discounted relative error between the actual amount of work and the predicted amount of work (Eq. 4).

$$\text{Imp}_{t,norm}^i = \frac{\text{Imp}_t^i}{\sum_{i=1}^N \text{Imp}_t^i},\ \text{where}\ \text{Imp}_t^i = \frac{max[V_t^i, \epsilon]}{max[\sum_{i=1}^N V_t^i, \epsilon]}\frac{max[U_t^i, \epsilon]}{max[\sum_{i=1}^N U_t^i, \epsilon]} \quad (3)$$

$$\text{Error}_t^i = \frac{W_t^i - V_{t-1}^i}{max[V_{t-1}^i, \epsilon]} + \gamma\text{Error}_{t-1}^i,\ \text{where}\ \gamma\ \text{is a discounting factor.} \quad (4)$$

Algorithm 2. Resource allocation among regions per time t

Input   : volume of resource to be allocated M, projected volume of hazard V_t^i,
            projected level of importance U_t^i, error ϵ, discounting factor γ
Output: amount of allocated resources per each sub-region l_t^i
for $i = 0, 1, 2, \ldots$ **do**

> Estimate level of importance for each sub-region using Eq. 3;
> Normalize the importance parameter among all sub-regions;
> Calculate the cumulative error for each agent using Eq. 4;
> **if** $\sum_{i=1}^{N} \text{Imp}_{t,norm}^i (1 + \text{Error}_t^i) \leq 1$ **then**
>> Resource allocation at time t per sub-region i:
>>
>> $$L_t^i = \left\lfloor M \cdot \text{Imp}_{t,norm}^i (1 + \text{Error}_t^i) \right\rfloor \tag{5}$$
>
> **else**
>> Resource allocation at time t per sub-region i:
>>
>> $$L_t^i = \left\lfloor M \cdot X_t^i \right\rfloor, \text{ where } X_t^i = \frac{\text{Imp}_{t,norm}^i (1 + \text{Error}_t^i)}{\sum_{i=1}^{N} \text{Imp}_{t,norm}^i (1 + \text{Error}_t^i)} \tag{6}$$
>
> **end**

end

---

4   Experiments

For our experiment we applied our framework to the 2019 snowstorm disaster in the Western New York area. We collected daily snow observations from the Global Historical Climatology Network (GHCN) stations for New York, USA, for the period of January-February 2019[1]. During the period there was a heavy snowfall with high winds, which resulted in massive accumulations of snow across the area. Emergency managers deployed 1,602 large plow trucks with a total of 3,900 operators. As an importance parameter, we utilized the daily average volume of traffic, using traffic count data provided by the Greater Buffalo-Niagara Regional Transportation Council[2] that contains more than 32,000 entries.

The training used a 10-day period, taking one day as a time step. The total volume of plow trucks that our framework needed to allocate was 500. For the purposes of our experiment we sampled stochastic normal distributions to simulate a traffic dynamic at each time step for the sub-regions. At each time step, the lead agent makes a resource allocation which is distributed across six subregions. To show the impact of real-time traffic and snow depth on optimal resource allocation, we tabulate the root means square error between the predicted volume of work and the actual volume of work. Table 1 shows that the

---

[1] https://www.ncdc.noaa.gov/snow-and-ice/daily-snow/NY/snow-depth/20190131.
[2] https://data.buffalony.gov/Transportation/Annual-Average-Daily-Traffic-Volume-Counts/y93c-u65y.

real-time traffic information decreases the error by 60%, and the real-time snow information decreases the error by 75%.

Table 1. Root mean square error between predicted and actual volume of work, with and without traffic and snow information.

Model	MSE	RMSE
Random allocation	18108	134
Without traffic importance	3377	58
Without snow depth importance	1335	36

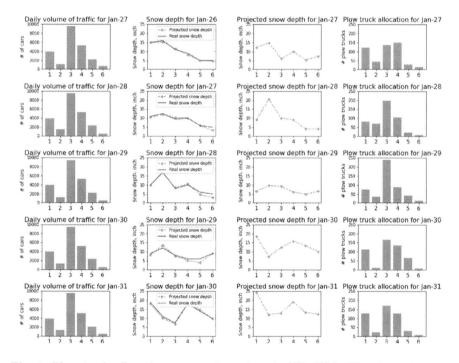

Fig. 4. Plow truck allocation among six regions in NY, USA. The six regions are Buffalo-Niagara, East Aurora, Hamburg, Kenmore, Tonawanda, Lockport

Figure 4 illustrates the daily traffic volume, the predicted and actual snow depth, and the resource allocation. The data is shown for a 5-day period across six regions in Western New York, USA. This figure has 4 columns of panels. The left most column shows that the average daily traffic over six regions is relatively stable during the 5-day period. The middle two columns show the actual snow depth on a day and the predicted snow depth on the next day. The worker agents can predict snow depth reasonably well with a policy gradient algorithm. The right-most column shows the resource allocation among the 6 regions during the 5 days. As shown in the plot, the prediction of snow depth from real time information by the worker agents is critical to make optimal resource allocation, and day-to-day fluctuations of traffic volume are also important for determining effective resource allocation.

As a result, a hierarchical multi-agent reinforcement learning framework that incorporates the distributed depth of snow and traffic information can help to achieve better resource allocation through real-time optimization.

5   Conclusions

We presented a novel approach to achieve optimal resource allocation during natural disasters by using a hierarchical multi-agent reinforcement learning. In this approach, a group of autonomous local agents with partial observations of the environment predict the time-series dynamics through deep reinforcement learning, and then a lead agent makes a high level decision on limited resource allocation based on the predictions by local agents.

A direction of improvement for the current framework would be to incorporate additional parameters, for example the cost of moving the resources from one region to another. We can also utilize a model based approach for training agents on the lower level, thus, while adding a new agent, it will allow the reuse of past experiences of the similar agents. This can improve the learning speed for new tasks.

Acknowledgements. We would like to thank Nathan Margaglio for insightful discussions of the resource allocation problem and SBP-BRiMS reviewers for feedback that has improved this work.

References

1. United Nations Office for Disaster Risk Reduction (UNISDR) and Centre for Research on the Epidemiology of Disasters (CRED), Economic losses, poverty and disasters, pp. 1998–2017 (2018)
2. Frias-Martinez, E., Williamson, G., Frias-Martinez, V.: An agent-based model of epidemic spread using human mobility and social network information. In: 2011 IEEE Third International Conference on Privacy, Security, Risk and Trust and 2011 IEEE Third International Conference on Social Computing, pp. 57–64. IEEE (2011)
3. Wesolowski, A., et al.: Quantifying the impact of human mobility on malaria. Science **338**(6104), 267–270 (2012)
4. Bengtsson, L., Lu, X., Thorson, A., Garfield, R., Von Schreeb, J.: Improved response to disasters and outbreaks by tracking population movements with mobile phone network data: a post-earthquake geospatial study in Haiti. PLoS Med. **8**(8), e1001083 (2011)
5. Abbasi, M.-A., Kumar, S., Filho, J.A.A., Liu, H.: Lessons learned in using social media for disaster relief - ASU Crisis Response Game. In: Yang, S.J., Greenberg, A.M., Endsley, M. (eds.) SBP 2012. LNCS, vol. 7227, pp. 282–289. Springer, Heidelberg (2012). https://doi.org/10.1007/978-3-642-29047-3_34

6. Chandan, S., et al.: Modeling the interaction between emergency communications and behavior in the aftermath of a disaster. In: Greenberg, A.M., Kennedy, W.G., Bos, N.D. (eds.) SBP 2013. LNCS, vol. 7812, pp. 476–485. Springer, Heidelberg (2013). https://doi.org/10.1007/978-3-642-37210-0_52

7. Petrovic, N., Alderson, D.L., Carlson, J.M.: Dynamic resource allocation in disaster response: tradeoffs in wildfire suppression. PloS one **7**(4), e33285 (2012)

8. Estuar, M.R.J.E., Rodrigueza, R.C., Victorino, J.N.C., Sevilla, M.C.V., De Leon, M.M., Rosales, J.C.S.: Agent-Based modeling approach in understanding behavior during disasters: measuring response and rescue in *eBayanihan* disaster management platform. In: Lee, D., Lin, Y.-R., Osgood, N., Thomson, R. (eds.) SBP-BRiMS 2017. LNCS, vol. 10354, pp. 46–52. Springer, Cham (2017). https://doi.org/10.1007/978-3-319-60240-0_5

9. Albrecht, S.V., Stone, P.: Autonomous agents modelling other agents: a comprehensive survey and open problems. Artif. Intell. **258**, 66–95 (2018)

10. Yang, F., Liu, B., Dong, W.: Optimal control of complex systems through variational inference with a discrete event decision process. In: Proceedings of the 2019 International Conference on Autonomous Agents and Multiagent Systems, International Foundation for Autonomous Agents and Multiagent Systems (2019)

11. Yang, F., Dong, W.: Integrating simulation and signal processing in tracking complex social systems. Comput. Math. Organ. Theor. 1–22 (2018). Special Issue: SBP-BRIMS2017

12. Yang, F., Dong, W.: Integrating simulation and signal processing with stochastic social kinetic model. In: Lee, D., Lin, Y.-R., Osgood, N., Thomson, R. (eds.) SBP-BRiMS 2017. LNCS, vol. 10354, pp. 193–203. Springer, Cham (2017). https://doi.org/10.1007/978-3-319-60240-0_23

13. Peng, P., et al.: Multiagent bidirectionally-coordinated nets for learning to play starcraft combat games, arXiv preprint arXiv:1703.10069 (2017)

14. Matignon, L., Jeanpierre, L., Mouaddib, A.-I.: Coordinated multi-robot exploration under communication constraints using decentralized Markov decision processes. In: Proceedings of the Twenty-Sixth AAAI Conference on Artificial Intelligence, pp. 2017–2023. AAAI Press (2012)

15. Ye, H., Li, G.Y., Juang, B.-H.F.: Deep reinforcement learning based resource allocation for V2V communications, arXiv preprint arXiv:1805.07222 (2018)

16. Mirzaei, H., Sharon, G., Boyles, S., Givargis, T., Stone, P.: Enhanced delta-tolling: traffic optimization via policy gradient reinforcement learning. In: 2018 21st International Conference on Intelligent Transportation Systems (ITSC), pp. 47–52. IEEE (2018)

17. Albrecht, S.V., Ramamoorthy, S.: A game-theoretic model and best-response learning method for ad hoc coordination in multiagent systems. In: Proceedings of the 2013 International Conference on Autonomous Agents and Multi-agent Systems, International Foundation for Autonomous Agents and Multiagent Systems, pp. 1155–1156 (2013)

18. Peters, J., Schaal, S.: Policy gradient methods for robotics. In: 2006 IEEE/RSJ International Conference on Intelligent Robots and Systems, pp. 2219–2225. IEEE (2006)

19. Schulman, J., Wolski, F., Dhariwal, P., Radford, A., Klimov, O.: Proximal policy optimization algorithms, arXiv preprint arXiv:1707.06347 (2017)

Parallelizing Convergent Cross Mapping Using Apache Spark

Bo Pu$^{(\boxtimes)}$, Lujie Duan, and Nathaniel D. Osgood

Department of Computer Science, University of Saskatchewan,
Saskatoon, SK, Canada
{bo.pu,lujie.duan,nathaniel.osgood}@usask.ca

Abstract. Identifying the causal relationships between subjects or variables remains an important problem across various scientific fields. This is particularly important but challenging in complex systems, such as those involving human behavior, sociotechnical contexts, and natural ecosystems. By exploiting state space reconstruction via lagged embedding of time series, convergent cross mapping (CCM) serves as an important method for addressing this problem. While powerful, CCM is computationally costly; moreover, CCM results are highly sensitive to several parameter values. While best practice entails exploring a range of parameter settings when assessing casual relationships, the resulting computational burden can raise barriers to practical use, especially for long time series exhibiting weak causal linkages. We demonstrate here several means of accelerating CCM by harnessing the distributed Apache Spark platform. We characterize and report on results of several experiments with parallelized solutions that demonstrate high scalability and a capacity for over an order of magnitude performance improvement for the baseline configuration. Such economies in computation time can speed learning and robust identification of causal drivers in complex systems.

Keywords: Causality · Convergent cross mapping · Spark · Parallelization · Performance evaluation

1 Introduction

Identification of causal relations between variables in many domains has traditionally relied upon controlled experimentation, or investigation of underlying mechanisms. The first of these requires a heavy investment of time and financial resources, and can pose ethical concerns. The limits of such controlled studies – such as Randomized Controlled Trials (RCTs) – are particularly notable in the context of complex systems, which exhibit reciprocal feedbacks, delays and non-linearities [6]. While ubiquitous in science, progress in the latter approach is commonly measured in decades. In recent years, a growing number of researchers have applied Pearl's causal inference framework [11] to identify causal linkages, but such methods can be challenging to apply in the presence of observability

© Springer Nature Switzerland AG 2019
R. Thomson et al. (Eds.): SBP-BRiMS 2019, LNCS 11549, pp. 133–142, 2019.
https://doi.org/10.1007/978-3-030-21741-9_14

constraints, and when the variables of interest are coupled but distant within the system. Most critically, in the context of complex systems, such inference techniques encounter challenges in the context of reciprocal causality.

Convergent cross mapping [13] is an algorithm based on Takens' embedding theorem [14] that can detect and help quantify the relative strength of unidirectional and bidirectional causal relationships between variables X and Y in coupled complex systems. Within the CCM algorithm, in order to assess if variable Y is causally governed by variable X, we attempt to predict the value of X on the basis of the state space reconstructed from Y (see below); for statistical reliability, this must be done over a large number r of realizations. To assess causality, we examine whether these results "converge" as we consider a growing numbers L of datapoints within Y within our reconstruction (also see below).

Overall, the results of CCM are sensitive to the parameters below:

- r The number of random subsamples, commonly 250 or larger.
- τ The embedding delay used in the shadow manifold reconstruction.
- E The embedding dimension of the dynamic system. For simplex projection, E will range from 1 to 10.
- L The size of the library extracted from time series.

Running CCM with a wide range of different parameter settings imposes a high computational overhead. But, as for many data science tasks, we believed that the performance could be elevated via parallel and distributed processing by implementing the CCM algorithm atop Apache Spark [19] (henceforth, "Spark") and distributing computations across a Yarn cluster [15].

In this paper, following additional background on CCM and the related literature, we describe a CCM parallel implementation which utilizes the MapReduce framework [2] provided by Spark. The paper then presents a performance evaluation and comparison of the framework. We can conclude from the experiments that, with the parallel techniques and cloud computing support, researchers can use CCM to confidently infer causal connections between larger time series in far less time than previously required.

2   Background

2.1   Convergent Cross Mapping Basics

In 2012, Sugihara et al. [13] built on ideas from Takens' Theorem [14] to propose convergent cross mapping (CCM) to test causal linkages between nonlinear time series observations. This approach has enjoyed varied applications. For example, Luo et al. [7] successfully revealed underlying causal structure in social media and Verma et al. [16] studied cardiovascular and postural systems by taking advantages of this algorithm.

We provide here a brief intuition for why and how CCM works. Consider two variables X and Y, each associated with eponymous time-series and – further – where Y depends on X. For example, consider a case where for each timepoint

X measures the count of hares, and Y that of lynx. In this situation, if Y (lynx) causally depends on X (hares), observing the values of Y over time (e.g., a steep drop or a plateauing in lynx numbers) tells us about the state of governing factors, including X (here, the fact that the number of hares is too small to effectively feed the lynx population, or that they are roughly in balance with lynx, respectively). A implication of this – captured by Takens' Theorem – is that information on the state of X is encoded in the state space reconstructed from Y, meaning that points that are located nearby within Y's reconstructed state space will be associated with similar values for X, and can thus be used to make accurate (skillful) prediction of the value of X. In most cases, such prediction of one variable (e.g., X) within the state space of another (Y) can be achieved by nearest neighbor forecasting using simplex projection [3]. Pearson correlation between observed and predicted values can be applied to measure prediction skill.

2.2 Past Work in CCM Performance Improvement

Despite the fact that CCM is increasingly widely applied, there remain pronounced computational challenges in applying the tool for moderate and large time series. In order to secure confidence in inferences regarding causality, use of appropriate parameter values and a relative longer input are required for the original CCM [10]. As such, since its first appearance in 2012, a number of modifications and improvements have been proposed to handle this drawback. In 2014, Ma et al. [8] developed cross-map smoothness (CMS) based on CCM which has the advantage of requiring a shorter time series. Compared to original CCM, CMS can be used for time series in the order of $n = 10$, whereas CCM requires time-series at least in the order of $n = 10^3$ to yield reliable results. Additionally, works such as [1,4,5] investigated and introduced mathematical methods to properly estimate parameters required by CCM (embedding dimension E, time delay τ and time subsample L). Such work expanded CCM-related research and also provided methods for quickly inferring causality in certain circumstances.

The previous improvements on CCM typically trade off potential accuracy for relatively fast execution, and the assumptions in some methods cannot be safely maintained with noisy time series observations. However, the original CCM can be improved by introduction of parallel computing techniques. In recent years, numerous studies such as [9,12] have been conducted using distributed computing frameworks such as MPI or Spark. Such parallel techniques can dramatically improve the algorithmic performance by effectively exploiting the cluster-based computational capacity. It is worthwhile to implement a parallel version of CCM to allow researchers to rapidly and robustly evaluate the existence and strength of causal connections between measured time series.

3 Methodology

To achieve a Spark parallel version of CCM, we introduce two core concepts: the Spark Resilient Distributed Dataset (RDD) [18] and Pipeline. The former is the immutable data structure that can be operated in a distributed manner, which brings significant benefits for concurrently draw r subsamples of time series to assess Cross-Mapping convergence. As for the pipeline, it is specified as a sequence of stages, and each stage transforms the original RDD to another RDD accordingly. In summary, the definition of pipeline supports an elegant design for a parallel CCM algorithm manipulating RDDs in Spark.

3.1 CCM Transform Pipeline

Consider applying CCM to test if the variable associated with time series Y is being driven by the variable associated with time series X. In the corresponding transform pipeline, the parallel version of CCM is implemented as several stages to transform the RDD of r random subsamples of the time series to the RDD of prediction skills for a given (τ, E, L) tuple. To start the transformation, an input RDD is created which includes a pair of subsamples of lengths L of each of the time series, and values for each of two parameters (τ, E). The output of the CCM transform pipeline is an RDD of sequences of prediction skills. In the whole procedure, Spark operates the whole transformation in parallel without extra coding as shown in Fig. 1.

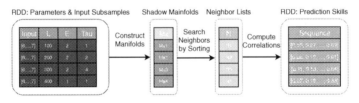

Parallel CCM Transform Pipeline

Fig. 1. A diagram of CCM RDD transformation which takes multiple realizations as input and outputs prediction skills.

3.2 Distance Indexing Table Pipeline

The CCM transform pipeline above achieves the aim of running CCM concurrently on multiple subsamples r. However, there is still a considerable potential for further optimization for this pipeline. As mentioned before, the most time-consuming part in the original CCM is finding the $E + 1$ nearest neighbors for every lagged-coordinate vector (τ) in the shadow manifold. For every point in the input RDD, the CCM transform pipeline computes the distances to all lagged-coordinate vectors of subsamples, sorts them and finally takes the top $E + 1$ as the nearest neighbors. This process is inefficient because of its repeated sorting and calculation for all the subsamples. It is particularly notable that as

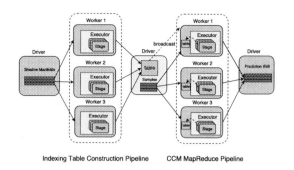

Indexing Table Construction Pipeline   CCM MapReduce Pipeline

Fig. 2. An illustration of the dependencies of two pipelines. After the distance indexing table is constructed in parallel, spark will broadcast it to all nodes and in the next pipeline, the executors can look up the table and fetch $E+1$ nearest neighbors quickly.

the length of subsamples L used for computation increases, the running time will grow superlinearly.

The best approach is to break down the nearest neighbors searching of CCM transformations into two parts: distance indexing table construction and nearest neighbors searching based on the constructed table. The first part can be achieved by setting another pipeline as a preprocessing step before applying the CCM transform pipeline. After building the distance indexing table, Spark can broadcast this table to each worker node on the cluster at one time rather than ship a copy of it every time they need it, as shown in Fig. 2. The pipeline of constructing the distance indexing table will be executed concurrently on the entire input time series, and it also reduces a significant amount of repeated calculation in the CCM transform pipeline. From the experiment results, the total computation time decreases in a pronounced fashion. As the library size L grows, the time spent on searching for the nearest neighbors increases correspondingly, and pre-building the distance indexing table secures increasing benefit.

3.3 Asynchronous Pipelines

After the pipeline is created to run CCM, a job is generated in the master node and then submitted to the cluster and partitioned into many tasks running in the executors of worker nodes. This setting is defined in the job submission and is, in general, constant. If we perform two pipelines one after other, they will always be executed sequentially. As such, we can adopt the asynchronous mechanisms to increase the parallelism and execute different pipelines concurrently. *Future-Action* is the Spark API to undertake asynchronous job submission. It provides a native way for the program to express concurrent pipelines without having to deal with the detailed complexity of explicitly setting up multiple threads. In this way, we can achieve running various combinations of the parameters (L, τ, and E) in parallel by executing multiple concurrent pipelines.

4   Scaling Analysis

The pseudocode of the CCM parallel algorithm is presented in Algorithm 1. In the analysis of parallel algorithms, the number of processes n $(n > 1)$ is generally introduced for the scaling analysis. Also, there are several additional parameters involved in the CCM test, which are listed below. L is the number of subsequence lengths set, while l is the length of subsequence; E represents the number of embedding dimensions set, and e is the embedding dimensions of shadow manifolds; Tau indicates the number of τ set, while τ is the embedding delay used in the shadow manifold; I is the length of input time series; r refers to the number of realizations.

Algorithm 1. Spark CCM parallel algorithm

1: INITIALIZE $\rho = \emptyset$
2: **for** e in E, τ in Tau, separately **do**
3:     Construct shadow manifold M_x for embedding dimension e and delay τ on X
4:     **for** i in I in parallel **do**
5:         $GDist_{Seq} \leftarrow$ calculate Euclidean distances of i to q for $R \in I$.
6:         $GSeq_{sort} \leftarrow$ sort based on $GDist_{Seq}$.
7:     **end for**
8:     **for** l in L **do**
9:         **for** $sample = 1,...,r$ in parallel **do**
10:            $Seq \leftarrow$ randomly draw sample from M_x with replacement for l times.
11:            **for** query point q in Seq **do**
12:                $K \leftarrow$ find top $e + 1$ indices for q in $GSeq_{sort}$.
13:                $\rho \leftarrow$ calculate correlation $\rho_{sample}^{e,\tau,l}$ between Y and X with indices $\in K$.
14:            **end for**
15:        **end for**
16:    **end for**
17: **end for**

As introduced before, the first preprocessing step takes $O(\frac{E \times Tau \times I^2 \times logI}{n})$. Then the time complexity of the state space searching with the preprocessed data is $O(\frac{E \times Tau \times L \times r \times l^2}{n}) < O(\frac{E \times Tau \times r \times I \times l}{n})$ as $l < I < L \times l$. The total input time series length is larger than the subsample sequence l but always smaller than the average factor of subsample sequences and the number of the testing set for l: $L \times l$. In these cases, the time complexity is determined by the preprocessing step. As such, the overall time complexity $O(\frac{E \times Tau \times I^2 \times logI}{n})$, which is smaller than the serial version $O(E \times Tau \times L \times r \times l^2 \times logl)$ even when n is 1. When we apply the parallel version in the cluster, the power of multiple processors (n) will dramatically decrease the overall execution time of CCM.

5   Experiment Results

The baseline scenario of parameters, with input time series size of 4000, r of 500, L with values $[500, 1000, 2000]$, E and τ both with $[1, 2, 4]$, is set for the comparison in the experiments. In the following experiments, the Spark parallel version of CCM will be run three times on the Google Cloud Platform to obtain the average computation time. The cluster setting is 1 master node and 5 worker nodes with 4 cores CPU and 15 GB Memory.

5.1   Overview of Improvements

This experiment compares the performance improvement of different implementations on the baseline scenario. These implementations in Table 1 are submitted on Yarn Mode and Local Mode, separately. Yarn Mode, or cluster mode will exploit all the worker nodes existing in the cluster while Local Mode only runs applications on the master node (Single Machine).

Table 1. Implementation levels

	Implementation level
Case A1	Single-threaded CCM (no RDD & pipeline)
Case A2	Synchronous CCM transform pipelines
Case A3	Asynchronous CCM transform pipelines
Case A4	Synchronous distance indexing table & CCM transform pipelines
Case A5	Asynchronous distance indexing table & CCM transform pipelines

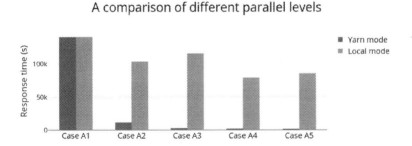

Fig. 3. Yarn mode utilizes all worker nodes in the cluster, while local mode only run experiments on the master node. Yarn mode significantly diminishes the average computation time of the parallel version of CCM with the help of worker nodes.

The results are shown in Fig. 3. Several conclusions can be drawn from the experimental results for different levels of parallel implementation. Firstly, the single-threaded version of CCM imposes a heavy computational cost, and there is no difference between two modes as they do not utilize the worker nodes in the cluster. Next, asynchronous pipelines can only reduce computation time in Yarn mode. After the comparison of the CPU utilization rates, it indicates that the asynchronous pipelines cannot offer more parallelization when the CPU utilization already reaches full throttle. However, when run with Yarn, the worker nodes still have rooms to improve the utilization rates. Also, as seen from the results, the spark full parallel version (*Case A5*) is approximately 1.2% the running time of the single-threaded version. Ultimately, the most significant improvement of marginal computation performance lies in adding the distance indexing table pipeline based on the CCM Transform pipeline. It reduces the computation time cost by over 80% relative to the baseline. Such considerable improvement shows the parallel version of CCM benefits strongly from establishing the distance indexing table globally for nearest neighbors searching pipeline.

When comparing current existing public CCM implementation, rEDM R package, which created by the Ye et al. [17] using lower level language C++, our Spark parallel implementation (*Case A5*) is approximately 15x faster than rEDM for baseline scenario on current cluster setup on Google Compute Platform. Obviously, the parallel version can perform more favorably with a more powerful cluster (vertical scaling) or adding more workers (horizontal scaling).

5.2   Elasticity Analysis

As a range of parameter settings been looped over for the best results to infer causality, testing the elasticity of running time concerning a given parameter value is necessary. Two versions of CCM (Parallel Version is the implementation *Case A5* which has all degrees of parallelism, while Single-threaded Version is the implementation *Case A1* which has not implemented any pipeline) will be tested with the parameter settings as shown in Table 2. Intuitively, these cases vary only one parameter from the baseline for comparison. When doubling parameter L, the average run time increases to 4.06x using the Spark single-threaded version, and 1.11x using the Spark parallel version. Similarly, doubling parameter τ and E almost has no impact on running time in the parallel version. However, doubling τ indeed increases the running time to 1.13x in the single-threaded version as it increase the dimension of shadow manifolds M_x and M_y.

Table 2. Elasticity analysis

Parameter varied	Parameter	Case B1	Case B2	Case B3
L	L	500	1000	2000
	Others	The same as baseline scenario		
E	E	1	2	4
	Others	The same as baseline scenario		
τ	τ	1	2	4
	Others	The same as baseline scenario		

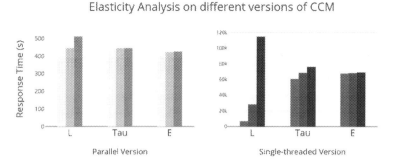

Fig. 4. The difference is the utility of worker nodes. The parallel version uses all of the optimization methods with five 4-core workers in the cluster, while the single-threaded version is only executed on the master node without any parallel optimization.

In summary, the values of these parameters, especially for L, do influence execution time for both the single-threaded and parallel versions; however, with the current optimization of the parallel methods by introducing indexing table before nearest neighbor searching, the relative impact shrinks, which make testing relatively large parameters for the causality assessment a reality.

6 Conclusion

The Spark framework provides relatively convenient APIs to exploit parallelism in algorithms such as CCM. This work conducted experiments demonstrating the performance benefits of exploiting the parallelism in CCM algorithm using Spark. The scalability of Spark offers considerable benefits in accelerating the execution with the support of clusters, allowing for a significant reduction in running time when adding more worker nodes into the cluster. Of critical importance for robust application of Spark, these performance gains make this algorithm a valuable modeling tool to assess causality with confidence in an abbreviated time. Such gains are particularly important in the context of high-velocity datasets involving human behavior and exposures, such as are commonly collected in human social and sociotechnical systems.

While it demonstrated potential for marked speedups, this work suffers from some pronounced limitations. Construction of the distance indexing table trades off higher space consumption for savings in computation time; for large shadow manifolds from a large value of L, the indexing table can require a large quantities of system memory. However, as previous study [8] shows, CCM can produce reliable results when input time-series length is around in the order of $n = 10^3$, for which the required memory space is well under what most current hardware can offer.

References

1. Cao, L.: Practical method for determining the minimum embedding dimension of a scalar time series. Phys. D: Nonlinear Phenom. **110**(1–2), 43–50 (1997)
2. Dean, J., Ghemawat, S.: Mapreduce: simplified data processing on large clusters. Commun. ACM **51**(1), 107–113 (2008)
3. Heylen, R., Burazerovic, D., Scheunders, P.: Fully constrained least squares spectral unmixing by simplex projection. IEEE Transact. Geosci. Remote Sens. **49**(11), 4112–4122 (2011)
4. Kantz, H., Schreiber, T.: Nonlinear Time Series Analysis, vol. 7. Cambridge University Press, Cambridge (2004)
5. Kugiumtzis, D.: State space reconstruction parameters in the analysis of chaotic time series–the role of the time window length. Phys. D: Nonlinear Phenom. **95**(1), 13–28 (1996)
6. Luke, D.A., Stamatakis, K.A.: Systems science methods in public health: dynamics, networks, and agents. Annu. Rev. Public Health **33**, 357–376 (2012)
7. Luo, C., Zheng, X., Zeng, D.: Causal inference in social media using convergent cross mapping. In: 2014 IEEE Joint Intelligence and Security Informatics Conference, pp. 260–263. IEEE (2014)
8. Ma, H., Aihara, K., Chen, L.: Detecting causality from nonlinear dynamics with short-term time series. Sci. Rep. **4**, 7464 (2014)
9. Maillo, J., Ramírez, S., Triguero, I., Herrera, F.: kNN-is: An iterative spark-based design of the k-nearest neighbors classifier for big data. Knowl.-Based Syst. **117**, 3–15 (2017)
10. Mønster, D., Fusaroli, R., Tylén, K., Roepstorff, A., Sherson, J.F.: Causal inference from noisy time-series data–testing the convergent cross-mapping algorithm in the presence of noise and external influence. Future Gener. Comput. Syst. **73**, 52–62 (2017)
11. Pearl, J., et al.: Causal inference in statistics: an overview. Statist. Surv. **3**, 96–146 (2009)
12. Reyes-Ortiz, J.L., Oneto, L., Anguita, D.: Big data analytics in the cloud: spark on hadoop vs MPI/OpenMP on Beowulf. Proc. Comput. Sci. **53**, 121–130 (2015)
13. Sugihara, G., et al.: Detecting causality in complex ecosystems. Sci. **338**(6106), 496–500 (2012)
14. Takens, F.: Detecting strange attractors in turbulence. In: Rand, D., Young, L.-S. (eds.) Dynamical Systems and Turbulence, Warwick 1980. LNM, vol. 898, pp. 366–381. Springer, Heidelberg (1981). https://doi.org/10.1007/BFb0091924
15. Vavilapalli, V.K., et al.: Apache Hadoop YARN: yet another resource negotiator. In: Proceedings of the 4th annual Symposium on Cloud Computing, p. 5. ACM (2013)
16. Verma, A.K., Garg, A., Blaber, A., Fazel-Rezai, R., Tavakolian, K.: Analysis of causal cardio-postural interaction under orthostatic stress using convergent cross mapping. In: 2016 38th Annual International Conference of the IEEE Engineering in Medicine and Biology Society (EMBC), pp. 2319–2322. IEEE (2016)
17. Ye, H., Clark, A., Deyle, E., Sugihara, G.: rEDM: an R package for empirical dynamic modeling and convergent cross-mapping (2016)
18. Zaharia, M., et al.: Resilient distributed datasets: a fault-tolerant abstraction for in-memory cluster computing. In: Proceedings of the 9th USENIX Conference on Networked Systems Design and Implementation, p. 2. USENIX Association (2012)
19. Zaharia, M., Chowdhury, M., Franklin, M.J., Shenker, S., Stoica, I.: Spark: cluster computing with working sets. HotCloud **10**(10–10), 95 (2010)

Continuous-Time Simulation of Epidemic Processes on Dynamic Interaction Networks

Rehan Ahmad and Kevin S. Xu$^{(\boxtimes)}$

EECS Department, University of Toledo, Toledo, OH 43606, USA
`Rehan.Ahmad@rockets.utoledo.edu`, `Kevin.Xu@utoledo.edu`

Abstract. Contagious processes on networks, such as spread of disease through physical proximity or information diffusion over social media, are continuous-time processes that depend upon the pattern of interactions between the individuals in the network. Continuous-time stochastic epidemic models are a natural fit for modeling the dynamics of such processes. However, prior work on such continuous-time models doesn't consider the dynamics of the underlying interaction network which involves addition and removal of edges over time. Instead, researchers have typically simulated these processes using discrete-time approximations, in which one has to trade off between high simulation accuracy and short computation time. In this paper, we incorporate *continuous-time network dynamics* (addition and removal of edges) into continuous-time epidemic simulations. We propose a rejection-sampling based approach coupled with the well-known Gillespie algorithm that enables *exact simulation* of the continuous-time epidemic process. Our proposed approach gives exact results, and the computation time required for simulation is reduced as compared to discrete-time approximations of comparable accuracy.

Keywords: Stochastic epidemic model · SIR model ·
Continuous-time network · Dynamic network · Rejection sampling ·
Gillespie algorithm

1 Introduction

Epidemic modeling has been an area of significant interest to the network science community, with applications including the spread of rumors or viral content over social media and the spread of infectious disease over face-to-face contact networks. The distribution and duration of contacts crucially affect the transfer of infection between individuals because it may increase or decrease the chances for infection to occur. Hence, the underlying network topology and properties are very important in epidemic modeling.

Epidemic models are mainly classified into two classes: deterministic and stochastic. We consider stochastic epidemic models, which can be used to simulate a range of possible outcomes for any given set of parameters. The infection process depends upon the instantaneous topology of the network, which

© Springer Nature Switzerland AG 2019
R. Thomson et al. (Eds.): SBP-BRiMS 2019, LNCS 11549, pp. 143–152, 2019.
https://doi.org/10.1007/978-3-030-21741-9_15

is changing continuously over time. Researchers often discretize time into short "snapshots" where the network topology is considered fixed during the time period of a snapshot. This is an approximation of the actual network where its topology changes only after a fixed interval of time. The epidemic process is then simulated in discrete time over this snapshot-based representation [2,4]. There is a trade-off between accuracy and computation time that is attached to discrete-time simulation models, depending on the length of the time snapshots.

Continuous-time stochastic epidemic models have traditionally been used for tractability of mathematical analysis [3]. Fennell et al. [6] propose an approach for simulating continuous-time stochastic epidemic models directly using the Gillespie algorithm [8] rather than by using discrete time intervals. They also demonstrated the inaccuracies that can result by using longer intervals for discrete-time simulations. However, the Gillespie algorithm-based approach is not applicable when the network changes over time because the infection rates change based on both the network dynamics (addition and removal of edges) and the infection dynamics. Vestergaard and Génois [12] address this limitation by proposing a method for exact continuous-time simulation on dynamic networks; however, it applies only to discrete-time dynamic networks.

In this paper, we propose an algorithm for simulating continuous-time epidemic processes on *continuous-time dynamic networks* that can change at arbitrary times unlike [12]. Our algorithm combines the Gillespie algorithm-based approach with a *rejection sampling* procedure that rejects inter-event times that occur after a change in the network, i.e. addition or deletion of an edge. We demonstrate that our approach is exact—that is, it correctly simulates the event times in the presence of network changes. We also demonstrate that our rejection sampling Gillespie algorithm results in faster simulations than comparable discrete-time approximations on two real dynamic social network data sets.

2   Background

2.1   Dynamic Interaction Networks

Real networks are generally time-varying in which the edges (interactions) between nodes (individuals) are not fixed or static. Therefore, the underlying network for any dynamic process like infection spreading or information diffusion changes with time, e.g. due to changing patterns of human interactions. The dynamics of this change in network topology may have a non-trivial impact over the processes. Holme [9] discusses several representations used for temporal networks. An exact continuous-time representation includes a sequence of interactions in the form (u, v, t, d), where u and v denote the two nodes involved, t denotes the timestamp of the start of the interaction, and d denotes the duration. The typical discrete-time representation is a sequence of aggregated graphs that represents how the topology of a temporal network changes with time. Each graph in the sequence is an aggregated representation over a time interval.

2.2   Stochastic Epidemic Models

The spread of infectious disease over a population is frequently modeled by a compartmental model in which the population is divided into a set of disjoint compartments. Some of the most common compartment models are the SIR, SIS, and SEIR models, where S stands for Susceptible, I for Infectious, R for Recovered, and E for Exposed. Any individual in this population exists in one compartment (or group) at a time and assumes similar properties of that group.

In this paper, we consider the SIR model, although our approach generalizes to the other compartmental models as well. In the SIR model, a susceptible individual may get infected after coming in contact with an infectious individual. This individual will recover after a certain amount of time. There are two approaches for modeling the transitions between these groups: deterministic and stochastic. In deterministic models, the transitions between these states are governed by differential equations. In stochastic models, which we consider in this paper, transitions between states occur with certain probabilities [3]. An infectious individual can spread the infection to a susceptible individual with infection probability β. Similarly, an infectious individual can transition to the recovered state with recovery probability μ. Stochastic models are sometimes also simulated to validate analytical results from deterministic models [13].

Discrete-Time Models. Discrete-time epidemic simulations consider time progressing in constant intervals of length Δt. In a single time interval or snapshot, any individual may make a single transition between compartments. For example, a susceptible individual may transition to the infectious compartment depending upon its contact with other infectious people, or an infectious individual may recover. Each of these transitions happens synchronously at a fixed time. Discrete-time epidemic models are quite convenient for simulation and have been used both with static [1,10] and dynamic networks evolving over discrete time steps [11], but mathematical analysis is much more difficult in the discrete-time setting, especially for dynamic networks.

Continuous-Time Models. Despite the simplicity of simulating discrete-time epidemic models, the accuracy and efficiency of discrete-time models is dependent upon the length of time interval being considered. In the continuous-time setting, infection and recovery probabilities are replaced with infection and recovery *rates*, respectively, which denote probabilities per unit time. Allen [3] considers analysis and simulation of continuous-time epidemic models for a fully-connected static network. Fennell et al. [6] consider simulation of continuous-time epidemics over a static (but not necessarily fully connected) network and investigate the effects of discretization of time and its limitations. In both studies, the underlying continuous-time epidemics are simulated using the well-known Gillespie algorithm [8]. Each individual in a population is considered to have an instantaneous rate $r_i(t)$ to transition from one state to another. The Gillespie algorithm works as per the following two properties:

1. The time that the network remains in the same state (no node transitions between compartments) is an exponentially distributed random variable with parameter $\lambda(t) = \sum_i r_i(t)$, the sum of the rates of all nodes in the network.
2. The probability that the next node i to transition from one compartment to another depends on the relative rate of the node $r_i(t)/\lambda(t)$.

Each edge (interaction) with an infectious node (individual) in the network at a certain time increases the infection rate for a susceptible node by an amount β that denotes the rate per infectious neighbor. Similarly, an infectious node will recover at a rate μ. The instantaneous rate r_i for node i to transition between compartments is given by

$$r_i(t) = \begin{cases} \beta m_i(t) & \text{if node } i \text{ is susceptible} \\ \mu & \text{if node } i \text{ is infectious} \end{cases}, \tag{1}$$

where $m_i(t)$ denotes the number of infectious individuals connected to node i [6].

The Gillespie algorithm is not applicable to dynamic networks because $m_i(t)$ can change over time independently from the epidemic process, i.e. when edges with an infectious node are added or deleted. Vestergaard and Génois [12] propose a temporal Gillespie algorithm to simulate continuous-time epidemics over a discrete-time dynamic network. In this paper, we propose a different modification to the Gillespie algorithm to deal with continuous-time dynamic networks.

3   Rejection Sampling Gillespie Algorithm

3.1   Inadequacy of Gillespie Algorithm

Fennell et al. [6] propose a method for simulating continuous-time stochastic epidemic models on a static network using the Gillespie algorithm. The exact Gillespie algorithm-based approach is also shown to be much faster than discrete-time approximations that achieve reasonable accuracy. The approach works on a static network because the network topology remains the same over time, hence the number of infectious neighbors an individual can have will remain the same until the next transition (infection or recovery). Thus, the instantaneous transition rate for each node, denoted by $r_i(t)$ in (1), is *constant* until the next transition happens at the simulated event time.

This assumption of static network topology does not hold in case of a dynamic network where edges can be added or removed because the network topology may change before the simulated event time, in which case the simulated event time no longer follows the correct distribution. For example, if a new edge is added with an infectious node i at time t', then $m_i(t')$ in (1) increases by 1, and thus the instantaneous transition rate $r_i(t')$ also increases. Thus, the inter-event time is no longer exponentially distributed as assumed in property 1 of the Gillespie algorithm. Instead, the cumulative distribution function (CDF) of the inter-event time has a knot (instantaneous change in slope) at time t' when $r_i(t')$ increases.

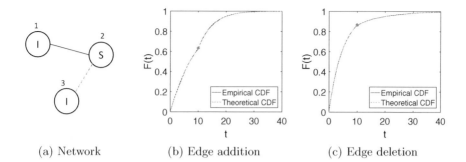

(a) Network   (b) Edge addition   (c) Edge deletion

Fig. 1. Comparison of theoretical CDF and empirical CDF of inter-event distribution for (b) addition and (c) deletion of an edge at time $t = 10$ in 3-node network shown in (a). The red dot in both theoretical CDFs at time $t = 10$ denotes the knot (instantaneous change in slope) when an edge is added or deleted. The two CDFs are almost identical, validating the correctness of our rejection sampling approach. (Color figure online)

3.2 Theoretical Inter-Event Distribution

The CDF of the inter-event time can be derived analytically as a continuous function with a series of knots at times when edges are added or removed. Without loss of generality, we can consider 2 cases: the addition of a single edge at a given time and the removal of a single edge at a given time.

Consider a simple network of three nodes and one edge between node 1 and node 2 as shown in Fig. 1(a). Nodes 1 and 3 are initially infectious at time $t = 0$. For a static network where the edge between nodes 2 and 3 is not added, the CDF for the inter-event time, which is the time to infection in this case, is the CDF of an exponential distribution with rate λ given by $P(T \leq t) = 1 - e^{-\lambda t}$, where λ is the infection rate parameter, i.e. the rate at which an infectious individual infects a susceptible individual when they are connected by an edge.

Assume now that, at time t', an edge is being added between an infectious node (node 3) and a susceptible node (node 2). Beginning from the Law of Total Probability and exploiting the memoryless property of the exponential distribution, the CDF for the inter-event time can be shown to be

$$F(t) = P(T \leq t) = \begin{cases} 1 - e^{-\lambda t}, & t \leq t' \\ 1 - e^{-2\lambda t + \lambda t'}, & t > t' \end{cases}$$

The CDF for the case of deletion of an edge between an infectious node and a susceptible node can be derived in a similar manner. Consider again the network in Fig. 1(a), but assume now that the edge between nodes 2 and 3 exists at time $t = 0$. At time t', the edge between node 2 and node 3 is deleted. The CDF for the inter-event time is given by

$$F(t) = P(T \leq t) = \begin{cases} 1 - e^{-2\lambda t}, & t \leq t' \\ 1 - e^{-\lambda t - \lambda t'}, & t > t' \end{cases}$$

In the general case where multiple edges are added and removed over time, each addition or removal of an edge with an infectious node creates a new knot in the CDF. This makes it difficult to analytically express the CDF.

3.3 Rejection Sampling Gillespie Algorithm

To overcome the problem with the Gillespie algorithm in a dynamic network setting, we employ the idea of *rejection sampling*[1]. Up to the first knot at time t' in the inter-event time CDF, the CDF matches that of an exponential random variable. Thus, we can sample from the exponential distribution, and if the sampled inter-event time occurs prior to the first knot, then we accept the sample. Otherwise, we reject the sample because the exponential distribution is no longer valid after the knot. We then re-set the current time in our simulation to the time of the knot. Then, the inter-event time will again be exponentially distributed, so we can once again sample the inter-event time from an exponential distribution and decide to accept or reject based on the time of the next knot. We repeat this process until we accept a sample. This approach is valid because the CDF of the inter-event time after each knot is exponentially distributed until the next knot, and the exponential distribution is memoryless. Figure 2 shows the entire algorithm formulated and used for simulation of epidemics in this work.

Input: Interaction network with timestamps of edge additions and deletions.
1 Initialize starting time as $t = 0$.
2 Randomly select k Infectious nodes to initialize epidemic.
3 Compute transition rate r_i for each Susceptible or Infectious node.
4 Sample the inter-event time for the next transition (infection or recovery) s from an exponential distribution with rate parameter $\lambda = \sum_i r_i$.
5 Update current event time by setting $t = t + s$.
6 **if** $t \le$ *time of the next added or removed edge* **then** {Sampled time t is valid}
7   Select node i to transition with probability $r_i / \sum_i r_i$.
8   **if** *node i is Susceptible* **then**
9   Change the status of node i to Infectious.
10   **else**
11   Change the status of node i to Recovered.
12 **else** {Sampled time t is invalid}
13   Reject the sampled time t and set $t =$ time of next added or removed edge.
14 If Infectious nodes still exist, go to step 3.

Fig. 2. Exact simulation of continuous-time stochastic epidemic process on a dynamic network using proposed rejection sampling Gillespie algorithm.

We evaluate the correctness of our rejection sampling approach on the 3-node network shown in Fig. 1(a) by simulating 5000 epidemics for both the edge

---

[1] Simulating a continuous-time epidemic model using a discrete-time approximation is also sometimes referred to as rejection sampling, e.g. in [6,12]. We refrain from such terminology in this paper as our proposed rejection sampling approach is exact.

addition and deletion scenarios and recording the inter-event time for each simulation. Figure 1(b)–(c) show the comparisons between the derived theoretical CDFs from Sect. 3.2 and the empirical CDFs computed from the simulations. The two plots match almost exactly, confirming the validity of our rejection sampling approach.

4   Datasets

In this paper, we consider two real-world datasets on face-to-face interactions in a high school setting. The datasets are collected among students of classes in a high school in Marseilles, France using wearable RFID sensors with a proximity range of roughly 1 to 1.5 m [7]. 126 individuals (118 students and 8 teachers) from 3 classes wore the sensors for a period of 5 days in 2011. 180 students from 5 classes wore them for a period of 7 days in 2012. Every 20 s, each sensor scanned and recorded the IDs of other sensors in proximity. We convert these 20-s scans to dynamic interaction networks by considering two nodes u and v to have interacted for s seconds if the pair (u, v) shows up in $s/20$ consecutive scans. Since 20 s is such a short interval compared to the dynamics of an epidemic process spreading over a dynamic network, these datasets are a good fit for our continuous-time approach, and thus we treat time as varying continuously. The dynamic networks are very sparse—the average instantaneous number of active edges is 5.1 in the 2011 data and 4.0 in the 2012 data.

5   Experiments

To evaluate our proposed rejection sampling Gillespie algorithm, we simulate epidemics on both high school networks using our exact rejection sampling Gillespie algorithm and discrete-time approximations for snapshot lengths $\Delta t \in \{10, 20, 50, 100\}$ seconds. In order to test the robustness of our method, we use a range of values for the infection rate $\beta \in \{0.1, 0.02, 0.002\}$. For each network and each value of infection rate considered, we simulate 1000 epidemics with our exact continuous-time epidemic model and 1000 epidemics with a discrete-time epidemic model for each value of Δt. The recovery rate is fixed at $\mu = 2 \times 10^{-5}$. Each simulation is run until the number of infectious individuals becomes zero, indicating that the epidemic has ended. Figure 3 shows the mean number of Susceptible (S), Infectious (I) and Recovered (R) individuals for continuous- and discrete-time epidemics simulated over the High School 2012 network with infection rate $\beta = 0.1$. Notice that the discrete-time approximations vary significantly in accuracy depending on the snapshot length Δt.

To compare the disease dynamics in the networks we use the area between the normalized mean continuous-time simulation curve and the normalized mean discrete-time simulation curve as our metric, summed over each of the 3 compartments (Susceptible, Infectious, and Recovered). We refer to this as the *normalized error*. This is a variant of an error metric used in [1], with the addition

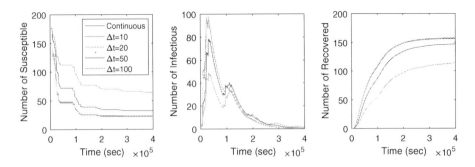

Fig. 3. Mean Susceptible (S), Infectious (I), and Recovered (R) plots over 1000 simulated epidemics for the exact continuous-time model and discrete-time approximations of varying lengths on the High School 2012 dataset with infection rate $\beta = 0.1$. The double peak in the number of infectious over time results from taking the mean of the 1000 simulated epidemics, each with different peak infectious times.

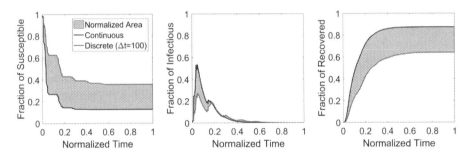

Fig. 4. Computation of normalized error metric by summing the normalized area between the mean curve for simulations with the continuous-time model and the discrete-time model over all three compartments. smaller normalized error denotes a more accurate discrete-time approximation.

of a normalization step. Since the mean simulation end time for each of the different models may be different, we normalize the computed area with respect to time by dividing it with the minimum mean end time of the models being compared. A lesser value of a normalized area between the plots indicates a better approximation to the simulation outcomes of the exact continuous-time model. The maximum normalized error is 3, consisting of a normalized area of 1 for each compartment. An illustration of the computation of the normalized error metric is shown in Fig. 4 for the High School 2012 data with $\beta = 0.1$.

6   Results

The normalized error for the discrete-time models is shown in Fig. 5. To compute the standard error, we use 100 bootstrap replicates [5]. Notice that the overall trend matches what one would expect—as the discrete-time snapshot length

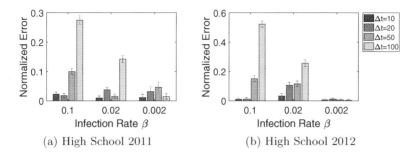

Fig. 5. Normalized errors between the mean continuous-time simulated SIR curves and mean discrete-time simulated SIR curves with different Δt values, with error bars denoting standard errors computed using the bootstrap. Error tends to increase with increasing Δt, indicating poorer approximations with longer time snapshots.

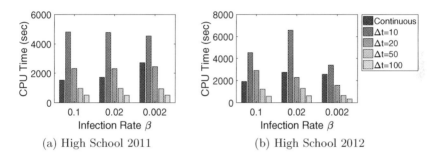

Fig. 6. CPU time taken to execute 1000 simulations for each scenario. CPU time for our exact continuous-time simulation is slightly lower than for discrete-time simulation with $\Delta t = 20$. Longer snapshots can be simulated faster but incur higher approximation error as shown in Fig. 5.

increases, the approximation gets worse. This is particularly true for $\Delta t = 100$ s, where the error is significantly larger than for smaller snapshot lengths, except for the $\beta = 0.002$ when the infection is not spreading rapidly. We also observe that the error increases as the infection rate β and the snapshot length Δt increase. This observation extends the observation by Fennell et al. [6] in the static network setting to dynamic networks.

The CPU time required to simulate 1000 epidemics using our proposed continuous-time model and each of the discrete-time models is shown in Fig. 6. Notice that our proposed continuous-time model is significantly faster than a discrete-time simulation with $\Delta t = 10$ s and also slightly faster than with $\Delta t = 20$ s in most cases. On the other hand, discrete-time simulations with $\Delta t = 50$ or 100 s are faster than our continuous-time simulation, but at the cost of higher approximation error as shown in Fig. 5, particularly for $\Delta t = 100$ s.

There is a trade-off between computation time and accuracy for discrete-time models. However, since our continuous-time model is exact and faster than discrete-time approximations with extremely short snapshot lengths, we argue that there is no benefit to using discrete-time models with such short lengths. This is because one could use our rejection sampling Gillespie algorithm to simulate the *exact* continuous-time epidemic process in a shorter amount of time! Thus, the only reason to use discrete-time approximations would be due to constraints on computation time, in which case one would use longer snapshots and have to tolerate the loss of accuracy. Otherwise, our proposed continuous-time approach is superior both in accuracy and computation time and is thus well-suited for general use in simulating epidemic processes over dynamic networks.

Acknowledgements. This material is based upon work supported by the National Science Foundation grant IIS-1755824.

References

1. Ahmad, R., Xu, K.S.: Effects of contact network models on stochastic epidemic simulations. In: Ciampaglia, G.L., Mashhadi, A., Yasseri, T. (eds.) SocInfo 2017. LNCS, vol. 10540, pp. 101–110. Springer, Cham (2017). https://doi.org/10.1007/978-3-319-67256-4_10
2. Allen, L.J.: Some discrete-time SI, SIR, and SIS epidemic models. Math. Biosci. **124**(1), 83–105 (1994)
3. Allen, L.J.: A primer on stochastic epidemic models: formulation, numerical simulation, and analysis. Infect. Dis. Model. **2**(2), 128–142 (2017)
4. Dong, W., Heller, K., Pentland, A.S.: Modeling infection with multi-agent dynamics. In: Yang, S.J., Greenberg, A.M., Endsley, M. (eds.) SBP 2012. LNCS, vol. 7227, pp. 172–179. Springer, Heidelberg (2012). https://doi.org/10.1007/978-3-642-29047-3_21
5. Efron, B., Tibshirani, R.J.: An Introduction to the Bootstrap. CRC Press, Boca Raton (1994)
6. Fennell, P.G., Melnik, S., Gleeson, J.P.: Limitations of discrete-time approaches to continuous-time contagion dynamics. Phys. Rev. E **94**, 052125 (2016)
7. Fournet, J., Barrat, A.: Contact patterns among high school students. PLoS One **9**(9), e107878 (2014)
8. Gillespie, D.T.: Exact stochastic simulation of coupled chemical reactions. J. Phys. Chem. **81**(25), 2340–2361 (1977)
9. Holme, P.: Modern temporal network theory: a colloquium. Eur. Phys. J. B **88**(9), 234 (2015)
10. Kim, L., Abramson, M., Drakopoulos, K., Kolitz, S., Ozdaglar, A.: Estimating social network structure and propagation dynamics for an infectious disease. In: Kennedy, W.G., Agarwal, N., Yang, S.J. (eds.) SBP 2014. LNCS, vol. 8393, pp. 85–93. Springer, Cham (2014). https://doi.org/10.1007/978-3-319-05579-4_11
11. Stehlé, J., et al.: Simulation of an SEIR infectious disease model on the dynamic contact network of conference attendees. BMC Med. **9**(1), 87 (2011)
12. Vestergaard, C.L., Génois, M.: Temporal Gillespie algorithm: fast simulation of contagion processes on time-varying networks. PLOS Comput. Biol. **11**(10), 1–28 (2015)
13. Volz, E.M., Miller, J.C., Galvani, A., Ancel Meyers, L.: Effects of heterogeneous and clustered contact patterns on infectious disease dynamics. PLOS Comput. Biol. **7**(6), 1–13 (2011)

Characterizing Bot Networks on Twitter: An Empirical Analysis of Contentious Issues in the Asia-Pacific

Joshua Uyheng$^{(\boxtimes)}$ ⓘ and Kathleen M. Carley ⓘ

CASOS Institute for Software Research, Carnegie Mellon University,
Pittsburgh, PA 15213, USA
{juyheng, carley}@andrew.cmu.edu

Abstract. This paper empirically analyzes bot activity in contentious Twitter conversations using case studies from the Asia-Pacific. Bot activity is measured and characterized using a series of interoperable tools leveraging dynamic network analysis and machine learning. We apply this novel and flexible methodological framework to derive insights about information operations in three contexts: the senatorial elections in the Philippines, the presidential elections in Indonesia, and the relocation of a military base in Okinawa. Varying levels of bot prevalence and influence are identified across case studies. The presented findings demonstrate principles of social cyber-security in concrete settings and highlight conceptual and methodological issues to inform further development of analytic pipelines in studying online information operations.

Keywords: Information operations · Social cyber-security · Asia-Pacific

1 Introduction

Online social networks (OSNs) have provided citizens worldwide with a powerful platform to participate in political discourse and exchange views at an unprecedented scale [1, 2]. However, targeted information operations on these platforms curb such democratizing potentials by introducing artificial influences into online conversations [3]. Utilizing artificial agents like automated bots, information operations exploit the network structure of the social media platform (e.g., who speaks to whom) to manipulate dominant topics of discussion (e.g., who speaks about what). As political discourse increasingly encompasses the digital sphere, it is of prime significance for researchers and policymakers to develop integrated and effective frameworks for detecting and characterizing information operations across a variety of contexts [4].

This paper proposes a novel and flexible methodological framework for understanding bot-driven information operations on Twitter. In the context of three contentious issues in the Asia-Pacific, we utilize a series of interoperable tools leveraging machine learning and dynamic network analysis to (a) detect user accounts that exhibit bot-like characteristics, (b) examine the topics of discussion in which such agents are involved, and (c) assess their overall impact on the online conversation [5, 6]. On a basic level, our findings contribute valuable insights into information operations in a

© Springer Nature Switzerland AG 2019
R. Thomson et al. (Eds.): SBP-BRiMS 2019, LNCS 11549, pp. 153–162, 2019.
https://doi.org/10.1007/978-3-030-21741-9_16

highly significant geopolitical region. More broadly, we illustrate the value of inter-operable pipelines for analyzing bot activity from the perspective of social cyber-security [4, 7]. Finally, we provide practical considerations for stakeholders involved in the cases examined, especially as regards mapping artificial influences on contentious political discourse.

1.1 Information Operations in Online Social Networks

Over the past decade, social media sites like Facebook and Twitter have become instrumental in facilitating online communication in both personal and large-scale contexts. In the political sphere, OSNs have introduced a highly democratized space for public discourse. The open lines of communication introduced by OSNs allow for diverse actors within and across societies to exchange ideas and information in a relatively unconstrained manner [2]. However, for the same reasons, OSNs are also susceptible to manipulation. In the present work, we are interested in empirically detecting and characterizing forms of manipulation such as targeted information operations.

The use of information operations to influence public opinion is not new. However, targeted communications in the context of OSNs acquire a vastly accelerated potential to spread malignant messages. Disinformation, polarization, and even terrorist radi-calization have all been associated with the unregulated proliferation of OSNs [8–10]. Extensive work shows how OSNs can be used to heighten intergroup conflict [11], recruit support for extremist organizations [12], or nudge electoral decisions [13]. By combining human and automated agents (i.e., bots), information operations leverage the dynamics of OSNs to affect real-world outcomes. Key stakeholders in such oper-ations may span individuals and organizations to nation-states.

1.2 Social Cyber-Security with Interoperable Tools

Scholarship in the burgeoning field of social cyber-security seeks to understand and combat digital technologies utilized for adversarial purposes using a 'multidisciplinary' and 'multimethodological' approach [4]. In this paper, we propose three research questions which operationalize foundational facets of social cyber-security inquiry. What is the level of bot activity in the online conversation? What messages are most targeted by bot activity? How does bot activity influence the social network? As we aim to show, these questions offer a general yet insightful framework for characterizing information operations across a variety of contentious political issues.

We further propose a novel and flexible methodological pipeline of interoperable tools to empirically address these questions in concrete contexts. The tools embedded in our framework leverage various concepts and techniques in network science and machine learning. Network science studies how entities and their relationships can be modeled using a graph structure with nodes and edges [14]. Dynamic network analysis characterizes large, high-dimensional, and time-variant graphs to quantify the structure of online conversations, the importance of some actors over others, and the relation-ships between agents and the topics they discuss. Several machine learning algorithms are also embedded in our pipeline. First, Bothunter deploys individual- and network-level features to compute a probability that a given user is a bot [6]. The algorithm is a random forest trained on a labeled dataset of known bots. The second tool identifies whether a user is a news agency, a government account, a company, or a celebrity, among other pre-identified account types known to drive significant Twitter traffic. This tool is based on a neural network model trained on a labeled dataset of user-provided account descriptions and their latest 20 tweets.

At this juncture, we note that while these concepts are not new in the study of OSNs and bot activity, the key feature of our proposed framework is its integrated pipeline of analysis. Whereas numerous studies have produced novel means of analyzing social media data and bot behavior online, few demonstrate frameworks for synergistically and effectively using these tools to distill holistic and actionable findings. In a practical setting, diverse data types and research questions must be examined concurrently, and therefore ought to be compatible for the generation and triangulation of insights. As our findings illustrate in subsequent sections, this paper demonstrates an informative model of such an integrated framework.

1.3   Contentious Issues in the Asia-Pacific

To illustrate the utility of our proposed pipeline, we examine three case studies of contentious issues in the Asia-Pacific. We collect Twitter conversations surrounding the Philippine senatorial elections, the Indonesian presidential elections, and campaigns against the relocation of a military base to Oura Bay in Okinawa. In all cases, tweets were in English, the local language (e.g., Filipino), or a blend of both.

#Halalan2019: The Philippine Senatorial Elections. The Philippines holds midterm elections after three years in the sitting President's six-year term. In May 2019, mid-term elections will decide twelve senatorial positions, comprising half of the seats of the nationally elected Senate [15]. Set against the backdrop of the polarizing regime of President Rodrigo Duterte, the senatorial race constitutes a high-stakes contest for legislative power in the country. The race features major political figures representing administration-backed candidates and a coalition of opposition candidates.

#Pemilu2019: The Indonesian Presidential Elections. In April 2019, Indonesia will hold general elections to fill a vast majority of its high-ranking political posts [16]. Foremost among these contests will be the race for the presidency, for which incumbent Joko Widodo will reiterate his 2014 clash against former general Prabowo Subianto. Whereas substantial controversy has surrounded Widodo's term for its economic underperformance and unconvincing stance on China, Subianto's "Make Indonesia Great Again" rhetoric has likewise been subjected to scrutiny for its similarity to Donald Trump's populist rhetoric during the 2016 US presidential elections. Both candidates will campaign for a five-year term of the nation's leadership.

#StandWithOkinawa: Military Base Relocation in Oura Bay. The final case concerns the relocation of a US military base to Oura Bay in the Okinawa prefecture [17]. While relocation plans had been in progress since the mid-1990s, resistance from the local population delayed construction until a Supreme Court ruling in 2016. Concerns of the opposition include people's safety in areas linked to military facilities, the

purportedly dubious proceedings legalizing the relocation's approval, and environmental effects of the construction which may impact dugong habitats.

2   Data and Methods

To analyze information operations in the three Asia-Pacific contexts described, this
study employed an integrated methodological framework. Utilizing interoperable network science and machine learning tools, we characterized the users, topics, and overall
network influence associated with bot-like accounts on public Twitter data harvested
with the STREAM and REST APIs.

2.1   Datasets

For each case study, a set of initial hashtags was generated based on mainstream media
coverage of the issues. From this initial set, further search terms were iteratively
identified and incorporated into the search. For the Philippine and Indonesian elections,
search terms consisted of official election hashtags (e.g., #Halalan2019 in the Philippines; #Pemilu2019 in Indonesia) as well as the names and campaign slogans of
candidates running for office. Meanwhile, for the case in Japan, the initial hashtag was
#standwithokinawa, which encapsulated the campaign to resist military base relocation.
From this initial search term, new terms were added including hashtags in Japanese.
Table 1 summarizes the data collection methods used for each case study, the inclusive
dates of the tweets analyzed, and the total number of tweets examined.

Table 1. Description of case study datasets.

Case study	Sample hashtags	Inclusive dates	Tweets collected
Philippine elections	#Halalan2019	Oct 2018–Jan 2019	473993
Indonesian elections	#Pemilu2019	Dec 2018–Jan 2019	305959
Stand with okinawa	#standwithokinawa	Dec 2018–Jan 2019	260562

2.2   Analytic Procedure

The overall analytic scheme followed a sequence of three broad stages: (a) user
analysis, (b) topic analysis, and (c) network analysis. Implementation flow is summarized in Fig. 1.

   During user analysis, Bothunter and role identity algorithms were deployed in
parallel. The results of both algorithms were converged to construct a contingency table
of predicted bots and possible role identities. The latter analysis clarified bot predictions by providing an alternative explanation for the 'bot-like' behavior identified by
the former tool. Only bots concurrently classified as non-special actors by the role
identity algorithm were retained, providing conservative estimates of bot activity.

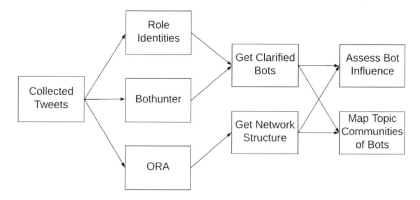

Fig. 1. Interoperable pipeline of social cyber-security tools.

Topic communities in the Twitter conversation for each case study were concurrently identified using ORA. ORA visualized Twitter users as nodes connected by weighted edges depending on the degree to which users had cited the same hashtags. Using Louvain clustering, we identified unique topics of discussion based on the hashtags used by the identified topic communities. The Louvain clustering algorithm optimizes modularity [18], which measures the network density within candidate clusters relative to vertices outside these clusters. Converged with the identified bots, we also identified which topics featured the highest level of predicted bot activity.

Finally, network analysis on ORA examined the higher-level Twitter interactions which took place in each online conversation. Simultaneously computing a host of network centrality metrics, ORA generated reports of Twitter users identified as 'super spreaders' and 'super friends' who wielded substantial influence in the collected dataset through their high-ranking centrality scores on communication networks (i.e., their messages are widely spread) or friendship networks (i.e., they are friends with many users), respectively. Such agents play an instrumental role in the online discussion. Interpreted in conjunction with the user analysis and topic analysis, this high-level approach provided a rich map of which messages (i.e., topics) are being propagated by whom (i.e., bots), thereby assessing the impact of potential information operations on the overall discussion (i.e., network influencers).

3   Results

Detected bots represented a non-negligible proportion of users across social networks, with notable country-level differences in general bot activity. Bot operations further varied across contexts, with bots attempting to drive intergroup opposition, rally intragroup support, or propel various facets of a contentious topic.

3.1   User Analysis: Bots and Special Actors

Bothunter produced probabilities that each user in the datasets was an automated bot. A 60% probability threshold was used for positive cases based on previous validation procedures with Bothunter. The role identity algorithm identified whether users belonged to special classes of actors with characteristically numerous tweets and followers. Table 2 cross-tabulates these results. Numbers in bold indicate the final bot estimates given that the role identity algorithm did not indicate they belonged to special actor classes. Percentages are relative to the total number of users in each dataset.

Table 2. Detected bots with predicted role identities.

Case study	Role identities	Unique users	Detected bots
Philippine elections	Normal	63260 (78.79%)	**9073 (11.30%)**
	Government	7424 (9.25%)	1189 (1.48%)
	News agency	3538 (4.41%)	510 (0.64%)
	News reporter	2950 (3.67%)	630 (0.78%)
	Company	802 (1.00%)	103 (0.13%)
	Celebrity	2015 (2.51%)	172 (0.21%)
	Sports	300 (0.36%)	44 (0.05%)
Indonesian elections	Normal	21987 (87.48%)	**2568 (10.22%)**
	Government	722 (2.87%)	95 (0.28%)
	News agency	471 (1.87%)	26 (0.10%)
	News reporter	60 (0.24%)	10 (0.04%)
	Company	1569 (6.24%)	215 (0.86%)
	Celebrity	230 (0.92%)	32 (0.13%)
	Sports	95 (0.28%)	6 (0.02%)
Stand with Okinawa	Normal	31428 (98.09%)	**8750 (27.31%)**
	Government	103 (0.32%)	30 (0.09%)
	News agency	171 (0.53%)	22 (0.07%)
	News reporter	23 (0.08%)	10 (0.03%)
	Company	261 (0.81%)	55 (0.17%)
	Celebrity	25 (0.08%)	3 (0.01%)
	Sports	29 (0.09%)	2 (0.01%)

A non-negligible proportion of captured users across the three datasets were detected as bots by our algorithms. Across the contentious issues analyzed, between 10% and nearly 30% of all captured Twitter users were bots or bot-like, second only to ordinary accounts in the datasets but outnumbering news agencies and government accounts. Country-level bot statistics appeared to be markedly different, with Japan having the highest number of bots and Indonesia the least among the case studies.

3.2   Topic Analysis: Bot-Active Hashtags

Topics in the Twitter conversations were identified by examining the most frequently used hashtags of Louvain clusters of agents. Top hashtags for the topic groups having the highest number of bots are presented in Table 3. Bot percentages are given relative to the total number of users in each topic group. Labels for topics were interpreted based on the top hashtags validated by manually reading sample tweets.

Table 3. Top hashtags in topics with most bot activity.

Case study	Topic	Bot activity	Hashtags
Philippine elections	Opposition campaign (focused on Roxas)	22.93%	OtsoDiretso, MarRoxas, RoxasTayo, OposisyonKoalisyon, TheLeaderIWant
	Opposition campaign (focused on Hilbay)	12.82%	HilbayForSenator, MarRoxas, TeamPhilippines, OtsoDiretso, OposisyonKoalisyon
	Anti-opposition, pro-Duterte	13.18%	NoToLiberalParty2019, NeverAgainToLP, ImeeSolusyon, NoToNeri, TeamDDS2019
Indonesian elections	Campaign for peaceful elections	11.11%	pilpres, pemiludamai, KPU, Pilpres2019, kampanye
	Campaign for opposition (Subianto)	6.60%	AkalSehatPilihPrabowoSandi, SumbarMemilihPrabowoSandi, 2019PrabowoPresidentRI
	Campaign for incumbent (Widodo)	1.40%	OrangBaikDukungJokowi, 01Indonesiamaju, jokowilagi, 2019TetapJokowi
Stand with Okinawa	White House petition against base	25.77%	辺野古の海を埋め立てないで, ホワイトハウスへ36万筆署名, Democracy
	Protesting Shinzo Abe	24.18%	ヤバすぎる安倍政権, ヤバすぎる緊急事態条項
	Referendum against military landfill	22.93%	県民投票, 辺野古県民投票, 普天間移設問題

For each topic above, the main hashtags (e.g., #Halalan2019, #standwithokinawa) were omitted as they appeared in all topics. In elections, both administration and opposition candidates were discussed by bot-like accounts, suggesting that information operations may be used by both parties, or that bots aim to influence messaging about all candidates. Further research may probe this question. In the Okinawa case, it appears that both local and international issues are amplified by such accounts.

3.3   Network Analysis: Influencer Assessment

Finally, top influencers were identified by network centrality. Super spreaders and super friends are identified in Table 4 with Bothunter scores. Asterisks mark verified accounts. Italics are for accounts suspended or moved since data collection. Bolded names may require deeper analysis due to high influence and bot-like behavior.

Table 4. Super spreaders and super friends.

Case study	Super spreaders	Super friends
Philippine elections	rapplerdotcom* (0.37)	ru6dy9 (0.50)
	MARoxas* (0.41)	raincyrainy (0.52)
	ATajum (0.44)	*mariagarciaah (0.42)*
	cnnphilippines* (0.20)	*MayDPoresBeWidU (0.54)*
	mariagarciaah (0.42)	MelLegaspi1 (0.44)
	AsecMargauxUson (0.51)	*AsecMargauxUson (0.51)*
	ru6dy9 (0.50)	jvejercito* (0.69)
	BembangBiik (0.37)	**BoyoKiss (0.65)**
Indonesian elections	Sandiuno* (0.47)	Addarul1 (0.57)
	CakKhum (0.25)	HotPepperminTea (0.51)
	Gerindra* (0.53)	abiid_d (0.56)
	Addarul1 (0.57)	abiyyu231299 (0.38)
	02Sandiaga (0.45)	Rusydi_riau40 (0.52)
		MangajatsCkp (0.55)
		Bagusalghazali (0.54)
Stand with Okinawa	**surumegesogeso (0.65)**	tkatsumi06j (0.30)
	robkajiwara (0.57)	affluencekana (0.46)
	ISOKO_MOCHIZUKI (0.65)	**sabor_sabole (0.74)**
	times_henoko (0.35)	robkajiwara (0.57)
	29ryukyu (0.33)	HempHere (0.34)
	mr_naha_das (0.49)	29_momechabo (0.57)
	BFJNews* (0.59)	HIROMI150303 (0.48)
		ActSludge (0.50)

In each case, major influencers appear to be a mix of politicians (e.g., MARoxas, Sandiuno), media outlets (e.g., rapplerdotcom, BFJNews), and bot-like agents (e.g., BoyoKiss, sabor_sabole). While the first two categories are expected, the last are noteworthy especially as they dominate super friends. While their messages are not entirely influential (e.g., retweeted), their vast connections (e.g., followers) might still make their messages visible to many. The present analysis thus flags accounts which may require deeper investigation of their activities and impact on public discourse.

4 Discussion

This paper analyzed three case studies of contentious Twitter conversations in the Asia-Pacific. Bots and bot-like accounts appeared to be non-negligible in terms of prevalence, participating in diverse activities such as promoting causes, antagonizing the opposition, and elaborating on various facets of an issue [7]. Country-level variations were observed, suggesting that information operations are more widely utilized in certain nations over others. Nonetheless, in each case, bots and bot-like accounts wielded significant influence, boasting high centrality measures on par with media outlets and government officials. Accounts requiring deeper investigation were flagged, among which some had already been suspended or deactivated, potentially providing partial validation of our predictions.

In the growing field of social cyber-security, our results illustrate the utility and need for integrated approaches to assess, characterize, and combat targeted information operations online. Pipelined interoperable tools triangulate insights beyond a single analytical approach. By providing a high-level map of potentially suspicious activity, our findings enable more in-depth inquiry into specific operations informed by specialized knowledge of local sociopolitical dynamics. Due to the ongoing nature of each campaign upon the writing of this paper, we note that new developments may arise beyond the scope of the present analysis. Within and beyond the Asia-Pacific, potential for covert intervention calls for rigorous vigilance from a diversity of fronts. We aimed to show how such vigilance might be implemented in practice.

Acknowledgments. This work is supported in part by the Office of Naval Research under the Multidisciplinary University Research Initiatives (MURI) Program award number N000141712675 Near Real Time Assessment of Emergent Complex Systems of Confederates, BotHunter award number N000141812108, award number N00014182106 Group Polarization in Social Media: An Effective Network Approach to Communicative Reach and Disinformation, and award number N000141712605 Developing Novel Socio-computational Methodologies to Analyze Multimedia-based Cyber Propaganda Campaigns. This work is also supported by the center for Computational Analysis of Social and Organizational Systems (CASOS). The views and conclusions contained in this document are those of the authors and should not be interpreted as representing the official policies, either expressed or implied, of the ONR or the U.S. government.

References

1. Kahne, J., Bowyer, B.: The political significance of social media activity and social networks. Polit. Commun. **35**, 470–493 (2018). https://doi.org/10.1080/10584609.2018.1426662
2. McGarty, C., Thomas, E.F., Lala, G., Smith, L.G.E., Bliuc, A.-M.: New technologies, new identities, and the growth of mass opposition in the arab spring. Polit. Psychol. **35**, 725–740 (2014). https://doi.org/10.1111/pops.12060
3. Bandeli, K.K., Agarwal, N.: Analyzing the role of media orchestration in conducting disinformation campaigns on blogs. Comput. Math. Organ. Theory (2018). https://doi.org/10.1007/s10588-018-09288-9

4. Carley, Kathleen M., Cervone, G., Agarwal, N., Liu, H.: Social Cyber-Security. In: Thomson, R., Dancy, C., Hyder, A., Bisgin, H. (eds.) SBP-BRiMS 2018. LNCS, vol. 10899, pp. 389–394. Springer, Cham (2018). https://doi.org/10.1007/978-3-319-93372-6_42

5. Benigni, M., Joseph, K., Carley, K.M.: Mining online communities to inform strategic messaging: practical methods to identify community-level insights. Comput. Math. Organ. Theory **24**, 224–242 (2018). https://doi.org/10.1007/s10588-017-9255-3

6. Beskow, D.M., Carley, K.M.: Its all in a name: detecting and labeling bots by their name. Comput. Math. Organ. Theory (2018). https://doi.org/10.1007/s10588-018-09290-1

7. Beskow, D.M., Carley, K.M.: Social cybersecurity: an emerging national security requirement. Mil. Rev. **99**(2), 117 (2019)

8. Bail, C.A., et al.: Exposure to opposing views on social media can increase political polarization. Proc. Natl. Acad. Sci. **115**, 9216–9221 (2018). https://doi.org/10.1073/pnas.1804840115

9. Tucker, J.A., et al.: Social media, political polarization, and political disinformation: a review of the scientific literature. Hewlett Foundation (2018)

10. Ong, J.C., Cabanes, J.V.A.: Architects of networked disinformation: behind the scenes of troll accounts and fake news production in the Philippines. Newton Tech4Dev Network (2018)

11. Babcock, M., Cox, R.A.V., Kumar, S.: Diffusion of pro- and anti-false information tweets: the Black Panther movie case. Comput. Math. Organ. Theory (2018). https://doi.org/10.1007/s10588-018-09286-x

12. Klausen, J.: Tweeting the Jihad: social media networks of western foreign fighters in Syria and Iraq. Stud. Confl. Terror. **38**, 1–22 (2015). https://doi.org/10.1080/1057610X.2014.974948

13. Allcott, H., Gentzkow, M.: Social media and fake news in the 2016 election. J. Econ. Perspect. **31**, 211–236 (2017). https://doi.org/10.1257/jep.31.2.211

14. Al-Garadi, M.A., et al.: Analysis of online social network connections for identification of influential users: survey and open research issues. ACM Comput. Surv. **51**, 16:1–16:37 (2018)

15. Bueza, M.: Survey says: how 2019 senatorial bets are faring so far (2019). https://www.rappler.com/newsbreak/iq/220707-senatorial-candidates-survey-performance-2019-elections

16. Cochrane, J.: Indonesia's presidential race takes shape, in shadow of hard-line Islam (2018). https://www.nytimes.com/2018/08/11/world/asia/indonesia-presidential-election.html

17. Japan begins filling in Henoko Bay in Okinawa to make room for unpopular US base (2018). https://www.japantimes.co.jp/news/2018/12/14/national/japan-starts-landfill-work-move-unpopular-u-s-base-okinawa

18. Blondel, V.D., Guillaume, J.-L., Lambiotte, R., Lefebvre, E.: Fast unfolding of communities in large networks. J. Stat. Mech: Theory Exp. **2008**, P10008 (2008). https://doi.org/10.1088/1742-5468/2008/10/P10008

A Hybrid Cellular Model for Predicting Organizational Recruitment in a k-Dimensional Space

Nicolas L. Harder[(⊠)] and Matthew E. Brashears

University of South Carolina, Columbia, SC 29208, USA
nharder@email.sc.edu

Abstract. Ecological models are useful in modeling organizations and their competition over resources. However, the traditional approaches, particularly Blau space models, are restrictive in their dependence on a continuous space. In addition, these models are susceptible to indicating competition in sparsely populated areas of an ecology. To deal with these problems we propose a reconceptualization of Blau space that utilizes a cellular structure to model a wider number of variable types, and simple probabilistic urn models to evaluate competition between organizations. We briefly review the basic concepts of Blau Space, demonstrate the issues with traditional Blau space modeling, and present a new model referred to as the Hybrid model.

Keywords: Simulation · Ecological models · Blau space

1 Introduction

Ecological models have often been adapted to model competition and influence in social settings, particularly among social organizations [2, 3]. One popular option, Blau space, models competition between organizations for resources (particularly members) in a k-dimensional sociodemographic space [3]. Blau space has been used to model organizational competition [4], organizational time demand and its influence on recruitment [10], and the spread of behaviors from organization members to non-members [9].

These models are useful, but often force researchers to adopt unreasonable assumptions, or result in interpretations that are sensible in a biological setting but not in a social setting. We propose a reconstruction of Blau space via a hybrid model combining a cellular structure and urn models. This new model evades many of the existing issues, and is a potential replacement for traditional Blau space modeling.

Our new approach is implemented in the R programing environment and we use the 8th Convocation of the Ukrainian Parliament as an example case to demonstrate both the traditional Blau space approach and the new Hybrid Blau space Model.

This work is supported by the Office of Naval Research Multidisplinary University Research Initiative (MURI) under grant number N00014-17-1-2675.

© Springer Nature Switzerland AG 2019
R. Thomson et al. (Eds.): SBP-BRiMS 2019, LNCS 11549, pp. 163–172, 2019.
https://doi.org/10.1007/978-3-030-21741-9_17

2 Blau Space Theory and Traditional Methods

Blau space, named in honor of social theorist Peter M. Blau, is a k-dimensional space in which the dimensions represent sociodemographic variables [3, 4]. Values on each dimension define a set of coordinates in Blau space and individuals possessing these values are located at those coordinates.

Modeling and analysis in Blau space is useful because of the prevalence of homophily, or the tendency for similar individuals to associate with each other [1]. Because of homophily, those near each other in Blau space have a greater likelihood of association across many different types of networks (including work, friendship, and romantic). This results in individual social worlds that are restricted to small regions of the k-dimensional Blau space and encompass only a small cluster of similar others. Homophily is one of the most robust findings in social science [8], and therefore proximity in Blau space is almost always a useful proxy for network closeness.

Blau space is primarily used in the modeling of competition between social organizations[1]. Organizations require a variety of resources to survive, but all organizations require members. A large fraction of recruitment into organizations occurs via networks; we are more likely to join a group if we are already connected to one or more members [7]. Because we tend to know individuals who are similar to ourselves, organizations tend to recruit from among those who are nearby in Blau space. This recruitment area is the organization's "niche"[2], and while members can be drawn from outside, the niche is the primary source of members for an organization. Second, the amount of time an individual can devote to an organization, and the number of individuals available in an area of Blau space, are limited. Consequently, organizations whose niches overlap are competing over the same pool of human time.

Overlap between organizational niches in Blau space indicates competition, with greater overlap indicating higher levels of competition. In traditional Blau space methods, this is represented by a competition coefficient that is calculated using the area of overlap between the niches of at least two organizations. However, these coefficients only rely on the overlap in niche areas and ignore the number of individuals within the overlapping space. Consequently, competition coefficients can be high in areas of Blau space where there are no individuals present (i.e., no resources to be exploited), resulting in over-estimates of the competition between organizations.

Using overlap in organizational niches to compute competition coefficients also imposes significant distributional assumptions. When the observations are not distributed approximately normally, competition coefficient estimates are often misleading.

---

[1] Traditionally, Blau space has been used to study "voluntary associations", such as athletic clubs, social clubs, and religious organizations. However, the theory itself is applicable to all social organizations [3].

[2] An organization's niche is calculated as range which extends out a fixed amount above and below the mean value on each dimension. The niche width is often fixed at 1.5 Std. Dev., but this parameter is tunable. See also: McPherson [3], McPherson et al. [4], and Popielarz and McPherson [5].

For example, Fig. 1 is a niche plot, an image that shows organizational niches within a Blau space, constructed using data on the members of the Ukrainian Parliament[3].

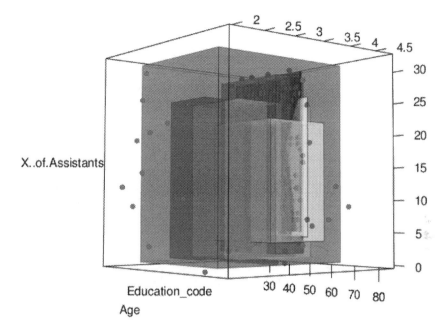

Fig. 1. Niche plot (Color figure online)

In Fig. 1, the dimensions are education, age, and number of assistants employed by a member of parliament. Individual members are shown as red dots positioned according to their attributes, and the shaded polygons are the niches for the eight factions present at this time. Because many individuals have a college degree there is a great deal of overlap in the niches, resulting in high competition coefficients. However, there is considerably more variation in the other dimensions and therefore the niches are spread broadly across a largely unpopulated space.

One solution would be to exchange the education dimension for a different one, but there are two complications. First, variables chosen as dimensions should be "salient", or meaningfully impact the likelihood of association between individuals. Excluding highly salient dimensions may not be an option. Second, variables used as dimensions typically must be continuous.[4] For highly salient variables such as sex, it is possible to define a dimension as a fraction of the organization that belongs to a particular category

---

[3] Produced using Blaunet Version 2.0.8.
  https://CRAN.R-project.org/package=Blaunet.

[4] The underlying mathematical framework for Blau space is a modification of the Lotka-Voltera Ecological Competition model [2], which assumes of populations with a continuous and somewhat normal distribution of resources that they are consuming.

(e.g., percentage female). However it is typically impossible to locate individuals within the resulting space, producing theoretical and methodological problems.

3   Hybrid Blau Space Model

To deal with the preceding difficulties, we reconceptualize Blau space's regular polygonal niches in a continuous space as irregular niches in a space composed of discrete ordered cells. Competition for members is captured using a simple probabilistic Urn model, sequentially updating randomly-chosen cells in a manner analogous to microstep updates used in other simulation-based network models (e.g., SAOM; [6]).

3.1   Cellular Structure

We divide Blau space into a series of cells comprising intersections between specific intervals in each dimension (e.g., males aged between 18 and 20 with a college degree). Individuals who belong to the same cell are within an interval distance of each other in all dimensions, while individuals in different cells differ in at least one dimension. Cells can be defined using equal intervals or can be scaled according to the social impact of particular differences.[5] While the cellular approach still imposes assumptions about the meaningfulness of differences, it can more easily be adjusted to match social reality than traditional Blau space approaches.

One advantage of this approach is that ordinal and categorical variables become much easier to model. Individuals can be placed in cells representing their unique combinations of attributes in the space so long as adjacency can be identified. Within this space, niches are no longer defined as regular polygons, but instead as the set of cells from which an organization has drawn members. This, in turn, means that the recruitment range of the organization is comprised of both the cells from which it draws members, as well as their adjacent cells. Individuals in these adjacent cells are at short network distances from organization members, and thus are more likely to be recruited. As a result, organizational niches can be highly irregular, and even discontinuous, thereby allowing a more flexible fit to non-normally distributed data.

The model proceeds by sequentially updating cells. When a cell is chosen for updating it is referred to as a "focal cell". Neighbor cells are those that are adjacent to the focal cell and therefore the neighborhood is the set of all such neighbor cells. An algorithm, described below, stochastically reassigns individual memberships in the focal cell based on the prevalence of organizational memberships in the focal cell and neighborhood. Over multiple updates this method allows the model to evolve in response to competitive pressure[6].

---

[5] This scaling might rely on MDS techniques or a Goodman RC-II model but is beyond the focus of this paper.

[6] The method proposed uses change in membership between two organizations as a proxy for competitive pressure, with the expectation that more organizations present in the neighborhood of a focal cell will result in more pressure for individuals to change their membership. See Fig. 2 for an example.

3.2   Urn Model

To simulate influence we turn to urn models, a class of stochastic model for discrete outcomes.[7] The urn model works in three steps. First, the number of members of each organization in the neighborhood, as well as the total number of individuals in the neighborhood regardless of membership, are stored as temporary variables. This is conceptualized as if each organization's members are balls of a specific color, and all balls present in the focal cell and neighborhood are pooled in a common urn. Second, the model uses the relative prevalence of these colored balls in the urn (i.e., the proportion of all individuals in the focal cell and neighborhood who belong to a particular organization) to determine the probability that a random draw from the urn would yield a ball of that color. These probabilities are stored as a set of temporary variables. Third, the probabilities are multiplied by the number of persons in the focal cell, generating the new predicted membership counts for each organization[8]. Finally, the simulation selects a new focal cell at random and repeats the process.[9,10]

Figure 2 is an example of this process in a two-dimensional Blau space. The urn model starts by selecting a random cell in the ecology and identifying the surrounding

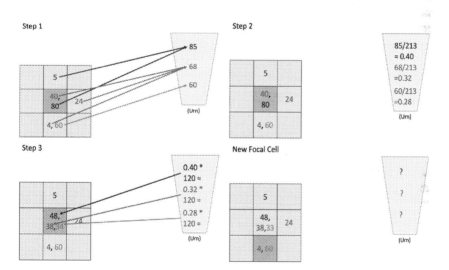

Fig. 2. Urn model example (Color figure online)

---

[7] We use the sampling with replacement implementation because the recruitment of one individual into an organization does not instantaneously make recruitment of another into the same organization *less* likely.

[8] The fraction of the focal cell that is updated in a given iteration is a tunable parameter.

[9] This process is similar to a cellular automata, where rule sets for updating a cell as "alive" or "dead" are based on the surrounding cells. In normal cellular automata models, it is the cell itself that is being updated, while in our model, it is entities within the cell that are being updated.

[10] All results reported herein use 10,000 iterations (i.e., 10,000 focal cell updates). This value is arbitrary as convergence criteria have not been defined, and thus analysts should engage in some degree of sensitivity analysis.

neighborhood. Because the neighborhood as a whole contributes to the relative probabilities, it is possible for individuals to join an organization that was not previously operating in the focal cell (e.g., the green organization in Fig. 2).

4 Metrics for Evaluating the Model

As the hybrid model dispenses with much of the machinery of Blau space, it is necessary to define new metrics appropriate to the approach. These metrics belong to two general types: metrics evaluating the breadth of the space from which an organization recruits (extensiveness), and intensity with which an organization exploits resources in a specific cell or cells (intensiveness). Combinations of these metrics can be used to identify specific organizational types, such as generalist organizations that recruit broadly, and specialist organizations that recruit heavily from a narrow range (e.g., [4, 5]).

4.1   Extensiveness

Extensiveness is an organization's breadth, or the range in the ecology from which the organization has recruited members. The corresponding metric (Eq. 1) defines a regular polygon whose top and bottom edges are the upper and lower bounds members of an organization have on each dimension (O_i). For example, if an organization's youngest member is 35, and their oldest 70, resulting in an edge from 35 to 70 on the age dimension. This value is then normalized to the unit interval by dividing the number of cells encompassed by the organization's polygon by a similar polygon defined by the minimal and maximal values of members of any organization in the same Blau space (R). The result is a fraction of the space occupied by any organization that is currently occupied by organization i. This fraction can change over time, indicating the expansion or contraction of an organizations recruitment and influence range.

$$Extensiveness = \frac{O_i}{R} \tag{1}$$

4.2   Intensiveness

Intensiveness is the success that an organization has in obtaining resources, such as members, in particular locations in Blau space. We compute two metrics for intensiveness, Cell Focused Intensiveness (CFI) and Range Focused Intensiveness (RFI). These metrics share a numerator, but differ in denominator, and thereby measure intensiveness relative to different baselines.

Cell Focused Intensiveness. CFI is the success of an organization in monopolizing resources in the cells they occupy. The metric is calculated (Eq. 2) by taking the number of members of an organization in a cell it occupies divided by the total number of individuals in that cell. This proportion is then summed up for all cells that the organization occupies and then divided by the number of cells from which the organization derives members (Q). Thus, an organization that is limited to a single cell, but possesses all the resources available in that cell, would have maximal CFI. And alternatively, if an organization has few individuals in multiple populous cells they will have a low CFI and may be at risk of membership loss as they lack the local influence to keep these "edge" members [5].

$$CFI = \frac{\sum \left(\frac{x_{ijk}}{C_{jk}}\right)}{Q} \qquad (2)$$

Range Focused Intensiveness. RFI is calculated similarly to CFI. However, the denominator is the number of cells within the range of the ecology in which the organization presently recruits members (O_i; See Eq. 1), instead of the number of individual cells in which an organization holds members (Eq. 3). RFI is interpreted as the mean recruitment success of an organization over its localized area of the ecology. RFI can be used as a general indicator of an organization's recruitment strength over the whole ecology.

$$RFI = \frac{\sum \left(\frac{x_{ijk}}{C_{jk}}\right)}{O_i} \qquad (3)$$

4.3   Interpretation of New Metrics

Table 1 computes intensiveness and extensiveness metrics for a subsection of the 8th Convocation of the Ukrainian Parliament. Table 2 shows the ranges and number of members for organizations in the ecology before and after model execution, as well as the differences between the two.

Table 1. Recruitment range evaluation metrics

	Extensiveness	CFI	RFI	Niche vol./space vol.
All-Ukrainian association (Fatherland)	0.1624	0.8615	0.0122	0.0986
Peter Poroshenko's Bloc	0.8305	0.8470	0.0207	0.0456
Opposition Bloc	0.4407	0.8083	0.0097	0.0587
People's front	0.6102	0.8227	0.0190	0.0904
People's will	0.1907	0.7843	0.0123	0.0747
Radical Party Oleg Lyashko	0.9216	0.8229	0.0025	0.3316
Economic development	0.1280	0.8009	0.0199	0.0777
Union (SAMOPOMICH) political party	0.3708	0.8857	0.0148	0.0952

Organizations that have larger recruitment areas, computed using the traditional Blau space approach, still have large recruitment areas in the new method. However, these numbers are insensitive to organizational membership, as the largest party, the Peter Poroshenko Bloc (PPB), and a much smaller party, the Radical Party of Oleg Lyashko (RP), both have high extensiveness scores. These same organizations also have similar CFI, indicating that they are roughly as successful in the cells where they are present, but vastly different RFI, with the PPB having the highest RFI.

These results allow us to discriminate generalist and specialist organizations [4, 5], as well as observe how they organize their recruitment space. Although the PPB has the greatest number of members in the ecology, they are not recruiting as extensively in the space as the RP, instead exploiting cells that are closer together, and with greater intensiveness (CFI and RFI). Comparing the recruitment ranges of the two political factions before and after model execution (Table 2), the PPB has a notably smaller range in age but matches the RP closely in all other dimensions. This small difference in the recruitment range increases the Extensiveness of the RP, but decreases their RFI and CFI. Distinctions like this are not picked up by traditional Blau space metrics, such as Niche Volume,[11] that only account for total space taken by the niche. As a result, traditional methods provide no way to identify where individuals are actually located, and or to see the effect in a localized area of Blau space (See Table 1; Niche Vol.).

We also note that changes in membership assignment are incremental rather than dramatic (ranging from 0–7), indicative that the model predicts smooth changes (see Table 2). Given our underlying data, gradual changes in factional alignment are more likely and so it is encouraging that the model duplicates this behavior. An example of member change between organizations can be seen in Fig. 3, showing pre and post-simulation distributions of members in the space for the PPB and the RP. PPB seems to disperse their memberships more widely over time, moving to areas that may have less competition. On the other hand, RP appears to concentrate into fewer but more widely spaced cells, keeping their extensiveness high and intensiveness low. Through different strategies are employed by PPB and RB, they manage to maintain fairly stable membership counts, preserving their place in the Ukrianian Parliament. But the differences in how they maintain membership sets the stage for future legislative and recruiting possibilities.

---

[11] Niche Volume is the total volume of Blau space taken by the niche or an organization. It is calculated by taking take volume of an organization's niche divided by the total volume of the k-dimensional space. All metrics are calculated using post-simulation results.

Table 2. Recruitment ranges

Initial Values (T=0)

	Members Orig.	Age-low Orig.	Age-high Orig.	Educ-low Orig.	Educ-high Orig.	Assist-low Orig.	Assist-high Orig.
All-Ukrainian Association (Fatherland)	14	30	69	4	4	0	29
Peter Poroshenko's Bloc	129	27	75	2	4	0	31
Opposition Bloc	35	35	68	2	4	4	28
People's Front	75	30	64	2	4	0	31
People's Will	19	38	80	4	4	0	28
Radical Party Oleg Lyashko	19	28	85	2	4	0	29
Economic Development	18	37	65	4	4	4	29
Union (SAMOPOMICH) political party	28	30	54	2	4	0	31

Modeled Values (T=10000)

	Members	Age-low	Age-high	Educ-low	Educ-high	# Assist-low	# Assist-high
All-Ukrainian Association (Fatherland)	13	30	69	4	4	7	29
Peter Poroshenko's Bloc	128	27	75	2	4	0	31
Opposition Bloc	30	37	68	2	4	3	28
People's Front	81	29	64	2	4	0	31
People's Will	18	45	80	4	4	0	29
Radical Party Oleg Lyashko	16	28	85	2	4	0	29
Economic Development	18	37	65	4	4	5	29
Union (SAMOPOMICH) political party	35	30	54	2	4	4	31

Differences in Range

	Diff. Members	Diff. Age-low	Diff. Age-high	Diff. Educ-low	Diff. Educ-high	Diff. # Assist-low	Diff. # Assist-high
All-Ukrainian Association (Fatherland)	1	0	0	0	0	7	0
Peter Poroshenko's Bloc	1	0	0	0	0	0	0
Opposition Bloc	5	2	0	0	0	1	0
People's Front	6	1	0	0	0	0	0
People's Will	1	7	0	0	0	0	1
Radical Party Oleg Lyashko	3	0	0	0	0	0	0
Economic Development	0	0	0	0	0	1	0
Union (SAMOPOMICH) political party	7	0	0	0	0	4	0

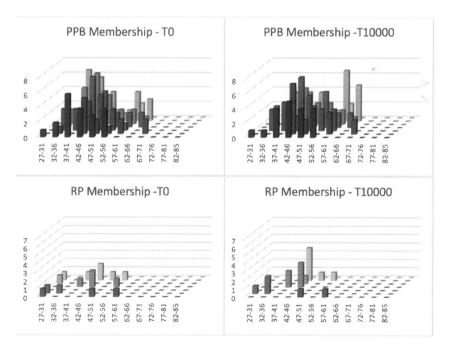

Fig. 3. Organization member shift

5 Conclusion

The hybrid model proposed here is a new and unique approach to modeling and simulation in Blau Space. Although it is not yet a replacement for existing models, it provides more detail on recruitment behavior and addresses several outstanding problems with prior methods. The strength of the new method is that it allows for a more granular and locally focused description of competition within Blau space, and supports a wider range of possible data types, including binary and categorical variables. The ability to distinguish between generalist and specialist recruitment strategies through metrics of extensiveness and intensiveness is more descriptive of the organization's general behavior and the influence each organization has on the ecology. In future work with this new Hybrid Blau space Model we hope to provide a metric of competition specialized for the new cellular structure, and visualization tools permitting easier exploration of the location of individuals and organizations within Blau space. We also plan to validate model predictions against observed system behavior from future sessions of the Ukrainian Parliament and to investigate convergence criteria for the model.

References

1. Lazerfield, P.F., Merton, R.K.: Friendship as a social process: a substantive and methodological analysis. Freedom Control Mod. Soc. **18**(1), 18–66 (1954)
2. Hannan, M.T., Freeman, J.: The population ecology of organizations. Am. J. Sociol. **82**(5), 929–964 (1977)
3. McPherson, J.M.: An ecology of affiliation. Am. Sociol. Rev. **48**, 519–532 (1983)
4. McPherson, J.M., Ranger-Moore, J.R.: Evolution on a dancing landscape: organizations and networks in dynamic Blau space. Soc. Forces **70**(1), 19–42 (1991)
5. Popielrz, P.A., McPherson, J.M.: On the edge or in between: niche position, niche overlap, and the duration of voluntary association memberships. Am. J. Sociol. **101**(3), 698–720 (1995)
6. Snijders, T.A.: Stochastic actor-oriented models for network change. J. Math. Sociol. **21**(1–2), 149–172 (1996)
7. Marsden, P.V., Gorman, E.H.: Social networks, job change, and recruitment. In: Sourcebook of labor market, pp. 467–502(2001)
8. McPherson, J.M., Smith-Lovin, L., Cook, J.M.: Birds of a feather: homophily in social networks. Annu. Rev. Sociol. **27**(1), 415–444 (2001)
9. Brashears, M.E., Genkin, M., Suh, C.S.: In the organizations shadow: how individual behavior is shaped by organizational leakage. Am. J. Sociol. **123**(3), 787–849 (2017)
10. Shi, Y., Dokshin, F.A., Genkin, M., Brashears, M.E.: A member saved is a member earned? The recruitment-retention trade-off and organizational strategies for membership growth. Am. Sociol. Rev. **82**, 407–434 (2017)

A Challenging Dataset for Bias Detection: The Case of the Crisis in the Ukraine

Andres Cremisini, Daniela Aguilar, and Mark A. Finlayson[✉]

School of Computing and Information Sciences, Florida International University,
Miami, FL 33199, USA
{acrem003,dagui082,markaf}@fiu.edu

Abstract. The use of disinformation and purposefully biased reportage to sway public opinion has become a serious concern. We present a new dataset related to the Ukrainian Crisis of 2014–2015 which can be used by other researchers to train, test, and compare bias detection algorithms. The dataset comprises 4,538 articles in English related to the crisis from 227 news sources in 43 countries (including the Ukraine) comprising 1.7M words. We manually classified the bias of each article as either *pro-Russian*, *pro-Western*, or *Neutral*, and also aligned each article with a master timeline of 17 major events. When trained on the whole dataset a simple baseline SVM classifier using doc2vec embeddings as features achieves an F_1 score of 0.86. This performance is deceptively high, however, because (1) the model is almost completely unable to correctly classify articles published in the Ukraine (0.07 F_1), and (2) the model performs nearly as well when trained on unrelated geopolitics articles written by the same publishers and tested on the dataset. As has been pointed out by other researchers, these results suggest that models of this type are learning journalistic styles rather than actually modeling bias. This implies that more sophisticated approaches will be necessary for true bias detection and classification, and this dataset can serve as an incisive test of new approaches.

Keywords: Media bias · Machine learning · Natural language processing

1 Introduction

Disinformation is the attempt to willfully deceive and sway public opinion to the benefit of some organization or state [19]. Bias is a more general phenomenon encompassing disinformation, in that bias is not necessarily strategic or deceitful, usually consisting of the articulation of a preference for a particular position on some issue [8]. In this regard, modeling disinformation likely requires modeling bias, and as such we investigate bias classification as a first step toward disinformation detection. Linguistically, the articulation of bias comprises a set of frames, which are combinations of words that seek to "promote a particular interpretation" of some concept or event [2]. Frames can be expressed in

© Springer Nature Switzerland AG 2019
R. Thomson et al. (Eds.): SBP-BRiMS 2019, LNCS 11549, pp. 173–183, 2019.
https://doi.org/10.1007/978-3-030-21741-9_18

any number of different ways, including "key-words, stock phrases, stereotype images, ... metaphors, exemplars, catchprases, depictions..." [2, p. 1473], all of which can be used in combination to create an interpretative "package." Thus, language exhibits bias in regard to some action or opinion, and framings are rhetorical tools used to support and express that bias; in other words, the framings of an argument are the surface signifiers for an underlying bias. For example, the frames *pro-choice* and *pro-life* support different biases towards the support for or abolishment of abortion.

Even for people, detecting bias is a difficult task. Recasens *et al.* reported that people achieve only 30% accuracy in identifying the bias-inducing word in a biased sentence (pairwise agreement of 40.73%) [17]. This may be due to the large repertoire of subtle framings available to authors when constructing their rhetoric. Take, for example, an event that took place during the writing of this paper, namely, the President of the United States declaring a national emergency to fund a wall at the U.S.-Mexico border. Examples (1) and (2) ref show headlines on the topic from articles published within hours of each other by left-leaning and right-leaning news outlets[1].

(1) Vox (left): *Trump will declare a national emergency to secure money for his border wall.*[2]

(2) Fox (right): *Trump declares national emergency to build border wall.*[3]

While lexically these sentences are nearly identical, within the context of recent American politics the pronoun *his* in (1) signals an anti-Trump bias. A few months before these articles were published, the President held a televised meeting with two Democratic representatives from the House and Senate. In this meeting, the President stated that he *owned* the controversial government shutdown—brought about by a disagreement over border wall funding—that ended a few weeks before the national emergency declaration. Subsequently Democratic politicians repeatedly referred to both the shutdown and the border wall as belonging to the President, painting both in a negative light. Building a model to automatically detect bias such as this is daunting, requiring not only sophisticated linguistic analysis but also a knowledge base of current events from which to extract the implication of the word *his*. While in this example a lexical model would not be adequate, there is some evidence that frames can be modelled lexically [6], which is the approach we adopt and investigate in our baseline classifier.

The primary contribution of this paper is a dataset that contains biased articles about a shared set of events. In particular, we have collected and annotated a set of 4,538 English language long form news articles from 227 news sources across 43 countries with three broad biases advocating for pro-Russian, Neutral,

---

[1] As classified by the news media site AllSides, https://www.allsides.com/unbiased-balanced-news.

[2] https://www.vox.com/2019/2/14/18222167/trump-border-security-deal.

[3] http://www.fox5dc.com/news/border-wall-national-emergency-government-funding-trump.

or pro-Western interests in the context of the 2014–2015 Ukrainian Crisis. All the articles report on one or more of a set of 17 events that occured during the crisis. We manually extracted a set of frames related to each event for each of the three biases, and then used these frames to determine what bias each article articulates. We created a simple supervised classification model (an SVM) with lexical features that is able to classify the bias of articles outside of the Ukraine, though fails when applied to articles from within the Ukraine. These results suggest that while some aspects of bias can be captured by lexical models, it seems that the lexicon of similar biases (say, pro-Russian from Russia vs. pro-Russian from Ukraine)—and thus their frames—vary across contexts. Further, the same model achieves comparable, though lower, performance when trained on an unrelated set of political articles and tested on the dataset, suggesting that it is actually capturing writing style more than it captures bias. These results suggest that a deeper level of regional and cultural awareness is necessary to detect and classify bias, and ultimately disinformation.

2   Data

We have collected and manually annotated 4,538 news articles that report on the situation in Ukraine of 2014–2015, with particular focus on the annexation of Crimea by Russia in 2014, the military conflicts in Southeastern Ukraine, and the Maidan protests. It has been noted by many commentators that the use of disinformation was prominent during the conflict [13, for example].

We began collection of the articles by crawling the reference lists of the twelve Wikipedia pages that discuss some facet of the 2014–2015 crisis. We preliminarily categorized the bias of each article based on its country of origin, placing each country into pro-Western, Neutral, or pro-Russian bias classes on the basis of known geopolitical alliances. As described below in Sect. 2.1 we developed a bias classification scheme using these same three classes. Our initial country-based categorization revealed that we had a disproportionate number of pro-Western articles, and therefore we augmented the dataset with more pro-Russian articles by crawling the Sputnik news website[4] and retrieving every article classified by Sputnik as dealing with the crisis.

The second author[5] manually annotated the bias of each article. After manually classifying the bias the dataset was still significantly imbalanced, though with a large number of both pro-Russian and pro-Western sources, as was our primary interest. Given the time consuming nature of identifying and classifying news articles, further balancing the dataset remains for future work. A final manual classification task involved classifying every Sputnik article into one of the events as described by one of our twelve Wikipedia pages on the Ukrainian Crisis. Similarly to Wikipedia, Sputnik organized the Ukrainian articles by event[6],

---

[4] https://sputniknews.com/.

[5] The second author is an undergraduate researcher majoring in International Relations and specializing in Russia.

[6] Sputnik uses the word "Topics" to refer to their article categories, though these serve the same organizing purpose as Wikipedia's events.

some of which aligned with the Wikipedia events. In order to merge these two event lists, we manually classified the Sputnik articles into one of the Wikipedia events. Those Sputnik articles that did not match any of the Wikipedia events resulted in a new Sputnik-only event in our timeline. This merging resulted in a total of 17 events. Therefore, at the end of the two annotation tasks, each article is classified by both bias and event. Table 1 shows a chronological list of the events, a breakdown of the number of articles that fall into each bias class and each major event, and the average and total word counts. Table 2 lists the number of articles for each of the top three publishers for each bias class.

Table 1. Breakdown of number of articles for each bias and event, in chronological order. Src $\equiv$ Source of the category: Wikipedia (Wk) or Sputnik (Sp); $N \equiv$ Number of articles in the category; $\overline{|W|} \equiv$ Average number of words per article in the category; $|W| \equiv$ Total number of words in the category. Some articles are classified into multiple categories.

Event	Src	pro-Russian			Neutral			pro-Western			Total		
		N	$\overline{|W|}$	$|W|$	N	$\overline{|W|}$	$|W|$	N	$\overline{|W|}$	$|W|$	N	$\overline{|W|}$	$|W|$
1 Ukraine-EU Association Agreement	Wk	0	0	0	10	310	3,098	36	610	21,958	46	545	25,056
2 Russia-Ukraine Gas Conflict	Wk	289	259	74,809	2	209	418	1	294	294	292	259	75,521
3 Euromaidan	Wk	1,302	275	357,595	84	247	20,764	126	641	80,813	1,512	304	459,172
4 Russian Photographer Stenin Killed	Sp	119	322	38,284	0	0	0	0	0	0	119	322	38,284
5 Annexation of Crimea	Wk	362	320	115,808	32	521	16,670	143	842	120,352	537	471	252,830
6 Crimea: New Life for Russia's Historic Resort	Sp	25	273	6,820	0	0	0	0	0	0	25	273	6,820
7 2014 Hrushevskoho Street Riots	Wk	0	0	0	15	232	3,478	30	577	17,312	45	462	20,790
8 Euromaidan (Post)	Sp	1,045	199	207,861	0	0	0	0	0	0	1,045	199	207,861
9 Russian Military Intervention	Wk	5	5,656	28,279	41	751	30,807	229	967	221,350	275	1,020	280,436
10 2014 Pro-Russian Unrest in Ukraine	Wk	14	1,809	25,322	30	560	16,794	128	879	112,479	172	899	154,595
11 International Sanctions During Ukrainian Crisis	Wk	3	125	376	6	469	2,813	43	584	25,092	52	544	28,281
12 War in Donbass	Wk	14	135	1,896	40	384	15,359	225	818	184,090	279	722	201,345
13 2014 Ukrainian Presidential Election	Wk	2	122	243	4	268	1,071	6	718	4,311	12	469	5,625
14 Russian Humanitarian Aid Convoys	Sp	281	234	65,792	0	0	0	0	0	0	281	234	65,792
15 2014 Donbass General Elections	Wk	6	485	2,909	3	392	1,175	10	782	7,821	19	627	11,905
16 2014 Ukrainian Parliamentary Election	Wk	7	501	3,507	1	730	730	0	0	0	8	530	4,237
17 2015 Ukrainian Local Elections	Wk	1	676	676	0	0	0	1	860	860	2	768	1,536
Total Unique Articles or Total Tokens		3,372	255	860,212	258	420	108,453	908	804	729,965	4,538	374	1,698,630

Table 2. Top three publishers per bias

	pro-Russian	Neutral	pro-Western	Total
Sputnik	**3,308**	3	0	3,311
TASS	**36**	1	0	37
Voice of Russia	**17**	0	0	17
Interfax	0	**73**	30	103
Euronews	0	**20**	11	31
OSCE	0	**18**	1	19
BBC News	0	9	**162**	171
Reuters	0	6	**125**	131
The Guardian	0	4	**74**	78

2.1 Bias Annotation Scheme

In order to lay the groundwork for validating the annotation on the data, the second author drew on her knowledge of the Ukrainian Crisis and Russian politics to identify sets of frames present in the dataset. We randomly selected a sample of 150 articles (roughly six per event) and she identified all of the frames she could, partitioning them by bias, resulting in a total of 51 sets of frames—one for each bias in each of the seventeen events. She then used these frames to manually classify every article in the dataset. As an example, the following frames are used in the *Annexation of Crimea* event:

pro-Russia: Crimea coming home; Russia welcomes Crimea; Crimea's accession to Russia; Russia welcomes Crimea; Admission of Crimea into Russia; Ukraine took over Crimea; Crimea wants to go back to its roots in Russia; Referendum website hit by cyber-attack; The U.S. plans to supply weapons to Ukraine.

pro-Western: Russia stealing land from a sovereign nation; Russian Separatists; Annexation by Moscow; Russia stages coup; Russia took over Crimea; Russia does not fear the West; Crimea has been isolated by Russia; Putin admits Russian actions to take over Crimea; Putin refuses to rule out intervention in Donetsk.

Neutral: Mention frames from both sides equally, reporting facts, or offer explanation for both pro-Russian and pro-Western frames. State factual information without any emotional, political or ideological charge.

2.2 Content Extraction

We archived all the articles and processed them into raw text using a tool which we specially built for this purpose called WART—the Web ARchiving Tool. While most web browsers have a function to archive webpages, the process often does not save a faithful snapshot of the page, nor do they automatically identify and extract the textual content. Webpages differ dramatically in how they are structured internally and there is no standard for identifying the text of a news article. Further, we wanted the ability to batch download articles, because manually saving thousands of web pages using browser functionality is inconvenient and inefficient. WART uses the `wget` [16] archiver on the backend to take an exact snapshot of a webpage and package it into a compressed archive. WART also provides the ability to batch download pages, view a page saved in an WART archive file, and automatically identify the content. For content extraction, we began with the Dragnet tool [15] and fixed a bug which improved content extraction F_1 score from 0.84 to 0.90.

3   Bias Classification

3.1   Preprocessing

We extracted the main content of each article using WART, and then tokenized the text with `nltk` [3], also removing capitalization. We also removed all mention of each article's publisher to ensure our model is not simply learning to relate publishers to biases.

3.2   Classifier

We trained a 1-vs-1 Support Vector Machine (SVM) with a linear kernel [14] with a 250-dimensional doc2vec model trained on our dataset [12]. Our decision to use document embeddings (rather than word or sentence embeddings) was motivated by various factors. From a performance standpoint, doc2vec has been shown to achieve state-of-the-art results in several tasks, including, specifically, text classification (of which bias detection can be thought of as a variant) [12]. From the standpoint of semantic richness of document representations, doc2vec embedding models capture quite a bit of semantic similarity between texts that are otherwise syntactically different; this is one main advantage of embeddings in general over, for example, bag of words approaches.

In addition to SVM, we also tested LDA and QDA, Multilayer Perceptron, k-Nearest Neighbors, and Random Forests [10]. The linear kernel SVM performed best on our dataset and as such we report on this model's performance. Importantly, all models performed similarly across all experiments.

To test the effectiveness of the embeddings, we carried out a retrieval test of embedding effectiveness. In this test, we computed an embedding for each document treating the document as previously unseen. We then used that computed embedding to find the closest embedding vector over all documents. This retrieval returned the original document 98% of the time, confirming a suitable doc2vec representation of our corpus.

4   Results and Analysis

We ran seven different experiments on our data. All experiments used 5-fold cross validation and were run twice through our data—once with the entire dataset and then using oversampling to compensate for the minority class imbalance in the dataset (the minority class was usually the Neutral bias, but in a few instances it was pro-Western or pro-Russian). We performed oversampling using the Borderline-SMOTE technique, which augments the minority class in the training data by generating synthetic samples near those minority samples closest to samples from other classes. This helps the model find a more general decision boundary less prone to overfitting to the majority classes [5,9]. We first split the data into one of the five fold-splits, performed oversampling on the training set, and then tested on the test set which was unseen by the SMOTE oversampling algorithm. In this way we ensured that the synthetic SMOTE interpolations were not seen by our model during both training and testing, providing a more realistic measure of performance. Table 3 shows the results of the experiments, along with the broad questions we were seeking to answer in each.

The first experiment, $\mathbf{A} \rightarrow \mathbf{A}$, seems to confirm that indeed a simple model can be used to model bias on news articles sampled from different countries. This is similar to the result in experiment $\{\mathbf{A{-}U}\} \rightarrow \{\mathbf{A{-}U}\}$, where the slight performance increase seems to suggest that the Ukrainian articles are more diffi-

Table 3. Experiments, results, and relevant questions. $\mathbf{A} \equiv$ *All articles from all countries;* $\mathbf{U} \equiv$ *All articles from Ukraine;* $\mathbf{WC}_{\bar{\mu}} \equiv$ *The average of train/test within a country over all countries;* $\mathbf{Aux} \equiv$ *Auxilliary dataset of non-Ukraine-Crisis geopolitical articles from Reuters and Sputnik*

Train	Test	F_1 Full Corpus	F_1 Over-sampled	Question
A	**A**	0.86	0.76	Naive experiment. Can a lexical model capture bias?
A–U	**A–U**	0.89	0.81	Given the domain of our data, is it easier to model bias outside of Ukraine, the central country?
U	**U**	0.57	0.57	Can our model capture bias only within Ukrainian articles?
A–U	**U**	0.07	0.34	Does bias generalize from non-Ukrainian to Ukrainian articles?
U	**A–U**	0.05	0.06	Does bias generalize from Ukrainian articles to non-Ukrainian articles?
$\mathbf{WC}_{\bar{\mu}}$	$\mathbf{WC}_{\bar{\mu}}$	0.74	-	Is bias more easily classified when trained and tested within a single country?
Aux	**A**	0.76	-	Is our model actually learning regional journalistic style?

cult to classify than the non-Ukrainian articles. The drop in performance on the third experiment, $\mathbf{U} \rightarrow \mathbf{U}$ suggests that the bias lexicon in Ukrainian articles is more difficult to learn, which is a reasonable interpretation given that Ukraine is the central country in the conflict (as well as the only country with articles classified into each of the three different biases). This seems to be supported by the penultimate experiment, $\mathbf{WC}_{\bar{\mu}} \rightarrow \mathbf{WC}_{\bar{\mu}}$, where we see a less significant drop in performance from $\mathbf{A} \rightarrow \mathbf{A}$ but an improvement over $\mathbf{U} \rightarrow \mathbf{U}$, suggesting that indeed modeling bias within the Ukraine is more difficult (at least with its smaller set of articles). The striking performance drops in experiments $\{\mathbf{A}\text{-}\mathbf{U}\}$ $\rightarrow \mathbf{U}$ and $\mathbf{U} \rightarrow \{\mathbf{A}\text{-}\mathbf{U}\}$ suggest that, at the very least, our lexical model does not generalize from the rest of the world to the Ukraine. The slightly higher performance in the oversampled $\{\mathbf{A}\text{-}\mathbf{U}\} \rightarrow \mathbf{U}$ experiment suggests that the larger sample size of the non-Ukrainian articles offers a richer lexicon for the Ukrainian articles than the other way around, as is to be expected.

These considerations naturally lead us to question if our model is learning bias at all or if, more likely, the model is simply learning journalistic style. The last experiment in Table 3 ($\mathbf{Aux} \rightarrow \mathbf{A}$) supports this view. The $\mathbf{Aux}$ set consists of a balanced dataset of 6,000 articles from Reuters and Sputnik dealing with geopolitics (removing all articles mentioning Ukraine). The articles were crawled from the websites using the publishers' *politics* news tags. We automatically annotated all Reuters articles as pro-Western and all Sputnik articles as pro-Russian, as these publishers are one of the top two majority publishers in our Ukrainian dataset for the pro-Western and pro-Russian biases, respectively (Table 2). The high performance of the model under these conditions, in light of our other experiments, suggests that we cannot be certain if a lexical model is learning bias rather than some other traits—such as regional or publisher style—that are correlated yet not causally linked to bias itself.

5   Related Work

Media bias has long been studied in the social sciences, but has historically received relatively little attention in computer science. Hamborg *et al.* gave a fairly broad history of media bias research both in Computer Science and in the Social Sciences, noting that the conceptual frameworks in Computer Science approaches tend to lack conceptual sophistication [8]. They attempted to bridge the gap between the two perspectives by compiling a sort of taxonomy of the forms of media bias as conceptualized by social scientists and a compilation of modeling frameworks devised to detect parts of the social science taxonomy by computer scientists. Grimmer and Stewart provide a helpful overview of some of the pitfalls in interpreting the results of automatic political text analysis, aligning with the general conclusion of our results [7].

We mention a few relevant computer science works here. Recasens *et al.* deconstructed bias into two components, epistemological bias and framing bias, the latter dealing with statements focusing on the truthfulness of a proposition and the former with subjective words or phrases associated with a particular point of view [17]. They constructed a large dataset of sentences flagged by Wikipedia users as violating the neutral point-of-view (NPOV) Wikipedia requirement, and trained a classifier to predict the bias inducing word in a sentence. They asked humans to perform the experiment and found that performance was surprisingly low, with around 30% prediction accuracy. Baumer *et al.* investigated how humans detect bias by asking annotators to highlight the parts of articles that contain instances of framing, in addition to building a classifier for the task [2]. They used lay annotators as opposed to subject matter experts, raising the important point that framing is only as effective as it is understandable, suggesting the need to study the effects of different framings on different subsets of populations. Field *et al.* studied framing and agenda setting in Russian news, finding that they could predict mentions of the United States in Russian news media by using fluctuations in Russian GDP [6], in addition to building a lexical classifier to predict an article's main frame using the Media Frames Corpus [4]. Krestel *et al.* used a similar lexical approach to detect bias in German Parliament speeches and German news [11].

Recent work in fake news detection consider the role of bias in fake news production, using bias detection as a feature in fake news detection models that tend to include some form of a knowledge database to judge the veracity of a news item [1,18]. While fake news can be seen as a more general form of disinformation—in that fake news does not necessarily have a discernible end goal tied to a state or organization—most fake news detection frameworks assume that bias is a good predictor, and so share a conceptualization similar to ours of the relationship between bias and disinformation.

6   Contributions

We have compiled and annotated a dataset of 4,538 articles from 227 news sources across 43 countries relating to the Ukrainian Crisis of 2014–2015 annotated with bias (pro-Russian, Neutral or pro-Western) and relevant events in the crisis timeline. We investigated the suitability of lexical models to capture bias in this dataset. We found that the lexical approach did not generalize from non-Ukrainian to Ukrainian publishers, suggesting that during the Ukrainian Crisis the lexicon of bias is more complex within the Ukraine itself. Further our results suggest that the lexical model is not in fact learning bias at all, but rather regional journalistic styles which are likely correlated with bias but not indicative of bias itself. Our results point both to the need for more sophisticated NLP techniques in building a general bias detector and simultaneously call into question the premise a general bias detector is possible given that the rhetorical tools used to express bias are so steeped in culture.

Acknowledgements. This work was supported by Office of Naval Research (ONR) grant number N00014-17-1-2983.

References

1. Baly, R., Karadzhov, G., Alexandrov, D., Glass, J., Nakov, P.: Predicting Factuality of Reporting and Bias of News Media Sources (2018)
2. Baumer, E.P.S., Elovic, E., Qin, Y.C., Polletta, F., Gay, G.K.: Testing and comparing computational approaches for identifying the language of framing in political news. In: ACL, pp. 1472–1482 (2015)
3. Bird, S., Klein, E., Loper, E.: Natural Language Processing with Python. O'Reilly Media, Newton (2009)
4. Card, D., Boydstun, A.E., Gross, J.H., Resnik, P., Smith, N.A.: The media frames corpus: annotations of frames across issues. In: Proceedings of the 53rd Annual Meeting of the ACM and the 7th International Joint Conference on Natural Language Processing (vol. 2: Short Papers) (2015). https://doi.org/10.3115/v1/p15-2072
5. Chawla, N., Bowyer, K.: SMOTE: synthetic minority over-sampling technique Nitesh. J. Artif. Intell. Res. **16**, 321–357 (2002). https://doi.org/10.1613/jair.953
6. Field, A., Kliger, D., Wintner, S., Pan, J., Jurafsky, D., Tsvetkov, Y.: Framing and Agenda-Setting in Russian News: a Computational Analysis of Intricate Political Strategies (2018)
7. Grimmer, J., Stewart, B.M.: Text as data: the promise and pitfalls of automatic content analysis methods for political texts. Polit. Anal. **21**(3), 267–297 (2013). https://doi.org/10.1093/pan/mps028
8. Hamborg, F., Donnay, K., Gipp, B.: Automated identification of media bias in news articles: an interdisciplinary literature review. Int. J. Digit. Libr. (2018). https://doi.org/10.1007/s00799-018-0261-y
9. Han, H., Wang, W.-Y., Mao, B.-H.: Borderline-SMOTE: a new over-sampling method in imbalanced data sets learning. In: Huang, D.-S., Zhang, X.-P., Huang, G.-B. (eds.) ICIC 2005. LNCS, vol. 3644, pp. 878–887. Springer, Heidelberg (2005). https://doi.org/10.1007/11538059_91
10. James, G., Witten, D., Hastie, T., Tibshirani, R.: An Introduction to Statistical Learning – with Applications in R. Springer Texts in Statistics, vol. 103. Springer, New York (2013). https://doi.org/10.1007/978-1-4614-7138-7
11. Krestel, R., Wall, A., Nejdl, W.: Treehugger or petrolhead? In: Proceedings of the 21st International Conference Companion on World Wide Web - WWW 2012 Companion, p. 547 (2012). https://doi.org/10.1145/2187980.2188120
12. Le, Q.V., Mikolov, T.: Distributed representations of sentences and documents. CoRR abs/1405.4053 (2014)
13. Nimmo, B.: Anatomy of an info-war: how Russia's propaganda machine works, and how to counter it. Technical report, Central European Policy Institute (CEPI) (2015)
14. Pedregosa, F., et al.: Scikit-learn: machine learning in python. J. Mach. Learn. Res. **12**, 2825–2830 (2011)
15. Peters, M.E., Lecocq, D.: Content extraction using diverse feature sets. In: Proceedings of the 22nd International Conference on World Wide Web, WWW 2013 Companion, pp. 89–90. ACM, New York (2013). https://doi.org/10.1145/2487788.2487828
16. Project, G.: Gnu Wget 1.20 Manual (2018). https://www.gnu.org/software/wget/manual/
17. Recasens, M., Danescu-Niculescu-Mizil, C., Jurafsky, D.: Linguistic models for analyzing and detecting biased language. In: Proceedings of the 51st Annual Meeting on ACM, pp. 1650–1659 (2013)

18. Sharma, K., Qian, F., Jiang, H., Ruchansky, N., Zhang, M., Liu, Y.: Combating Fake News: A Survey on Identification and Mitigation Techniques, vol. 37, no. 4 (2019). https://doi.org/10.1145/1122445.1122456
19. Zhou, X., Zafarani, R.: Fake News: A Survey of Research, Detection Methods, and Opportunities (2018). https://doi.org/10.13140/RG.2.2.25075.37926

Does Causal Coherence Predict Online Spread of Social Media?

Pedram Hosseini[1]([✉]), Mona Diab[1,2]([✉]), and David A. Broniatowski[1]([✉])

[1] The George Washington University, Washington DC, USA
{phosseini,broniatowski}@gwu.edu
[2] AWS, Amazon AI, Seattle, USA
diabmona@amazon.com

Abstract. Online misinformation is primarily spread by humans deciding to do so. We therefore seek to understand the factors making this content compelling and, ultimately, driving online sharing. Fuzzy-Trace Theory, a leading account of decision making, posits that humans encode stimuli, such as online content, at multiple levels of representation; namely, gist, or bottom-line meaning, and verbatim, or surface-level details. Both of these levels of representation are expected to contribute independently to online information spread, with the effects of gist dominating. Important aspects of gist in the context of online content include the presence of a clear causal structure, and semantic coherence – both of which aid in meaning extraction. In this paper, we test the hypothesis that causal and semantic coherence are associated with online sharing of misinformative social media content using Coh-Metrix – a widely-used set of psycholinguistic measures. Results support Fuzzy-Trace Theory's predictions regarding the role of causally- and semantically-coherent content in promoting online sharing and motivate better measures of these key constructs.

Keywords: Social media · Misinformation · Causal coherence · Gist · Coh-metrix

1 Introduction

The speed and scale at which false news spreads through online platforms have posed new challenges to society with implications in politics, economics, and mental health. Although automated accounts ("bots") have been implicated in dissemination of such misinformation—especially at early stages on social media networks [19]—recent studies demonstrate that human users are the primary vector for the spread of false content [25]. We therefore seek to understand what content features lead humans to spread false information. Although significant effort has focused on meta-information, such as social network structure [20], individual personality differences [16], source credibility [8], and prior exposure [15], relatively little attention has focused on online *content*.

© Springer Nature Switzerland AG 2019
R. Thomson et al. (Eds.): SBP-BRiMS 2019, LNCS 11549, pp. 184–193, 2019.
https://doi.org/10.1007/978-3-030-21741-9_19

We seek to determine what online content features make this content compelling, leading people to share it. Our analysis is based on Fuzzy-Trace Theory (FTT) [17], a leading theory of decision making under risk. According to FTT, there are multiple representations of online content encoded in parallel: (1) the gist representation is a developmentally-advanced, subjective, simplified, memorable, and contextualized representation of the bottom-line meaning of a stimulus, and (2) the verbatim representation is detailed, factual, brittle in memory, and decontextualized. In the context of online misinformation, gist tends to be semantically- and causally-coherent representation of the situation described in the online content, typically linking some events, actors, and outcomes— regardless of its truth value [3,18].

In practice, humans tend to rely on gist representations when making decisions, such as whether or not to share online content. However, prior work examining the role of content in online information spread has primarily focused on verbatim content features, e.g., [24], although see [5].

In this paper, we test the hypothesis that more causally- and semantically-coherent text documents—both necessary (but not sufficient) conditions for meaningful gist—are more likely to be shared by individuals on social media. We briefly review the literature on measures of semantic and causal coherence, and mental representation of text content. Based on this review, we use indices derived from Coh-Metrix [11] to compute a variety of indices including causality indices in text and run a linear regression analysis using these indices. In particular, we investigate the associations between Coh-Metrix indices related to causal and semantic coherence, and numbers of shares of a collection of news documents on Twitter.

The rest of paper is organized as follows: in Sect. 2 we review the existing metrics of causal and semantic coherence, mainly from the psychology literature. In Sect. 3, we delineate our analysis of a dataset of fake and real news articles to test our hypothesis. Next, we discuss our results from linear regression analysis in Sect. 4. Finally, we conclude the paper in Sect. 5 and outline next steps and future work.

2   Literature Review

In this section, we review methods of measuring causal and semantic coherence in text documents. It is important to emphasize that semantic coherence in text documents has been measured using different techniques, including topic models [12] and Latent Semantic Analysis [6]. However, coherence has different types and our intention here is to measure *causal coherence*. Since there are, to our knowledge, no studies on measuring causal coherence using Natural Language Processing (NLP) methods, we mainly review the psychology and psycholinguistics literature to find a metric for causal coherence that can be implemented using text analysis techniques.

At first glance, most of the methods of rating causal coherence are based on discovering *causal links* among *units* of information in content. As a result, there are two key questions to address when finding a metric for causal coherence.

First, what is a "unit" of information in a document? A unit of information could be in form of a sentence, clause, event, action, state or any combination of them. And second, how do we define and extract causal links between these units of information? In the following, we address these two questions.

In the majority of studies the definition of causal coherence is based on causal links in documents. In [2], causal coherence is measured by checking if there is a causal link between a pair of sentences. Here, causal relations were manually extracted. The definition of causal coherence in [10] is also based on the existence of causal connections across sentences, which are manually checked by participants by assigning a score between 1 to 7 to triplets of sentences.

In [13], the question regarding the unit of information in documents is addressed. A causal coherence relation, explicit or implicit, is not a relation between propositions. Rather, coherence between two sentences describing cause and effect, respectively, is the result of the fact that *events* described in these sentences are causally related in the world described by the text—not between the *sentences*. Thus, inferring causal coherence requires background knowledge.

In [9], three measures of causal coherence are introduced including: (1) meaning-making, (2) narrative complexity, and (3) the use of causal terms. Meaning-making refers to the tendency of subjects to describe lessons or insights that they gained connected to the story they told. Narrative complexity analyzes the recognition of multiple dimensions, perspectives or emotions in a narrative. And the coding scheme is for measuring narrative complexity. For measuring causal terms, fifteen causal terms were coded. The frequency of word occurrence was obtained using Linguistic Inquiry and Word Count (LIWC) [14] software, which reports the frequency of occurrence of categories of words in narrative units.

In [1], a summary of what we need to know about causal coherence and how it helps readers' comprehension of narratives is presented. Authors refer to Trabasso's [23] definition of causal coherence in their work: "causal coherence is a property of stories that readers mentally represent as a causal network, where events are interconnected through causal links". One interesting point mentioned here is that causal coherence is not strictly dependent on the use of elaborate syntactic structures or cohesion devices (such as causal conjunctions) but depends mostly on conceptual links of causality. Their unit of analysis in text is the "T-unit", or terminable unit, which refers to a main clause plus any subordinate clauses that may be attached to it. Finally, they distinguish between different causal relation types (enabling, physical, psychological, and motivational)—focus on the relations particularly relevant to the narration of human-like events: psychological and motivational—and between local vs. global coherence[1].

---

[1] Finding local and global coherence is based on the work by Trabasso and van den Broek [22] on causal network model. The idea is that causal connections between text units are established at three levels. First level: organizing events in episodes, or GAO (goal, action, outcome) structures. Second level: linking GAOs to adjacent GAOs in a linear sequence. And third level: linking GAOs, even distant ones, in a global causal network.

[4] uses Trabasso and Sperry's paradigm [22] to identify causal connections and a story's causal chain. Here, a causal connection refers to the direct causal relationship between two story events, which can be evaluated with a necessity test (i.e., event X is caused by event Y if event X would NOT have occurred if event Y had not occurred).

2.1 Summary

Measuring causal coherence is highly dependent on the **causal connections** between **events** in a text document. When we extract the causal relations between every pair of events in a document, we will be able to measure the causal coherence of the document in a variety of ways. For instance, we can build a causal chain and assess the level of causal coherence in a document based on the length of the causal chain. Or, we can build a causal network, based on Trabasso's model, where nodes are the events in the document and edges are the causal links between events and define a causal coherence score using such a network.

Extracting implicit causal relations is a challenging open task in NLP. As mentioned in [1], causal coherence is not strictly dependent on the use of elaborate syntactic structures or cohesion devices (such as causal conjunctions) but depends mostly on conceptual links of causality, which are mostly implicit in content. Even though there is still no publicly available tool for tagging both implicit and explicit causal relations in text with a high and reliable accuracy, we can utilize tools that at least provide us with an estimate of the degree to which a document contains causal relations.

In the next section, we explain how and using what tool we extract the most relevant information to causal connections assuming that these connections are the fundamental elements in measuring causal gist.

3 Analysis

In this section, we test[2] our hypothesis that more coherent documents are more likely to be shared on social media.

3.1 Measuring Causal Coherence

We use Coh-Metrix [11] to compute a suite of features for coherence and cohesion features of text documents. Definition of cohesion consists of characteristics of the explicit text that play some role in helping the reader mentally connect ideas in the text [7]. Coh-Metrix is a tool for producing the indices of linguistic and discourse representation of a text. Coh-Metrix indices can be used to investigate the cohesion of the text and the ease with which a coherent underlying mental representation may be generated. There are 108 different indices produced

---

[2] https://github.com/phosseini/SBP-BRiMS2019.

by Coh-Metrix which lie in several categories pertinent to text comprehension. Among these indices, we are looking for those which may have some association with the constructs of causal and semantic coherence. In the following subsections, we delineate our hypothesis testing process step by step.

Dataset and Preprocessing. We use FakeNewsNet [21], which is a dataset of fake and real news articles and their social context information. We chose this dataset because it contains an article's content (body text) and measures of social engagement and number of shares on Twitter. Furthermore, these articles have been fact-checked enabling us to know their truth value.

We filtered the dataset and removed those news documents that did not have a proper value in their "Body" feature. In particular, we removed those records that either had no text or included only one sentence potentially due to unsuccessful parsing of the corresponding html files. After the filtering process, a total of 414 news articles remained in the dataset[3]. We did *not* remove punctuation marks and stop words. Also we did not do any sort of stemming, lemmatizing, or lower-casing to avoid any possible impact on Coh-Metrix's output. And, all the paragraphs in input text documents are separated by new line characters.

3.2 Principal Component Analysis (PCA)

Coh-Metrix yielded 108 indices for each document; however, several of these features were correlated. To account for multicollinearity, we applied Principal Component Analysis (PCA) on this set of features before doing our regression analysis. PCA was conducted with *principal* package in R.

3.3 Linear Regression

After applying a logarithm transform to the dependent variable (distinct number of shares) to control for skew[4], and using the PCA factors, truth (fake/real) labels, and source of fact checking as inputs, we conducted a multiple linear regression analysis to determine which factors predicted online sharing.

4   Results and Discussion

Results of the regression analysis, shown in Table 1, indicate that there are 16 significant predictors. Of these, one is a measure of referential cohesion. Referential cohesion aids readers in making connections between propositions, clauses, and sentences, enabling semantic coherence and helping readers to create a coherent mental representation of text. A greater cohesion indicates the greater coherence in mental representation and gist comprehension [26]. As a result, presence of

---

[3] Our inputs to the Coh-Metrix are all plain text stored in .txt file format.

[4] Normal probability plots show that residuals follow a normal distribution, see Fig. 1.

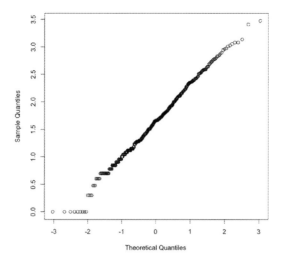

Fig. 1. Normal probability plots of logarithm of the number of shares of news documents on Twitter.

Table 1. Note. *** = p < 0.001, ** = p < 0.01, * = p < 0.05. "RCs" are the PCA components. "Description" contains the description of the Coh-Metrix index with the highest absolute loading value in the corresponding PCA component.

Components	Description	Estimate	Std. error	t statistic	
RC58	Number of syllables	0.11	0.03	3.56	***
Truth labels	Fake/Real labels	0.32	0.09	3.54	***
RC14	Preposition phrase density	−0.10	0.03	−3.27	**
RC16	Adverbial phrase density	−0.09	0.03	−3.02	**
RC2	Argument overlap, all sentences	0.09	0.03	2.86	**
RC24	First person plural pronoun incidence	0.09	0.03	2.80	**
RC72	Referential cohesion	−0.07	0.03	−2.42	*
RC7	Number of words	0.08	0.03	2.41	*
RC11	Number of letters	−0.07	0.03	−2.33	*
RC47	LSA overlap, adjacent paragraphs	−0.07	0.03	−2.22	*
RC31	Third person plural pronoun	−0.07	0.03	−2.18	*
RC66	Stem overlap, adjacent sentences	0.07	0.03	2.14	*
RC94	Hypernymy for nouns and verbs	−0.06	0.03	−2.05	*
RC30	Temporal connectives incidence	0.06	0.03	2.02	*
RC39	Word frequency for content words	−0.06	0.03	−2.00	*
RC37	Noun phrase density	−0.06	0.03	−1.98	*
(Intercept)	-	1.52	0.13	11.95	

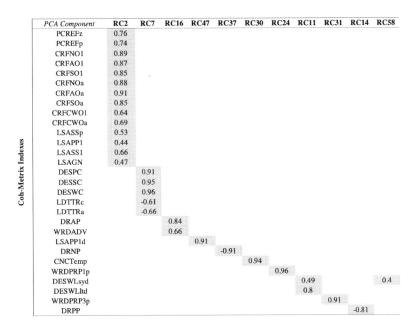

PCA Component	RC2	RC7	RC16	RC47	RC37	RC30	RC24	RC11	RC31	RC14	RC58
PCREFz	0.76										
PCREFp	0.74										
CRFNO1	0.89										
CRFAO1	0.87										
CRFSO1	0.85										
CRFNOa	0.88										
CRFAOa	0.91										
CRFSOa	0.85										
CRFCWO1	0.64										
CRFCWOa	0.69										
LSASSp	0.53										
LSAPP1	0.44										
LSASS1	0.66										
LSAGN	0.47										
DESPC		0.91									
DESSC		0.95									
DESWC		0.96									
LDTTRc		-0.61									
LDTTRa		-0.66									
DRAP			0.84								
WRDADV			0.66								
LSAPP1d				0.91							
DRNP					-0.91						
CNCTemp						0.94					
WRDPRP1p							0.96				
DESWLsyd								0.49			0.4
DESWLltd								0.8			
WRDPRP3p									0.91		
DRPP										-0.81	

Fig. 2. Results of running PCA. RCs are the PCA components. Significant components based on p-values in the linear regression model and factor loadings ≥ 0.40 are shown in the table. A detailed explanation of Coh-Metrix features is available on Coh-Metrix's website at http://www.cohmetrix.com/.

referential cohesion as one of the significant variables in our linear regression model demonstrates the importance of gist in predicting online sharing.

Also, as shown in Fig. 2, the *RC2* component is highly associated with the semantic features of text. The Latent Semantic Analysis (LSA) related features measure the semantic overlap between sentences or paragraphs.

Another significant predictor in our linear regression model is the truth label (fake/real) of news documents. In order to better understand what features predict and explain these truth labels, we conducted another experiment using the PCA scores in a stepwise backward elimination regression with truth label as dependent variable. A summary of the regression model is shown in Table 2. There are 32 significant components in the model. Based on the loading values of the PCA, there are three Coh-Metrix indexes, *PCDCz*, *PCDCp*, and *CNC-Caus*, highly associated with one of the significant predictors, *RC5*. PCDCz and PCDCp which are indexes of deep cohesion, demonstrate the degree to which the text contains causal and intentional connectives. According to the documentation of Coh-Metrix, these connectives help the reader to form a more coherent and deeper understating of causal events in text. Also, CNCCaus indicates the incidence score of causal connectives in documents. As a result, causal and deep cohesion connectives predict truth assessment (label) which itself is a significant predictor of online sharing.

Table 2. Note. *** = p < 0.001, ** = p < 0.01, * = p < 0.05. Summary of the regression model of predicting truth labels. "RCs" are the PCA components. "Description" contains the description of the Coh-Metrix index with the highest absolute loading value in the corresponding PCA component.

PCA component	Description	Sum sq	Mean sq	Statistic	
RC1	Number of letters	1.09	1.09	8.05	**
RC2	Argument overlap, all sentences	1.46	1.46	10.73	**
RC5	Text Easability PC Deep cohesion	1.05	1.05	7.77	**
RC7	Number of words	3.61	3.61	26.61	***
RC9	Additive connectives incidence	0.55	0.55	4.08	*
RC3	Sentence syntax similarity	0.71	0.71	5.22	*
RC15	Negative connectives incidence	0.89	0.89	6.55	*
RC17	Number of sentences	0.87	0.87	6.41	*
RC16	Adverbial phrase density	0.85	0.85	6.25	*
RC8	Number of words	1.75	1.75	12.88	***
RC27	Infinitive density	2.21	2.21	16.26	***
RC12	Word frequency in sentences	5.71	5.71	42.07	***
RC37	Noun phrase density	1.65	1.65	12.18	***
RC30	Temporal connectives incidence	0.60	0.60	4.44	*
RC31	Third person plural pronoun	2.07	2.07	15.28	***
RC19	Agentless passive voice density	0.62	0.62	4.56	*
RC89	Lexical diversity, type-token ratio	2.03	2.03	14.98	***
RC18	Text Easability PC Temporality	0.65	0.65	4.80	*
RC46	Age of acquisition for content words	1.77	1.77	13.06	***
RC60	Anaphor overlap, all sentences	2.22	2.22	16.37	***
RC39	Word frequency for content words	0.86	0.86	6.34	*
RC58	Number of syllables	1.36	1.36	10.02	**
RC50	Sentence syntax similarity	2.08	2.08	15.31	***
RC75	Number of paragraphs	0.70	0.70	5.16	*
RC79	Content word overlap, adjacent sentences	1.99	1.99	14.69	***
RC67	Number of letters	0.61	0.61	4.48	*
RC87	Text Easability PC Word concreteness	1.06	1.06	7.83	**
RC84	Text Easability PC Deep cohesion	0.65	0.65	4.80	*
RC86	Concreteness for content words	1.34	1.34	9.85	**
RC91	Stem overlap, all sentences	1.50	1.50	11.03	***
RC92	Positive connectives incidence	0.67	0.67	4.96	*
RC62	Noun overlap, all sentences	0.56	0.56	4.16	*

5   Conclusion and Future Work

In this paper, we use Coh-Metrix to test our hypothesis that documents with more semantic and causal coherence are more likely to be shared on social media by individuals. Our results show statistically significant associations between proxies of these constructs and the number of shares of news documents.

The deep cohesion score in Coh-Metrix is mainly based on the incidence of explicit causal connectives. Thus, there is reason to believe that this metric could be improved. For future work, we plan to also extract the implicit causal relations from text and build a causal network to better model the mental representation of readers. We are interested to see if including both implicit and explicit causal relations as additional features in our linear regression analysis provides us with better material to support our hypothesis. We will also run experiments on additional datasets to determine how well these results generalize.

Based on our review of casual coherence metrics, we intend to formulate and implement our measure of causal coherence using NLP techniques. As part of this task, we will build a corpus of causally coherent stories by the means of crowd-sourcing tools which will be used as a gold standard collection of documents with causal coherence scores.

Tagging causal relations in documents, specifically those of implicit type, is not an easy task in NLP. We believe that the focus in automatically modelling causal gist and mental representation of readers should be on extracting causal relations. Since there is a fairly clear idea of how causal coherence can be modeled from psychological perspective, having a reliable tool for extracting events and causal relations between them in content will enable us to better model causal coherence and, consequently, gist. This will enable us to test any hypothesis on the relation between this model and patterns of information sharing online.

Acknowledgement. We would like to sincerely thank the Coh-Metrix team for their invaluable advice and assistance with using their tool for our analysis.

References

1. Arfé, B., Boscolo, P.: Causal coherence in deaf and hearing students' written narratives. Discourse Process. **42**(3), 271–300 (2006)
2. Black, J.B., Bern, H.: Causal coherence and memory for events in narratives. J. Verbal Learn. Verbal Behav. **20**(3), 267–275 (1981)
3. Broniatowski, D.A., Reyna, V.F.: Combating Misinformation and Disinformation Online: The "Battle of the Narrative". White Paper, Submitted to the Committee on a Decadal Survey of Social and Behavioral Sciences and Applications to National Security, National Academies of Sciences, Engineering, and Medicine, Washington, DC, February 2017
4. Diehl, J.J., Bennetto, L., Young, E.C.: Story recall and narrative coherence of high-functioning children with autism spectrum disorders. J. Abnorm. Child Psychol. **34**(1), 83–98 (2006)
5. Dredze, M., Broniatowski, D.A., Smith, M.C., Hilyard, K.M.: Understanding vaccine refusal: why we need social media now. Am. J. Prev. Med. **50**(4), 550–552 (2016)
6. Foltz, P.W., Kintsch, W., Landauer, T.K.: The measurement of textual coherence with latent semantic analysis. Discourse Process. **25**(2–3), 285–307 (1998)

7. Graesser, A.C., McNamara, D.S., Louwerse, M.M.: What do readers need to learn in order to process coherence relations in narrative and expository text. In: Rethinking Reading Comprehension, pp. 82–98 (2003)
8. Grinberg, N., Joseph, K., Friedland, L., Swire-Thompson, B., Lazer, D.: Fake news on Twitter during the 2016 US presidential election. Science **363**(6425), 374–378 (2019)
9. Grysman, A., Hudson, J.A.: Abstracting and extracting: causal coherence and the development of the life story. Memory **18**(6), 565–580 (2010)
10. Kuperberg, G.R., Paczynski, M., Ditman, T.: Establishing causal coherence across sentences: an ERP study. J. Cogn. Neurosci. **23**(5), 1230–1246 (2011)
11. McNamara, D.S., Louwerse, M.M., McCarthy, P.M., Graesser, A.C.: Coh-metrix: capturing linguistic features of cohesion. Discourse Process. **47**(4), 292–330 (2010)
12. Mimno, D., Wallach, H.M., Talley, E., Leenders, M., McCallum, A.: Optimizing semantic coherence in topic models. In: Proceedings of the Conference on Empirical Methods in Natural Language Processing, pp. 262–272. Association for Computational Linguistics (2011)
13. Mulder, G., Sanders, T.J.: Causal coherence relations and levels of discourse representation. Discourse Process. **49**(6), 501–522 (2012)
14. Pennebaker, J.W., Boyd, R.L., Jordan, K., Blackburn, K.: The development and psychometric properties of LIWC2015. Technical report (2015)
15. Pennycook, G., Cannon, T.D., Rand, D.G.: Prior exposure increases perceived accuracy of fake news. J. Exp. Psychol.: Gen. (2018)
16. Pennycook, G., Rand, D.G.: Lazy, not biased: susceptibility to partisan fake news is better explained by lack of reasoning than by motivated reasoning. Cognition (2018)
17. Reyna, V.F.: A new intuitionism: meaning, memory, and development in fuzzy-trace theory. Judgm. Decis. Making (2012)
18. Reyna, V.F.: Risk perception and communication in vaccination decisions: a fuzzy-trace theory approach. Vaccine **30**(25), 3790–3797 (2012)
19. Shao, C., Ciampaglia, G.L., Varol, O., Yang, K.C., Flammini, A., Menczer, F.: The spread of low-credibility content by social bots. Nat. Commun. **9**(1), 4787 (2018)
20. Shu, K., Bernard, H.R., Liu, H.: Studying fake news via network analysis: detection and mitigation. In: Agarwal, N., Dokoohaki, N., Tokdemir, S. (eds.) Emerging Research Challenges and Opportunities in Computational Social Network Analysis and Mining. LNSN, pp. 43–65. Springer, Cham (2019). https://doi.org/10.1007/978-3-319-94105-9_3
21. Shu, K., Wang, S., Liu, H.: Exploiting tri-relationship for fake news detection. arXiv preprint arXiv:1712.07709 (2017)
22. Trabasso, T., Sperry, L.L.: Causal relatedness and importance of story events. J. Mem. Lang. **24**(5), 595–611 (1985)
23. Trabasso, T., et al.: Causal cohesion and story coherence (1982)
24. Tsur, O., Rappoport, A.: What's in a hashtag?: content based prediction of the spread of ideas in microblogging communities. In: Proceedings of the Fifth ACM International Conference on Web Search and Data Mining, pp. 643–652. ACM (2012)
25. Vosoughi, S., Roy, D., Aral, S.: The spread of true and false news online. Science **359**(6380), 1146–1151 (2018)
26. Wolfe, C.R., Widmer, C.L., Torrese, C.V., Dandignac, M.: A method for automatically analyzing intelligent tutoring system dialogues with coh-metrix. J. Learn. Anal. **5**(3), 222–234 (2018)

Detecting Disruption: Identifying Structural Changes in the Verkhovna Rada

Thomas Magelinski[1]([✉])[ID], Jialin Hou[2][ID], Tymofiy Mylovanov[2][ID],
and Kathleen M. Carley[1][ID]

[1] CASOS, Institute for Software Research, Carnegie Mellon University, Pittsburgh,
PA 15217, USA
tmagelin@andrew.cmu.edu
[2] Department of Economics, University of Pittsburgh, Pittsburgh, PA 15217, USA
http://www.casos.cs.cmu.edu

Abstract. We identify time periods of disruption in the voting patterns of the Ukrainian parliament for the last three convocations. We compare two methods: ideal point estimation (PolSci) and faction detection (CS). Both methods identify the revolution in Ukraine in 2014. The faction detection method also detects structural changes prior to the revolution (election of the president whose tenure was ended early by the revolution), while the ideal points method performs stronger after 2014, identifying a disruption around voting on constitutional changes to implement Minsk II agreements between separatists and Ukraine. The ideal point method is better at detecting position changes of the members of parliament, while the faction method is better at detecting changes in relationships between different MPs. The results suggest that after 2014, the Ukrainian parliament has become more consolidated, but the distribution of its political positions continues to evolve in response to changes in geo-political conditions.

Keywords: Change point analysis · Ideal points · Faction detection · Verkhovna Rada · Dynamic network analysis · Ukraine

1 Introduction

In 2014, there was a revolution in Ukraine triggered by the decision by the president to stop negotiations on the association agreement with EU. The president fled to Russia after violent protests in which a number of protesters and police

This material is based upon work supported by the Office of Naval Research Multidisciplinary University Research Initiative (MURI) under award number N00014-17-1-2675. Any opinion, findings, and conclusions or recommendations expressed in this material are those of the authors and do not necessarily reflect the views of the Office of Naval Research. Additionally, Thomas Magelinski was supported by an ARCS foundation scholarship.

R. Thomson et al. (Eds.): SBP-BRiMS 2019, LNCS 11549, pp. 194–203, 2019.
https://doi.org/10.1007/978-3-030-21741-9_20

were killed. Shortly after that, Russia annexed Crimea and Russia backed a security conflict in the East of Ukraine. The government of Ukraine lost control over the territories in parts of two regions: Donetsk and Lugansk. The international community have mediated two agreements between the government of Ukraine and separatists - Minsk I and Minsk II. Both agreements have not been successful at resolving the crisis.

The political power in Ukraine is centralized. Despite recent legislation implemented after 2014 that serves to decentralize some economic power to the regions (budgets, service provision), the majority of legislative and executive power is concentrated at the national level. Verkhovna Rada is the national parliament. It appoints the government (subject to nominations by the president), votes on national laws that can regulate economy, allocate national budget, and has the power to approve constitutional amendments. Minsk agreements between the separatists and the government of Ukraine require such amendments.

In this paper, we study disruptions in the voting patterns of the parliament of Ukraine (Verkhovna Rada) over the period of 2007–2018. This period covers three convocations, prior, during, and after the revolution. The prior convocation is important because the president who left the office because of the revolution was elected in the middle of this convocation. We employ two methods: a standard ideal points model from political science and a network science approach to identify factions in the Ukrainian parliament and structural changes in the composition and behavior of these factions.

The paper makes two contributions. First, we identify the time periods of substantive changes in the Ukrainian Parliament: (1) election of the president prior to the revolution, (2) the revolution, and (3) voting for constitutional amendments to implement Minsk II agreements. We find that during these moments, the Ukrainian parliament has become polarized. Polarization can be a sign of conflict, in-fighting, and weakness within the Ukrainian political elites. Second, we compare the performance of two different methods of identification of disruption in the parliament behavior. We show that the network science faction detection method picks up structural changes prior to the revolution (election of the president whose tenure was ended early by the revolution), while ideal points method performs stronger after 2014 identifying a disruption around voting on constitutional changes to implement Minsk II agreements between separatists and Ukraine. Ideal point method is better at detecting position changes of the members of parliament, while faction method is better at detecting changes in relationships between different MPs. These results suggest that the Ukrainian parliament has become more consolidated, but the distribution of its political positions continues to evolve in response to changes in geo-political conditions.

The ideal points model is the standard approach to understand a legislative body of government. It models the decision of individual voters, and estimates, for each voter, a point which represents the voter's relative degree of conservativeness/liberality in the government. Researchers have developed the classical static ideal points model [6] into several dynamic forms [5,11,14], and used this approach to analyze the US Congress [20] and legislation in other countries [3].

We contrast ideal point results with those of an alternate modeling perspective: network science. Under a network science framework, MPs are connected to each other based on how often they agree on bills. From this network underlying communities can be detected automatically in a variety of ways [13]. Finally, disruptions in the underlying structure of the parliament are measurable through changes in faction assignment.

These two approaches compliment each other. The ideal points method can see fine-grained change since it operates on individuals, allowing for earlier detection of changes in the legislature. Faction analysis, on the other hand, is typically more interpretable, as it simplifies the entire legislature into a small number of groups. In the following sections we discuss the data set, walk through the details of both ideal points and faction disruption, display the results of both methods, and finally put them in context with Ukraine's political landscape.

2   Data

We use the publicly available roll call votes in Ukraine's Parliament, the Verkhovna Rada [1]. The parliament is split into convocations, roughly terms for the parliamentarians (MPs). In this work we analyze the roll calls from convocations 6, 7, and 8. The data was analyzed on a month to month basis. Summary statistics for each convocation, including their start and end dates, can be found in Table 1. MPs that are not present throughout the entire convocation are not considered, since these members cannot be analyzed through ideal points. This keeps the number of MPs consistent through the methods of analysis.

Table 1. Summary of the data for each convocation.

Convocation	Start date	End date	Months	Bills	MPs	Parties
6	11/2007	12/2012	59	3974	539	8
7	12/2012	10/2014	22	1062	479	8
8	11/2014	07/2018*	21	1267	325	9

*Note: Convocation 8 is ongoing, but our data is truncated.

The Verkhovna Rada has unique voting rules. For each bill, MPs have 6 voting options: Vote For, Vote Against, No Vote, Absence, Do Not Vote, and Abstain. Votes in favor of bills are common, while vote against bills are not. Local experts suggest that unless an MP feels very strongly in opposition of a bill, they will use one of the other non-for voting options.

3   Ideal Points

3.1   Background

The ideal points model estimates a political space from the roll call votes data by directly modeling voters' decisions. We assume that each voter has an ideal point x_i in the policy space R^d with the dimensionality $d \in \{1, 2, ...\}$ representing the issues which are independent to each other. The issues do not necessarily have a direct interpretation, but they are, by the model set-up below, pairwise independent to each other. Commonly, $d = 1$ is used. The position of x_i indicates Voter i's political ideology, i.e. relative degree of conservativeness or liberality in the parliament for each identified issue.

Denote MPs $i \in \{1, ..., I\}$, Bills $j \in \{1, ..., J\}$. We code Vote For as 1, Vote Against as -1, and other votes as 0, denoted as $y_{i,j} \in \{1, -1, 0\}$. When abstention is modeled as a neutral attitude, we model instead a multi probit. The results of each roll call are either passed ξ_j or not passed ψ_j. ξ_j, ψ_j are vectors in R^d, representing the political consequences of a bill which vary across bills.

The utility of a voter i of having bill j passed is $U_i(\xi_j) = -||x_i - \xi_j||^2 + \eta_{i,j}$, and the utility of a voter i of not having bill j passed is $U_i(\psi_j) = -||x_i - \psi_j||^2 + \nu_{i,j}$. Given the setup, utility maximization of each voter implies

$$y_{i,j} = \begin{cases} 1 \text{ , if } U_i(\xi_j) > U_i(\psi_j) \\ -1 \text{ , if } U_i(\xi_j) \leq U_i(\psi_j) \end{cases} \tag{1}$$

The voters vote independently. Thus, the error terms are normalized to be $\eta_{i,j} - \nu_{i,j} \sim N(0, 1)$. The probability of Legislator i voting yea on Bill j is

$$P(y_{i,j} = 1) = P(U_i(\xi_j) > U_i(\psi_j)) = P(\nu_{i,j} - \eta_{i,j} < ||x_i - \psi_j||^2 - ||x_i - \xi_j||^2)$$
$$= \Phi(\beta_j' x_i - \alpha_j)$$

where x_i is the ideal point of MP i, and β_j is the parameter that describes the characteristics of Bill j. The likelihood across all MPs and all bills is

$$L = \prod_{i=1}^{n} \prod_{j=1}^{m} \Phi(\beta_j' x_i - \alpha_j)^{|y_{i,j} + \frac{1}{2}| - \frac{1}{2}} (1 - \Phi(\beta_j' x_i - \alpha_j))^{|y_{i,j} - \frac{1}{2}| - \frac{1}{2}} \tag{2}$$

where $y_{i,j}$ is the observed votes, and x_i, β_j, α_j are unknown parameters. We use Bayesian inference, assuming a normal (0,1) prior distribution of the parameters for each MP before estimation of each convocation.

The Ideal Point Scale (y-axis of Fig. 1) is a measure of relative liberty/ conservation with respect to an unpredictable, perfectly-neutral median voter whose measure is 0. A score of x on the Scale implies a difference from the median voter by a fraction $\Phi(x)$ of random voters, where $\Phi(\cdot)$ is the cumulative density function of the normal distribution with mean 0 and variance 1. We set the positive direction on y to mean more liberty, by manually putting the right-most parties which we know beforehand onto the positive half of the y-axis.

3.2   Methodology Used for Ukraine

We first describe the political space of the Ukraine Parliament from 2007-present. Figure 1 shows the random-walk ideal points of factions of the Parliament in Convocation 6, 7, and 8.

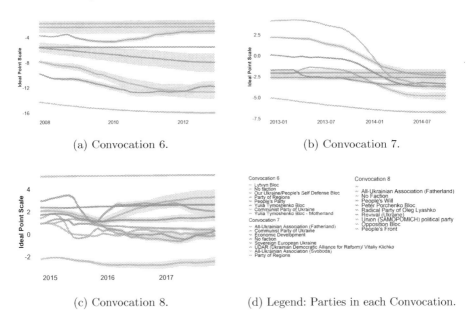

(a) Convocation 6.   (b) Convocation 7.

(c) Convocation 8.   (d) Legend: Parties in each Convocation.

Fig. 1. Ideal points by party.

On Fig. 1a we present the estimation result for parties in the Convocation 6. We estimate the ideal points dynamically assuming a random walk process for each party, while treating the members of each party as one voter. Through Convocation 6 we observe gradual divergence in parties, with the Communist Party of Ukraine at one end and the Yulia Tymoshenko Bloc at the other end. The scale on the vertical axis is the normalized standard deviation from 0 which is the ideal point of a perfect median voter.

Convocation 7 witnessed the Ukrainian revolution of 2014. The revolution was followed by a series of changes in Ukraine's political system, including the formation of an interim government and the restoration of the 2004 constitution. As Fig. 1b shows, our estimation identifies the revolution in the winter of 2014. The biggest party, Party of Regions, almost flipped their political standing from far left (pro-Russian) to slightly right. Other parties in the Parliament also moved to the right, showing the profound impact of the Euromaidan Revolution.

At the beginning of the Convocation 8, five of the parliament's pro-European parties, namely the Petro Poroshenko Bloc, People's Front, Self Reliance, Fatherland, and Radical Party, signed a coalition agreement. Per the coalition agreement, the current convocation is tasked with passing major reforms to ensure Ukrainian membership in European institutions such as EU and NATO, while dealing with the threat of further Russian aggression and intervene against Ukraine. As Fig. 1c shows, our estimation identifies a disruption in the Parliament in the late summer of 2015. We associate this disruption inside the Parliament with the historical event of the "Grenade Attack" that happened at that time. The attack occurred on the day parliament voted for the constitutional amendments required to implement the Minsk II agreements. The nationalist Radical Party of Oleg Lyashko was the most affected in the disruption.

4   Faction Change Detection

4.1   Background

An alternate method of discovering legislative change points is to first find groups of aligned MPs or *factions*. More formally, a faction is defined as a recognized political group with a defined political agenda and sometimes with formal membership requirements [8,9]. Large changes in faction membership, then, can be thought of as legislative change points.

Much of legislative analysis relies on co-voting, or the instances that MPs cast the same vote on the same bill [12,17]. One approach to faction detection, then, is through network analysis, where the network links MPs based on their co-voting frequency. After a network is created, a number of empirically verified community detection algorithms could be applied to find factions [4,7,16]. Comparison of network creation and grouping algorithm combinations has been completed in [13], which we use to inform our choice of method.

Once the factions are defined, a method of group-based change detection must then be applied. This problem is a simplified version of dynamic community detection, which seeks to understand how communities evolve in time [2,18]. The two most popular approaches to dynamic community detection are Generalized Louvain, and GraphScope [15,19]. GraphScope takes an *online* approach, meaning that it allows for real-time change detection and can account for MPs leaving and entering. However, it cannot leverage the full time-line, and as such is susceptible to noise. Generalized Louvain utilizes the full time-line, but does not account for MPs leaving or entering, and requires user-defined parameters. Instead, we use a simple alternative method described in Sect. 4.2.

4.2   Methodology Used for Ukraine

In [13], it was found that a Gower-Mean Shift method is both intuitive, and gives similar results to a many other possible methods. As such, we will apply it here. First, a brief summary of the method.

A network is constructed where nodes represent MPs. The nodes are linked by their Gower similarity: $S_{ij} = \frac{\sum_{k=1}^{N} w_k \delta(x_{ik}, x_{jk})}{\sum_{k=1}^{N} w_k}$, where w_k is a weighting on bill k, x_{ik} represents the vote cast on bill k by MP i, and δ is the Kronecker delta function [10]. A bills weight, w_k, is calculated based on the roll call's Shannon entropy: $w_k = \sum_v p_v \log(p_v)$, where v, is the type of vote cast (for, against, abstain, etc.), and p_v is the proportion of the parliament casting vote type v. Entropy, then, is a measure of how contentious a bill is. Bills that split the parliament in half, are weighted higher than bills for which everyone agrees. It should be noted that entropy weighting has a limitation which can be demonstrated through an example: If a bill gets 95% votes "for," it will receive a very low weight, however, the 5% in opposition are showing signs of strong ties. Finally, the Gower's similarity matrix is clustered using Mean Shift [7], classifying each of the MPs into a faction. This procedure was run on Convocations 6–8.

After defining factions, the change-point analysis can be performed. The basis of the analysis is that time segments should: one, not contain major changes in group membership, and two, have different group memberships from adjacent time segments. These competing goals balance the number of segments the timeline is split into.

More formally, a co-group network, G_t, is defined at each time segment, t. This network links nodes if they were placed in the same faction for time period t. Then, a pairwise similarity matrix is defined as $H_{t_1,t_2} = s(G_{t_1}, G_{t_2})$, where s is the Product-Moment Correlation between the graphs. The goals described, then, can be formalized as internal and external similarity:

$$s_{\text{internal}}(H, \mathbf{t}) = \sum_{p=1}^{P-1} H_{t_p:t_{p+1},t_p:t_{p+1}}; \ s_{\text{external}}(H, \mathbf{t}) = \sum_{p=1}^{P-2} H_{t_p:t_{p+1},t_{p+1}:t_{p+2}}, \quad (3)$$

where $\mathbf{t}$ is a vector indicating the time segments, and P is the number of partitions+2, for each of the ends. The value of P is initially 3. The partitions are placed such that s_{internal} is maximized. The number of partitions is then increased, and placement is repeated. If the gain in internal similarity outpaces that of external similarity, the process repeats. If not it is terminated. The second-to-last iteration is then taken as the time segmentation.

5   Results

5.1   Legislative Change Points

Ideal Points. From the ideal points perspective, divergence is identified by the variance of all MPs in the Parliament. We measure the polarization of the Ukraine Parliament by the auto-variance of party ideal points. Specifically, we calculate the following measure: $V_t = [\sum_i (x_{i,t} - x_{i,t-1})^2]^{\frac{1}{2}}$. From our dispersion measure, we identify two structural breaks in the Parliament, shown in Table 2.

Factions. The data for convocations 6, 7, and 8 were fed through the faction detection and network partitioning algorithms. The resulting partitions of the faction similarity matrix are shown in Fig. 3 and resulting breaks are given in Table 2. This figure shows the temporal structure of factions. In convocation 6, we see that there is a major change after April 2010, after which the factions are extremely stable. This stability is seen though the highly correlated block in the similarity matrix after the partition. Convocation 7 shows the opposite: stable faction structure until the break in February of 2014. We also see two partitions in convocation 8, on April 2015 and April 2016, though they are far less dramatic than those of prior convocations (Fig. 2).

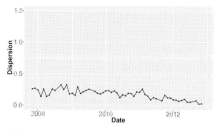

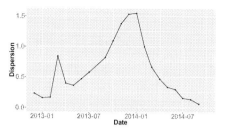

(a) C6. No spike in parties dispersion.   (b) C7. Spike is found on December 2013.

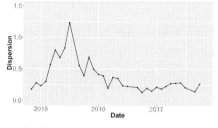

(c) C8. Spike is found on June 2015.

Fig. 2. Dispersion of parties ideal points.

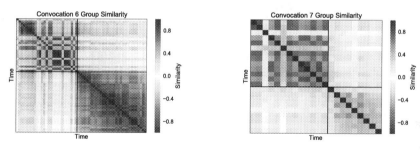

(a) C6. Partition on April 2010.   (b) C7. Partition on February 2014.

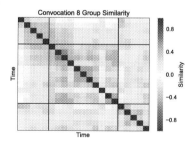

(c) C8. Partitions on April 2015 and 2016.

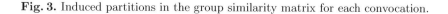

Fig. 3. Induced partitions in the group similarity matrix for each convocation.

Table 2. Structural breaks

Method	Date of Break	Event description
Ideal points	1/1/2014	Euromaidan Revolution
Ideal points	7/1/2015	Grenade Attack
Factions	4/1/2010	Presidential Election
Factions	2/1/2014	Euromaidan Revolution
Factions	4/1/2015	No Clear Event
Factions	4/1/2016	No Clear Event

5.2   Discussion

We first investigate polarization in Convocation 8 using the ideal points model. We find that polarization is most visible among new MPs to the convocation. Through the Convocation, most party leaders maintained their own stances (which are also consistent with their party ideology), while many independent MPs changed their position on the political spectrum over time.

In convocation 6, one large faction initially dominates the legislature. After April 2010, however, this faction splits roughly in half, creating a more balanced parliament. In convocation 7, we see large changes in membership from the presidential faction, to the opposing faction. In this shift, we see the opposing faction gaining a majority in the parliament. In convocation 8, the minority faction is jockeying for power. They gain a significant number of members after April 2015. In April 2016, there are just minor trades between factions.

6   Conclusion

Both methods identify the revolution in Ukraine in 2014. Faction detection method also detects structural changes prior to the revolution (election of the president whose tenure was ended early by the revolution), while ideal points method performs stronger after 2014 identifying a disruption around voting on constitutional changes to implement Minsk II agreements between separatists and Ukraine. Ideal point method is better at detecting position changes of the members of parliament, while faction method is better at detecting changes in relationships between different MPs. The results suggest that after 2014, the Ukrainian parliament has become more consolidated, but the distribution of its political positions continues to evolve in response to changes in geo-political conditions.

References

1. Verkhovna rada of ukraine, February 2018 . http://rada.gov.ua/en
2. Aynaud, T., Fleury, E., Guillaume, J.L., Wang, Q.: Communities in evolving networks: definitions, detection, and analysis techniques. In: Mukherjee, A., Choudhury, M., Peruani, F., Ganguly, N., Mitra, B. (eds.) Dynamics On and Of Complex Networks, Modeling and Simulation in Science, Engineering and Technology. Birkhäuser, vol. 2, pp. 159–200. Springer, New York (2013). https://doi.org/10.1007/978-1-4614-6729-8_9
3. Bailey, M.A.: Comparable preference estimates across time and institutions for the court, congress, and presidency. Am. J. Polit. Sci. **51**(3), 433–448 (2007)
4. Blondel, V.D., Guillaume, J.L., Lambiotte, R., Lefebvre, E.: Fast unfolding of communities in large networks. J. Stat. Mech.: Theor. Exp. **2008**(10), P10008 (2008)
5. Caughey, D., Warshaw, C.: Dynamic estimation of latent opinion using a hierarchical group-level irt model. Polit. Anal. **23**(2), 197–211 (2015)
6. Clinton, J., Jackman, S., Rivers, D.: The statistical analysis of roll call data. Am. Polit. Sci. Rev. **98**(2), 355–370 (2004)
7. Comaniciu, D., Meer, P.: Mean shift: a robust approach toward feature space analysis. IEEE Trans. Pattern Anal. Mach. Intell. **24**(5), 603–619 (2002)
8. Friedkin, N.E.: A structural Theory of Social Influence, vol. 13. Cambridge University Press, Cambridge (2006)
9. Friedkin, N.E., Johnsen, E.: Social influence networks and opinion chance. Adv. Group Process. **16**, 1–29 (1999)
10. Gower, J.C.: A general coefficient of similarity and some of its properties. Biometrics **27**(4), 857–871 (1971)
11. Jackman, S.: Multidimensional analysis of roll call data via Bayesian simulation: identification, estimation, inference, and model checking. Polit. Anal. **9**(3), 227–241 (2001)
12. MacRae Jr., D.: Dimensions of Congressional Voting: A Statistical Study of the House of Representatives in the Eighty-First Congress. University of California Press, California (1958)
13. Magelinski, T., Cruickshank, I., Carley, K.M.: Comparison of faction detection methods in application to ukrainian parliamentary data (2018)
14. Martin, A.D., Quinn, K.M.: Dynamic ideal point estimation via markov chain monte carlo for the us supreme court, 1953–1999. Polit. Anal. **10**(2), 134–153 (2002)
15. Mucha, P.J., Richardson, T., Macon, K., Porter, M.A., Onnela, J.P.: Community structure in time-dependent, multiscale, and multiplex networks. Science **328**(5980), 876–878 (2010)
16. Newman, M.E.: Fast algorithm for detecting community structure in networks. Phys. Rev. E **69**(6), 066133 (2004)
17. Poole, K.T., Rosenthal, H.: A spatial model for legislative roll call analysis. Am. J. Polit. Sci. **29**(2), 357–384 (1985)
18. Rossetti, G., Cazabet, R.: Community discovery in dynamic networks: a survey. ACM Comput. Surv. (CSUR) **51**(2), 35 (2018)
19. Sun, J., Faloutsos, C., Faloutsos, C., Papadimitriou, S., Yu, P.S.: Graphscope: parameter-free mining of large time-evolving graphs. In: Proceedings of the 13th ACM SIGKDD International Conference on Knowledge Discovery and Data Mining, pp. 687–696. ACM (2007)
20. Tausanovitch, C., Warshaw, C.: Measuring constituent policy preferences in congress, state legislatures, and cities. J. Polit. **75**(2), 330–342 (2013)

Lost in Online Stores? Agent-Based Modeling of Cognitive Limitations of Elderly Online Consumers

Justyna Pawlowska[✉], Radoslaw Nielek, and Adam Wierzbicki

Polish-Japanese Institute of Information Technology, Warsaw, Poland
{justyna.pawlowska,nielek,adamw}@pjwstk.edu.pl

Abstract. We have developed an agent-based model of e-commerce platform users' behavior with emphasis on reflecting decision-making characteristics of elderly adults. The model has been used to verify how cognitive deficits of older customers influence the effectiveness of collaborative filtering and content-based recommender systems. The results from our simulation suggest that the effectiveness of recommender systems in improving quality of elderly consumers choices is low for population of agents with strong cognitive deficits.

Keywords: Agent-based simulation · Cognitive modelling ·
Recommender system · Multi-attribute choice

1 Introduction

The global sales of B2C e-commerce products and services grow at a double-digit rate and is expected to hit 2.1 trillion USD in 2018. As the population in developed countries ages and the number of elderly Internet users increases, the clients over 60 become a more and more significant customer group. However, enjoying the benefits of growing e-commerce market can be challenging for elderly users. Their specific needs and preferences are not fully addressed by some service providers [1]. Moreover, as cognitive capabilities are deteriorating with age, older consumers' decision making can be suboptimal in an environment that favors the use of a more cognitively demanding strategy – such as multi-attribute choice on an e-commerce platform [2].

Taking into account that most of the main recommendation techniques are based on previous user decisions (or decisions of similar users) [3], the recommendations made by those systems can also be suboptimal and even create a vicious circle of wrong decisions. On the other hand, recommendations made based on observing behavior of a larger consumer population may be helpful for consumers with strong cognitive deficits.

To test how recommender systems influence decision making of elderly users we have developed an agent-based model mimicking the behavior of e-commerce platform users (agents) with emphasis on reflecting decision-making characteristics of elderly adults. Based on the analysis of the publications on the cognitive limitations of the elderly and taking into account properties of recommendation systems, we have stated the following hypotheses: (1) Collaborative recommender systems can improve

© Springer Nature Switzerland AG 2019
R. Thomson et al. (Eds.): SBP-BRiMS 2019, LNCS 11549, pp. 204–213, 2019.
https://doi.org/10.1007/978-3-030-21741-9_21

decision making of elders with cognitive limitations and biases, (2) Content-based recommendation systems can be detrimental for decision making of elders with cognitive limitations and biases. This research is the first approach towards understanding the impact of age-related cognitive deficits on recommender systems.

2 Related Work

The issue of adverse age-related changes in cognitive abilities, such as slowed attention allocation and speed of information processing, decreased working and long-term memory capacity or changes in reasoning is widely recognised and addressed in numerous studies [4, 5]. The combination of these cognitive changes leads to difficulties in navigating the e-commerce space. It has been found that seniors employ different heuristics in problem-solving tasks, e.g. rely more on deductive than inductive strategies or require less information prior to taking decision [2]. While some of the strategies may be useful, research has shown that overall, older subjects make less optimal decision compared with younger subjects. Although the decision-making strategies and performance among seniors are generally well described in scientific publications, a computational model of senior's cognitive strategies has not been created yet.

Obstacles faced by the elderly when navigating in the e-commerce space have also been discussed in many publications [1, 6, 7]. Another study [1] compares younger and older users of e-commerce platforms to find differences in their behavior. Usage of recommender systems by the elderly has been described in [8] where authors propose a tutoring component of e-commerce systems that would teach users how to navigate and use the recommender system. The publication most related to the main research problem is [9] where the authors analyzed how demographics such as age and gender influence the click-through rate of a recommender system and found out that elderly users clicked on the recommended items more often than younger users. The impact of suboptimal choices of elderly users on the recommender system's suggestions has not been researched yet.

3 Model Description

The model, build using Python 3.6 language[1], consist of 3 key elements: Agents, Items and a Recommender System. The purpose of the model is to allow an exploration of benefits and limitations of various designs of recommender systems. Agents are modelled based on knowledge about elder[2] consumer's decision making from psychological and cognitive research. Items are modeled based on observations of currently existing e-commerce stores, and the recommender system will be a target for detailed and realistic modeling.

---

[1] The source code is available on request.

[2] Agents referred to as 'elderly' have a specific level of cognitive deficiencies that can appear with very different intensity also at a different age.

3.1   Modelling Agents

To model mild cognitive deficits mainly associated with aging we have differentiated three elements that affect decision making: (a) how many attributes of the item the agent is able to assess (working memory), (b) how many items will she compare before making the purchase (combination of mental processing speed and endurance) and (c) how much the decision will be influenced by emotional reactions.

Each agent is characterized by her cognitive abilities: working memory size (WM) and mental processing speed (S). We also assign a variable describing the agent's endurance (E) for cognitive effort with an assumption that it depletes over time as she becomes more tired and confused.

We set working memory capacity to 10 pieces of information – 'chunks' for agents classified as 'young' on average, and 1–10 'chunks' for 'elderly' agents.

When an agent starts to browse items, she initially can keep values of up to WM features of the product in working memory. The agent calculates the utility of the first product and 'remembers' this number. When she browses a new product, she also calculates the product's utility. If the new product's utility is higher than the utility value of the product in agent's working memory, the new product 'pushes out' the old one from working memory. Mental processing speed (S) specifies how much time takes to assess expected utility for 1 feature.

Agent's endurance decreases linearly with the passage of time:

$$E_{t1} = E_{t0} - S$$

Preferences of agents are modeled using the reference point approach described in [11]. For each feature i of the product, every agent has a desired aspiration level qa_i and reservation level qr_i (which if not achieved, causes a sharp drop of utility). We assume that the features should be maximised so, given q_ilo and q_iup that are respectively lower and upper bounds:

$$q_ilo < qr_i < qa_i < q_iup$$

A way of aggregating the features into an order-consistent utility function starts by specifying partial utility functions σ_i:

$$\sigma_i(q_i, qr_i, qa_i) = \begin{cases} 1 + \alpha(q_i - qr_i)/(q_iup - qa_i), & qa_i \le q_i \le q_iup \\ (q_i - qr_i)/(qa_i - qr_i), & qr_i < q_i < qa_i \\ \beta(q_i - qr_i)/(qr_i - q_ilo), & qilo \le qi \le qri \end{cases}$$

The coefficients α, β should be chosen in such a way that partial utility functions are not only monotone, but also concave. The corresponding overall utility function would then have the form:

$$\sigma(q, qr, qa) = \left(\min_{1 \le i \le k} \sigma_i(q_i, qr_i, qa_i) \right) + \varepsilon \sum_{i=1}^{k} , \sigma_i(q_i, qr_i, qa_i)/(1 + k_\varepsilon)$$

Note that the above formula for the utility function can be calcualted using all features, or using a selected (remembered) subset of features (different for each agent).

We have chosen different preferences (reservation and aspiration levels) for agents using a pseudo-random number generator. The aspiration levels can they can take values in a range (0.5, 0.9) and reservation levels can they can take values in a range (0.1, 0.5).

Apart from the influence of cognitive factors, we model age-associated increased reliance on emotional reactions [12]. We include two heuristics in our model: (1) Sensitivity to negative reviews [13] and (2) Strong brand preferences [14].

In the first heuristic, spotting a single negative review causes perceived utility to be lower than the one resulting from the agent's utility function. The second heuristic has an opposite effect – if the product's brand is one of the agent's favorite, the perceived utility is significantly higher (Fig. 1).

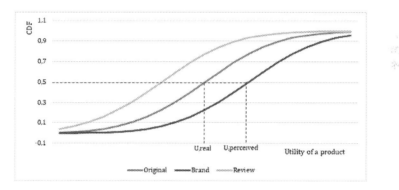

Fig. 1. Average utility of the items purchased by the agents

3.2 Modelling Items

Items are products or services of more sophisticated type, such as restaurants and hotel reservations, holidays or films and books. We assume that the number of item features of significance for agents is 10. We assume that an item's price is an additional feature that is always taken into account by the agent. The value of every non-price attribute of each item is determined using a pseudo-random number generator and the price is correlated with the sum of the values of non-price attributes.

In addition to the features, agents can see a product's brand and other agent's reviews that can trigger the heuristics.

We generate 3 sets of items, each containing 250 items, with various proportions of items triggering one or both heuristics[3]. Table 1 shows the percentages of items that trigger heuristics in the three item sets A, B, C.

---

[3] Attributes 'brand' = 1, 'negative review' = 1 or both, in which case only 'negative review' triggers the heuristc.

Table 1. Items sets.

Set	% of items with heuristics
A	10%
B	50%
C	90%

3.3 Modelling Recommender System

Recommendations are made based on purchases history data. The first recommender system tested in this research is a user-based nearest neighbor collaborative filtering (CF) strategy. The strategy is based on the premise that a user must like the favorite items of a user with similar taste. The CF algorithm uses the purchase history from each individual user to recommended items among similar users and is one of the most widely used methods for making recommendations in e-commerce [15].

The second recommender system is a content-based (CB) method where the algorithm identifies items similar to those a given user has purchased in the past.

In addition to the two more sophisticated recommender systems, we model a naive recommender system with an algorithm based on popularity voting. It recommends the most popular items based on purchase history of the entire population of agents.

3.4 Simulation

An agent enters the e-commerce platform with the aim to purchase a product. After entering her request into the search engine, she is presented with a set of items $(P_1, P_2, \ldots P_k)$ that may be arranged in order specified by the recommender system. Agent browses the items one by one. At the beginning of the 'shopping session' each agent makes a random choice which features will be considered while making decision. We assume that price is always taken into account.

After each action, the level of the agent's endurance decreases at a rate correlated with the complexity of the comparison task. When an agent's endurance level reaches 0, the agent chooses the best product she remembers. Knowing an agent's utility function and preferences, we can calculate the utility of the chosen product.

4 Experiments

Experiment 1. The goal of Experiment 1 is to analyse how cognitive limitations and biases influence the quality of agents' choices.

We run the simulation, changing working memory of agents from 1 to 10 and the strength of heuristics to one of the values (0, 0.25 and 0.5). The results of 90 scenarios (3 item sets × 3 heuristics levels × 10 WM levels) are compared to benchmark results from scenarios where all agents are modelled as 'young'[4].

---

[4] Agents with WM = 10 and h = 0.

Experiment 2. The goal of Experiment 2 is to verify if the CF recommendation system can improve decision making of elders with cognitive limitations and biases. Using data from Experiment 1 as a training set, we develop a CF recommendation system that provides agents with a suggested list of items based on their purchase history.

Experiment 3. The goal of Experiment 2 is to verify if the CB recommendation system can be detrimental for decision making of elders with cognitive limitations and biases. Using data from Experiment 1 as a training set, we develop a CB recommendation system that provides agents with suggested list of items based on their purchase history.

5 Results

Experiment 1. Figure 2 shows the average utility of the items purchased in scenario where items were presented to the user in a random order. The results confirm that the cognitive limitations significantly deteriorate the quality of choices of elderly agents. 'Young' agents achieved average utility of 0.80. 'Old' agents took a smaller number of attributes into consideration, which resulted in lower average utility of purchased items. The quality of the choices is additionally deteriorated by heuristics, especially when the strength of the heuristics is high. The effect is the more severe, the larger part of items can trigger heuristics. If the heuristics strength is moderate, or just a small part of the items can trigger heuristics, the overall quality of choices made by elders does not significantly decrease for higher levels of WM, but it is strongly deteriorated for low levels of WM.

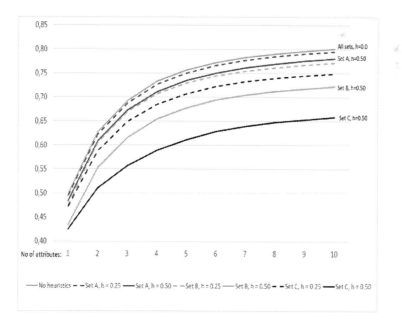

Fig. 2. Average utility of the items purchased by the agents, no recommender system

This leads to important conclusion that when the number of attributes taken into account by elderly agents is low, increasing it can significantly improve the quality of choice. This illustrates how choosing less cognitively demanding strategies in multi-attribute choice tasks can lead to significantly worse choices. Even small improvements in working memory capacity can significantly improve the quality of choices.

Experiment 2. The results (shown on Fig. 3) of using recommendations from the CF recommender system show that the system can significantly improve the quality of choice compared to benchmark[5]. This effect, however, weakens when the heuristic's strength is higher. In the first case, when heuristics strength is set to 0, the results from item sets A, B and C are the same, as the proportion of products triggering the heuristics does not influence the results. In the second case, when heuristics can change the perceived utility of products by as much as 25%, agent utilities in Experiment 2 are lower, especially for the lower numbers of attributes taken into account. This suggests that agents with more severe cognitive limitations are part of the group that suffers the most from the effect of heuristics. In the last case, when the strength of heuristics is set to 50%, the CF recommender system improves results significantly only on item sets A and B. The results from simulation on set C are worse than the benchmark. This suggests that for agents strongly relying on emotions when making decisions, the CF recommender system can only improve the quality of choices if the proportion of items triggering those emotional reactions is moderate.

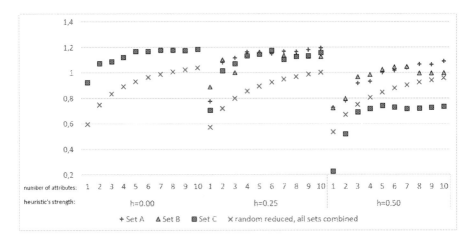

Fig. 3. Av. utility of the items purchased by the agents, recommendation from CF rec. system

Experiment 3. The results after recommendations from the CB recommender system show similar pattern (see Fig. 4). The quality of choices with the support of the CB

---

[5] Random reduced: results of simulation where items are presented in a random order, assuming that agent put a filter on the items set first. The items with attribute values below the reservation levels are excluded if given attribute is kept in agent's WM.

recommender system is deteriorating with the heuristics strength and level of cognitive deficiencies. Moreover, the choices supported by CB system are worse than the benchmark on all levels of heuristics strengths. Significant deterioration of the average utility from purchases supported by the CB recommendation system is evident even when heuristic strength set at 0%, as this system prompts agent to repeat a suboptimal choice once it has been done in the past.

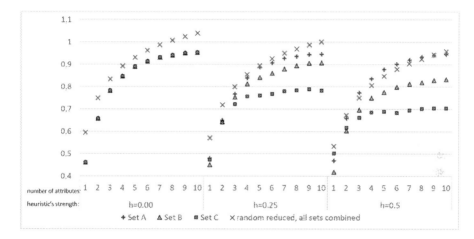

Fig. 4. Av. utility of items purchased by agents, recommendation from CB rec. system

5.1   Comparison of Recommender Systems

We compared combined results from Experiments 2 and 3 with the group that was recommended the most frequently purchased items[6]. In all cases, the CF recommender system is more effective in improving agents' decision than the CB system. Moreover, naive popularity voting recommender system is more effective than both more sophisticated recommender systems. This suggests that systems recommending items based on larger pool of agents are more effective when part of the population makes suboptimal choices due to cognitive limitations. The more the recommender system is fitted to individual agent past choices, the lower the chance that it can correct suboptimal choices by suggesting more diverse items (Fig. 5).

---

[6] Popularity voting recommender system.

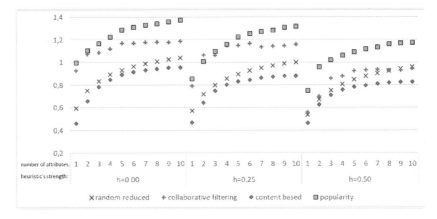

Fig. 5. Av. utility of the items purchased by the agents, all sets combined

6 Model Limitations

The model introduces many simplifications and does not fully reflect the complex process of human decision making.

The results of experiments are strongly dependent on the way we model populations of items and agents.

We have developed just one representative version of each recommender system. CF and CB systems can be designed and calibrated in many different ways, and the results obtained from our versions may not be universal for all. Those limitation will be basis for future analysis and tests.

7 Conclusions

We have developed an agent-based model mimicking the behavior of e-commerce platform users (agents) with emphasis on reflecting decision-making characteristics of elderly adults. We have used the model to verify how cognitive deficits of e-commerce customers influence the effectiveness of collaborative filtering and content based recommender systems.

The results from our simulation confirm the hypothesis (1): we have demonstrated that a CF recommender system improved the average utility of items purchased by agents with cognitive limitations, in most scenarios. Only in the extreme case, when the heuristic strength was highest and cognitive limitations were strongest, the decisions supported by the recommender system were less optimal than the benchmark.

The second hypothesis was also confirmed. We demonstrated that the CB recommender system can create a vicious circle of wrong decisions, deteriorating, instead of improving, the quality of decisions.

Results suggest that systems recommending items based on larger pool of agents, like CF and naive popularity voting model are more effective when part of the population makes suboptimal choices due to age-related cognitive limitations.

This conclusions can be basis for future work on designing recommender systems in a way that best supports specific needs of growing number of older e-commerce platform users.

Acknowledgements. This research was supported by the Polish National Science Center grants 2015/19/B/ST6/03179 and 2018/29/B/HS6/02604.

References

1. Lian, J.W., Yen, D.C.: Online shopping drivers and barriers for older adults: age and gender differences. Comput. Hum. Behav. **37**, 133–143 (2014)
2. Besedeš, T., Deck, C., Sarangi, S., Shor, M.: Decision-making strategies and performance among seniors. J. Econ. Behav. Organ. **81**, 524–533 (2012)
3. Ricci, F., Rokach, L., Shapira, B.: Recommender systems: introduction and challenges. In: Ricci, F., Rokach, L., Shapira, B. (eds.) Recommender Systems Handbook, pp. 1–34. Springer, Boston, MA (2015). https://doi.org/10.1007/978-1-4899-7637-6_1
4. Belleville, S.: Cognitive training for persons with mild cognitive impairment. Int. Psychogeriatr. **20**(1), 57–66 (2008)
5. Hess, T.M., Strough, J., Löckenhoff, C. (eds.): Aging and Decision Making: Empirical and Applied Perspectives. Academic Press, Cambridge (2015)
6. Tauringana, M., Good, A., Fitch, C.J.: E-commerce adoption among the elderly people. J. Electron. Commer. Organ. (2015)
7. Wagner, N., Hassanein, K., Head, M.: Computer use by older adults: a multi-disciplinary review. Comput. Hum. Behav. **26**(5), 870–882 (2010)
8. Savvopoulos, A., Virvou, M.: Tutoring the elderly on the use of recommending systems. Campus-Wide Inf. Syst. **27**(3), 162–172 (2010)
9. Beel, J., Langer, S., Nürnberger, A., Genzmehr, M.: The impact of demographics (age and gender) and other user-characteristics on evaluating recommender systems. In: Aalberg, T., Papatheodorou, C., Dobreva, M., Tsakonas, G., Farrugia, Charles J. (eds.) TPDL 2013. LNCS, vol. 8092, pp. 396–400. Springer, Heidelberg (2013). https://doi.org/10.1007/978-3-642-40501-3_45
10. Miller, G.A.: The magical number seven, plus or minus two: some limits on our capacity for processing information. Psychol. Rev. **63**, 81–97 (1956)
11. Wierzbicki, A.P.: Reference point approaches. In: Gal, T., Stewart, T.J., Hanne, T. (eds.) Multicriteria Decision Making, vol. 21, pp. 237–275. Springer, Boston (1999). https://doi.org/10.1007/978-1-4615-5025-9_9
12. Mikels, J.A., Shuster, M.M., Thai, S.T.: Aging, emotion, and decision making. In: Aging and Decision Making, pp. 169–188. Elsevier (2015)
13. von Helversen, B., Abramczuk, K., Kopeć, W., Nielek, R.: Influence of consumer reviews on online purchasing decisions in older and younger adults. Decis. Support Syst. **113**, 1–10 (2018)
14. Lambert-Pandraud, R., Laurent, G.: Why do older consumers buy older brands? The role of attachment and declining innovativeness. J. Mark. **74**, 104–121 (2010)
15. Yang, X., Guo, Y., Liu, Y., Steck, H.: A survey of collaborative filtering based social recommender systems. Comput. Commun. **41**, 1–10 (2014)

Identifying Toxicity Within YouTube Video Comment

Adewale Obadimu$^{(\boxtimes)}$, Esther Mead$^{(\boxtimes)}$,
Muhammad Nihal Hussain$^{(\boxtimes)}$, and Nitin Agarwal$^{(\boxtimes)}$

University of Arkansas at Little Rock, Little Rock, AR 72204, USA
{amobadimu, elmead, mnhussain, nxagarwal}@ualr.edu

Abstract. Online Social Networks (OSNs), once regarded as safe havens for sharing information and providing mutual support among groups of people, have become breeding grounds for spreading toxic behaviors, political propaganda, and radicalizing content. Toxic individuals often hide under the auspices of anonymity to create fruitless arguments and divert the attention of other users from the core objectives of a community. In this study, we examined five recurring forms of toxicity among the comments posted on pro- and anti-NATO channels on YouTube. We leveraged the YouTube Data API to collect video and comment data from eight channels. We then utilized Google's Perspective API to assign toxic scores to each comment. Our analysis suggests that, on average, commenters on the anti-NATO channels are more likely to be more toxic than those on the pro-NATO channels. We further discovered that commenters on pro-NATO channels tend to use a mixture of toxic and innocuous comments. We generated word clouds to get an idea of word use frequency, as well as applied the Latent Dirichlet Allocation topic model to classify the comments into their overall topics. The topics extracted from the pro-NATO channels' comments were primarily positive, such as "Alliance" and "United"; whereas, the topics extracted from anti-NATO channels' comments were more geared towards geographical locations, such as "Russia", and negative components such as "Profanity" and "Fake News". By identifying and examining the toxic behaviors of commenters on YouTube, our analysis lends aid to the pressing need for understanding this toxicity.

Keywords: Social network analysis · Topic modeling · Toxicity analysis

1 Introduction

Technology is changing the social media landscape at a very fast pace. Despite the myriad of advantages of utilizing this medium to educate and connect to like-minded individuals, a consensus is emerging suggesting the presence of malicious actors, otherwise known as trolls [1]. These users (hereafter referred to as toxic users) and the content they post (hereafter referred to as toxic behaviors) thrive on disrupting the norms of a given platform and causing emotional trauma to other users [2]. The proliferation of smart devices and mobile applications has further exacerbated the social side effect of these nefarious acts on various social media platforms [3]. However, despite the rich vein of academic research on identifying toxic behaviors online,

© Springer Nature Switzerland AG 2019
R. Thomson et al. (Eds.): SBP-BRiMS 2019, LNCS 11549, pp. 214–223, 2019.
https://doi.org/10.1007/978-3-030-21741-9_22

tackling these behaviors at scale remains surprisingly challenging [3]. In this study, similar to extant literature, we give an operational definition of toxicity as "the usage of rude, disrespectful, or unreasonable language that will likely provoke or make another user leave a discussion [1, 3–5]. Therefore, in this regard, toxicity analysis is different from sentiment analysis, which is the attempt to assign sentiment scores of positive, neutral, and negative to text data.

A report by Pew Research Center indicated that 73% of adult internet users have seen someone harassed online, and 40% have experienced it personally [1, 6]. Another survey by [6] highlighted that 19% of teens reported that someone has written or posted mean or embarrassing things about them on social networking sites. Due to the growing concerns about the impact of online harassment, many platforms are taking several steps to curb this phenomenon [1, 3, 7, 8]. For instance, on YouTube, a user can simply activate the safety mode to filter out offensive languages [7]. Wikipedia has a policy of "Do not make personal attacks anywhere in Wikipedia", and notes that attacks may be removed and the users who wrote them may be blocked [3]. Similarly, Facebook allows users to add comma-separated keywords to a "moderation blacklist." [7]. Twitter was forced to revise its hate speech guidelines when Zelda, the daughter of Robin Williams, was bullied after posting a memorial to her Father [8]. On CNN.com, over one in five comments are removed by moderators for violating community guidelines [1]. In 2013, Facebook came under fire for hosting pages which were hateful against women [8]. The aforementioned are a few examples that highlight the negative impact toxic behavior can have on a community and a company. Toxic behavior, if not curbed at the initial stage, can have a ripple effect. It can dissuade other people from joining a community [5] by perceiving the community as a hostile or unfriendly environment.

OSNs such as YouTube provide a means for users to upload digital content and allow other users to interact with this content via comments. A report by [7] highlights that YouTube is used by at least 758 million users around the world every month, with each visitor watching an average of 79 videos each month. In 2017, eighteen percent of Americans reported obtaining news from YouTube [6]. Research shows that YouTube is believed to contain widespread toxicity within comments [7]. Due to the widespread exploitation of these platforms by toxic users, automatic detection of toxicity has become a pressing need [8]. As a result, the Conversation AI team and Google collaborated to build a multi-headed model–an API called Perspective–that is capable of detecting various types of toxicity [9].

There is a dearth of systematic research concerning toxicity within the user-generated content on YouTube. This, along with the immensity of the amount of data and the speed with which the data is generated and shared, motivated us to conduct this analysis. Specifically, we examine the comment text data on pro- and anti-NATO YouTube channels. These channels were deemed suitable due to the high level of distinction in the type of digital content they contain. The remainder of the article is set out as follows. First, a few extant literatures on toxicity behavior are reviewed. Next, the empirical study is described, and the findings are discussed. Lastly, we discuss conclusions, limitations, and ideas for future work.

2 Related Work

2.1 Toxicity on Social Media

Threads of extant literature on antisocial behavior suggest that toxicity, in its various forms, oftentimes disrupts constructive discussions in an online community [1, 3–5, 8]. Prior studies have shown that online users participate in toxic behaviors out of boredom [10], others participate to vent [2] or simply to have fun [11]. A comprehensive examination of various forms of toxicity was conducted by [12]. Another study by [13] suggests that toxic users have unique personality traits and motivation, however, [1] noted that given the right condition, anyone can exhibit toxic tendencies. A study conducted by [14] shows that toxic users become worse over time, in terms of the toxic comment they post, and they are more likely to become intolerant of the community. One of the major obstacles in understanding toxic behavior is balancing freedom of expression with curtailing harmful content [7]. Closer to our work is a study by [5] that analyzed toxicity in multiplayer online games. The authors indicated that a competitive game might lead to an abuse of a communication channel. Toxic behavior, also known as cyberbullying [5], or online disinhibition [13], is bad behavior that violates social norms, inflicts misery, continues to cause harm after it occurs, and affects an entire community. Since toxic behavior has an offline impact, a deeper understanding of its properties is needed.

2.2 Identifying Toxicity on Social Media

Several researchers have tried to identify and suggest ways to mitigate hate speech in a community [3, 7]. Using data collected via crowdsourcing, the authors in [3] employed machine learning techniques such as linear regression and multilayer perceptron to analyze personal attacks at scale. They concluded that the problem of identifying and categorizing hate speech at scale remains surprisingly difficult. A study by [5] utilized Natural Language Processing techniques to detect the emergence of undesired and unintended behavior in online multiplayer games. [15] leveraged a set of regular expressions, n-grams, and supervised learning techniques to detect an abusive language. Another research by [16] combined lexical and parser features to identify an offensive language in YouTube comments. [17] presented a dataset with three kinds of comments: hate speech, offensive but non-hateful speech, and neither. The authors in [18] demonstrated the vulnerability of most state of the art toxicity detection tools against adversarial inputs. After experimenting with a transfer learning approach, [19] concluded that hate speech detection is largely independent of model architecture. They showed that results are mostly comparable among models but do not exceed the baselines.

3 Methodology

3.1 Data Collection and Visualization

Data collection is a critical task for analyzing information flow in social media. Our data collection approach employs a two-pronged effort. We started by consulting domain experts to obtain lists of pro- and anti-NATO channels on YouTube. We then utilized the YouTube API to collect videos' and commenters' data for each respective channels. Due to the noisy nature of the data, several data processing steps such as data standardization, noise elimination, and data formatting were conducted. Our final dataset consists of 1,424 pro-NATO videos with 8,276 comments, and 3,461 anti-NATO videos with 46,464 comments. Specifically, our final dataset consists of videos and comments scraped from eight YouTube channels. Four are pro-NATO channels: JFC Napels, SHAPE NATO, OTAN, and NATO; and four are anti-NATO channels: POTUS   BREAKING   NEWS   NETWORK,   DEFENSE   FLASH   NEWS, VOSSING KOG, and LATEST NEWS 360. Pro-NATO channels contain narratives that support NATO's objectives. Whereas; anti-NATO channels contain narratives that are counter to NATO's objectives. The network in Fig. 1 was built using the Organization Risk Analyzer (ORA) [20] visualization tool. The nodes represent commenters in each of the respective channels. The edge denotes the comments by the commenters. There is a presence of what is termed a bridge in the network, i.e., commenters that commented on both pro- and anti-NATO channels.

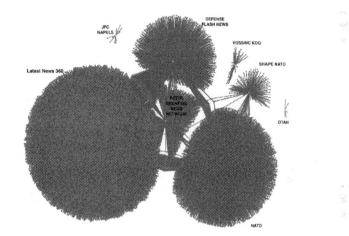

Fig. 1. Commenter-channel network for pro- and anti-NATO channels.

3.2   Assigning Toxic Scores

Obtaining a toxic score for each comment is imperative to our analysis. To this end, we leveraged a classification tool developed by Google's 'Project Jigsaw' and 'Counter Abuse Technology' teams - Perspective API [9] - to assign a score on the perceived toxicity of a given comment. This model uses a Convolutional Neural Network (CNN) trained with word-vector inputs to determine whether a comment could be perceived as "toxic" to a discussion [9, 21]. Specifically, the tool analyzes five (5) frequently recurring toxic attributes on YouTube viz: threats, insults, hate, sexually explicit, and identity-based attack. The API returns a probability score between 0 and 1, with higher values indicating a greater likelihood of the attribute label. Upon manual inspection, we observed that comments with a toxic score that is less than 0.1 have no apparent toxicity. Hence, we regard these comments as not toxic (assigned a value of "NONE") (see Algorithm 1).

Algorithm 1 Toxicity Computation

Input: C : Comments from pro- and anti-NATO channels
Output: T : Toxic scores for each comment
  for each $c \in C$ **do**
    $result \leftarrow$ COMPUTETOXICITY(c)
    if $result < 0.1$ **then**
      $toxicity \leftarrow NONE$
    else
      $toxicity \leftarrow result$
    end if
  end for
  function COMPUTETOXICITY(c)
    $attributes \leftarrow [SexuallyExplicit, Threat, IdentityAttack, Profanity, Insult]$
    for each $e \in attributes$ **do**
      $result \leftarrow$ PERSPECTIVEAPIREQUEST(c, e)
      $maxTocity_c \leftarrow$ MAX($result$)
    end for
  end function

Table 1 gives examples of attribute scores for some of the comments in our dataset. For instance, the first comment has a very high score for profanity and insult. The reason for this is that the comment contains toxic words that will likely escalate or stop the progression of the subsequent conversation.

Table 1. Convenience sampling of five (5) toxic comments from pro- and anti-NATO channels. * represents the attributes with the highest toxic score for the comment.

Comment	Threat	Sexual	Profanity	Insult	Attack
Youre a god damn idiot!!	0.18	0.07	0.97	*0.98	0.48
Russia is #1 terrorist nation	0.34	0.08	0.29	0.52	*0.72
Death to liberals	*0.98	0.11	0.57	0.63	0.88
America suck my Dick	0.26	*0.98	0.95	0.76	0.63
Eat shit. You white racist sucker	0.22	0.76	*0.99	0.98	0.94

3.3 Toxicity Analysis

Considering the diametric nature in the types of digital content being pushed by pro- and anti-NATO channels, obtaining quantitative insights about the comments' toxicity will give us an idea of how these content consumers react to various narratives. To this end, we compared the proportion of toxic comments in the pro- and anti-NATO channels. Our result shows that, when compared to comments on the anti-NATO channels, the majority of the comments on the pro-NATO channels are not toxic (Fig. 2). Thirty-five (35) percent of the comments on the pro-NATO channels contain words that are not toxic (were assigned a value of "NONE"); whereas, thirteen (13) percent of the comments on the anti-NATO channels fall into the not toxic category. This shows that the comment sections on anti-NATO channels are riddled with various forms of toxicity.

In terms of their toxicity, around thirty-seven (37) percent of the comments in the pro-NATO channels contain threats, nineteen (19) percent contain a form of identity attack; four (4) percent insult; three (3) percent sexual; and just two (2) percent profanity. Upon examination of some of these comments, we discovered that they contain subtle innuendo; for instance, one of the comments on a pro-NATO channel read 'NATO will crush Russia', and is regarded as a threat. The implication of this finding is that, even though most of the comments on the pro-NATO channels are a threat, they may not necessarily be against NATO.

In the same vein, we observed that a large proportion of words on the anti-NATO channels are made up of inflammatory, or negative comments towards a person or a group of people - including swear words, curse words, or other obscene or profane language. The implication of this finding is that commenters in the anti-NATO channels may be more inclined to post toxic comments about a person or group of people based on personal identity criteria. Further, threat, insult, and identity attack are the highest forms of toxicity being posted in the anti-NATO comment sections. We observe very little usage of words that are sexually explicit or profane on both channels. This finding suggests that commenters in both channels do not use words that contain references to sexual acts or body parts in a sexual way.

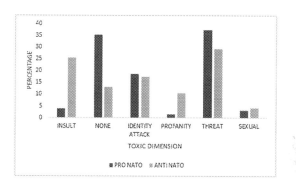

Fig. 2. The proportion of toxic comments on the pro-NATO channels and anti-NATO channels.

3.4 Topic Analysis

The proportion of toxic comments revealed in our toxicity analysis gave us a sneak-peek into reactions of commenters to the contents being pushed by pro- and anti-NATO channels. To further elaborate on this, in this section, we take a deeper dive into the specific words used by these commenters. We built a word cloud for all the comments in each respective channel type (see Fig. 3). The relative size of the words represents the frequency of the words used in the channel. We observed that prominent keywords among the pro-NATO channels comments are related to NATO, Russian, Ukraine, and Afghanistan. We also observed a high presence of positive words such as *Thank, Love,*

Nice, Good, and *Hope.* This observation is consistent with our earlier analysis that shows a high proportion of non-toxic comments on the pro-NATO channels. We posited that the high presence of these positive words is responsible for the general low proportion of toxicity in the pro-NATO channel. On the contrary, for the anti-NATO channels, the prominent keywords are *Fake, China,* and *News.* Further, there is a presence of profane words such as *shit* and *fuck* in the word cloud. This also validates our earlier assessment of the proportion of toxicity in anti-NATO channels.

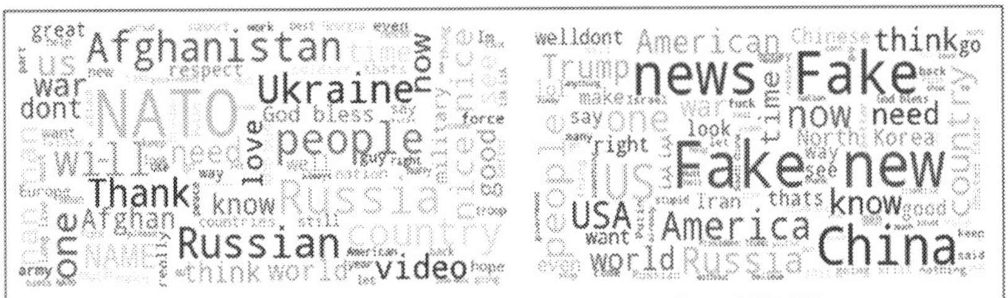

(a) Pro-NATO   (b) Anti-NATO

Fig. 3. Word cloud of comments posted on (a) pro-NATO videos and (b) anti-NATO videos.

To help us further explore the semantic structures within our collection of unstructured text bodies from these channels, we conducted topic modeling using Latent Dirichlet Allocation (LDA) [22]. This approach allowed us to discover the set of abstract topics that exist in the comments. We used the gensim python library to perform the topic modeling. Each comment was treated as a document, and was pre-processed, tokenized, and vectorized. WordNetLemmatizer was used to extract word roots, and to compute bigrams and trigrams of the tokens. The Natural Language Toolkit (NLTK) suite of python libraries was used to filter out English stopwords.

We visualized the topics extracted from our topic modeling by using pyLDAvis, Bokeh, and t-Distributed Stochastic Neighbor Embedding (t-SNE). t-SNE [23] is a non-linear dimensionality reduction algorithm used for exploring high-dimensional data. It maps multi-dimensional data to two or more dimensions suitable for human observation. By inspecting the topics of both channels, we observed that most of the topics on pro-NATO channels are relatively positive towards NATO, such as "Alliance", "United", and "Respect" (see Fig. 4). On the other hand, prominent topics on the anti-NATO channels are primarily geared towards some geological location, such as "Russia", coupled with "Soldier". Additionally, there are pockets of comments related to negative components, such as "Fake News", or to some sort of toxicity, such as "Profanity", "Insult", and "Attack" (see Fig. 5).

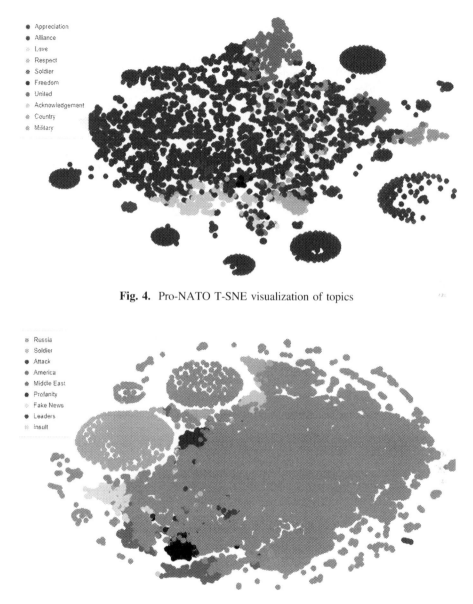

Fig. 4. Pro-NATO T-SNE visualization of topics

Fig. 5. Anti-NATO T-SNE visualization of topics

4 Conclusions and Future Work

This study outlined a methodology for identifying and scoring toxicity within the user-generated content posted on a specific OSN, YouTube. Specifically, this analysis was able to shed light on how the existence and magnitude of toxicity changes when there are shifts in the narrative, i.e. pro- or anti-NATO videos and their respective comments.

Although allowing for the ability to infer that anti-NATO videos attract more user engagement and result in the creation of more user-generated content in the form of comments, a seeming limitation of this study lies in the dataset being greatly skewed toward the anti-NATO narrative. For example, even though four YouTube channels were chosen from each of the two categories (pro-NATO, anti-NATO), more videos and more comments were posted on the anti-NATO channels, therefore skewing our final dataset so that it contained 2.5 times more anti-NATO videos than pro-NATO videos, and 5.6 times more comments from the anti-NATO channels than comments from the pro-NATO channels. Future work should address whether or not additional sampling techniques should be applied to determine the implications this skew had on analysis. Future work should also determine whether applying the concept of topic coherence and perplexity can provide any valuable insights from the topics. Additionally, and more complex, future work should attempt to establish the causal inference of the spread of toxicity on OSNs and to develop preventive measures.

Acknowledgements. This research is funded in part by the U.S. National Science Foundation (IIS-1636933, ACI-1429160, and IIS-1110868), U.S. Office of Naval Research (N00014-10-1-0091, N00014-14-1-0489, N00014-15-P-1187, N00014-16-1-2016, N00014-16-1-2412, N00014-17-1-2605, N00014-17-1-2675), U.S. Air Force Research Lab, U.S. Army Research Office (W911NF-16-1-0189), U.S. Defense Advanced Research Projects Agency (W31P4Q-17-C-0059), Jerry L. Maulden/Entergy Endowment at the University of Arkansas at Little Rock and the Arkansas Research Alliance. Any opinions, findings, and conclusions or recommendations expressed in this material are those of the authors and do not necessarily reflect the views of the funding organizations. The researchers gratefully acknowledge the support.

References

1. Cheng, J., Bernstein, M., Danescu-Niculescu-Mizil, C., Leskovec, J.: Anyone can become a troll: causes of trolling behavior in online discussions. In: Proceedings of the 2017 ACM Conference on Computer Supported Cooperative Work and Social Computing - CSCW 2017, pp. 1217–1230. ACM Press, Portland (2017)
2. Lee, S.-H., Kim, H.-W.: Why people post benevolent and malicious comments online. Commun. ACM **58**, 74–79 (2015)
3. Wulczyn, E., Thain, N., Dixon, L.: Ex machina: personal attacks seen at scale. arXiv:1610. 08914 [cs] (2016)
4. Nobata, C., Tetreault, J., Thomas, A., Mehdad, Y., Chang, Y.: Abusive language detection in online user content. Presented at the (2016)
5. Martens, M., Shen, S., Iosup, A., Kuipers, F.: Toxicity detection in multiplayer online games. In: 2015 International Workshop on Network and Systems Support for Games (NetGames), pp. 1–6. IEEE, Zagreb (2015)
6. Online Harassment | Pew Research Center (2014). http://www.pewinternet.org/2014/10/22/online-harassment/
7. Chen, Y., Zhou, Y., Zhu, S., Xu, H.: Detecting offensive language in social media to protect adolescent online safety. In: 2012 International Conference on Privacy, Security, Risk and Trust and 2012 International Conference on Social Computing, pp. 71–80. IEEE, Amsterdam (2012)
8. Cao, Q., Yang, X., Yu, J., Palow, C.: Uncovering Large Groups of Active Malicious Accounts in Online Social Networks. Presented at the (2014)

9. Perspective. http://perspectiveapi.com/#/
10. Varjas, K., Talley, J., Meyers, J., Parris, L., Cutts, H.: High school students' perceptions of motivations for cyberbullying: an exploratory study. West J. Emerg. Med. **11**, 269–273 (2010)
11. Shachaf, P., Hara, N.: Beyond vandalism: Wikipedia trolls. J. Inf. Sci. **36**, 357–370 (2010)
12. Warner, W., Hirschberg, J.: Detecting hate speech on the world wide web. In: Proceedings of the Second Workshop on Language in Social Media, pp. 19–26. Association for Computational Linguistics, Montréal (2012)
13. Suler, J.: The online disinhibition effect. Cyberpsychol. Behav. **7**(3), 321–326 (2004)
14. Cheng, J., Danescu-Niculescu-Mizil, C., Leskovec, J.: Antisocial behavior in online discussion communities, 10
15. Yin, D., Xue, Z., Hong, L.: Detection of harassment on Web 2.0., 7
16. Sood, S.O., Antin, J., Churchill, E.F.: Using crowdsourcing to improve profanity detection, 6
17. Davidson, T., Warmsley, D., Macy, M., Weber, I.: Automated hate speech detection and the problem of offensive language. In: Proceedings of the Eleventh International AAAI Conference on Web and Social Media (ICWSM 2017) (2017)
18. Hosseini, H., Kannan, S., Zhang, B., Poovendran, R.: Deceiving Google's perspective API built for detecting toxic comments. arXiv:1702.08138 [cs] (2017)
19. Gröndahl, T., Pajola, L., Juuti, M., Conti, M., Asokan, N.: All you need is "love": evading hate-speech detection. arXiv:1808.09115 [cs] (2018)
20. Carley, K.M., Reminga, J.: ORA: Organization Risk Analyzer: Defense Technical Information Center. Fort Belvoir, VA (2004)
21. Responding to Cognitive Security Challenges | StratCom. https://www.stratcomcoe.org/responding-cognitive-security-challenges
22. Blei, D.M.: Latent Dirichlet allocation, 30
23. van der Maaten, L., Hinton, G.: Visualizing data using t-SNE. J. Mach. Learn. Res. **9**, 2579–2605 (2008)

Examining Intensive Groups in YouTube Commenter Networks

Mustafa Alassad$^{(\boxtimes)}$, Nitin Agarwal, and Muhammad Nihal Hussain

University of Arkansas, Little Rock, AR, USA
{mmalassad, nxagarwal, mnhussain}@ualr.edu

Abstract. Focal structures are the sets of individuals in social networks that are not influential on their own but are influential collectively. These individuals, when coordinating, can be responsible for massive information diffusion, influence operations, or could coordinate (cyber)-attacks. These communities have high tension than other communities in the social network and can mobilize crowds. In this research, we propose a two-level decomposition optimization method for identifying these intensive groups in the complex social networks by constructing a two-level optimization problem for maximizing the local individual's degree centrality values and the global modularity measures. We also demonstrate the assembled centrality modularity method by applying to a network of YouTube users commenting on conspiracy theory videos to identify coordinating commenters. The dataset consisted of 9,661 users commenting on 4,145 conspiracy theory videos and the derived commenter network contained more than 4.4 million edges. Focal structure analysis was applied to this network to identify sets of users that are coordinating to promote disinformation dissemination. Our proposed model identifies smallest atomic units having high influence, interactions, higher reachability for information propagation. A multi-criteria optimization problem is also employed to rank the identified sets for further investigations.

Keywords: Focal structure analysis · Degree centrality · Modularity measure · Multi-criteria optimization · Bi-level problem

1 Introduction

Social media has become an important medium and provides a platform for people to voice their opinions during various events, protest or crises. During major protest campaigns like "The Egyptian Revolution", "Occupy Wall Street", and "Gezi Park", several groups harnessed the capability of social media to organize and mobilize crowds. Social media was integral during these protests and it would be impossible for the influential groups to stage successful protests without the power of social media.

Identifying the higher level of organizations (coordinating groups) in a complex network is an essential task. Focal Structure Analysis (FSA) enables us to identify these coordinating structures that usually go undetected by community detection algorithms. These coordinating groups have significantly higher interacting/influencing power than other communities. A variety of studies have been conducted by researchers to identify

© Springer Nature Switzerland AG 2019
R. Thomson et al. (Eds.): SBP-BRiMS 2019, LNCS 11549, pp. 224–233, 2019.
https://doi.org/10.1007/978-3-030-21741-9_23

influential groups during various global events including Saudi Women to drive campaign [1] have found focal structures had higher interactions and were able to mobilize crowd to stage protests. The users that formed focal structures had a significantly higher number of retweets, mentions, and replies than a random set of users in the network.

In this content, FSA is the identification of these sets of individuals in social networks that may or may not be influential on their own but are influential collectively [1]. Focal Structures was first identified by Şen et al. [1], where they applied a solely greedy algorithm for identifying the influential sets of individuals. FSA algorithm was based on local and global structures and was able to find the smallest intensive sets of groups in the social networks. However, the model suffers from some major drawbacks such as chain groups where their average clustering coefficient values were equal to zero, and the model assigned each node to only one group limiting the many central node's activities.

In this paper, we propose a bi-level optimization modification to overcome the shortcoming in FSA algorithm by maximizing degree centrality values and network modularity measures. We also demonstrate the efficacy of our algorithm by applying it to a network of commenters on YouTube disseminating disinformation [2].

The rest of the paper is organized as follows. Section 2 summarizes the current state of literature. Proposed methodology is explained in Sect. 3. In Sect. 4, we apply proposed model to a network and demonstrate its efficiency. We conclude with intended future work in Sect. 5.

2 Literature Review

The fields of graph theory, network science and network analysis have been extensively studied. Researchers have found various node level measures as well as network level measures to identify various aspects of the network.

Various studies [3–11] identified prominent nodes by measuring their power and opportunities based on their centrality values and positions in the network/graph. Researchers investigated how influential a node is, and who can maximize the information spread supporting their claims through developed methods such as PageRank [12] and HITS [13]. Researchers have also applied these measures to various domains like blogs [5], marketing [8–10] to identify influential nodes that maximize information spread [9, 11] in the network.

Although it is important to study various attributes of a node, it is also important to understand the structure of the network. Researchers have proposed various methods to detect communities in the network [14–21]. Newman [19, 27] proposed modularity to effectively communities in a complex network. Larger modularity values indicate the existence of denser communities. Optimizing modularity is an NP-hard problem.

3 Proposed Methodology

To overcome the drawbacks of the FSA algorithm, we propose a bi-level max-max approach. In the first level, we conduct node level analysis to identify candidate nodes that could form the FSA. In the next level, we use the sphere of influence of candidate nodes to identify candidate focal structure.

3.1 Data Set

We have applied the proposed bi-level max-max approach to many social network such as Zachary's karate network [24], Mis Miserable [25], and Dolphins social network [26]. Where the model was able to overcome the current FSA algorithm drawbacks as discussed in the next section. The selected dataset in this research is an undirected YouTube network consist of 4,145 videos along with 16,493 comments by 9,661 users [2]. To collect this dataset, a YouTube channel posting conspiracy theory videos was identified. We later used YouTube Data API [27] to Using the obtained dataset, a commenter network was constructed by connecting two commenters if they commented on the same video. Table 1 is summarizing the network's statistics.

3.2 Local Analysis-Node Level

To conduct node level analysis, we utilized the degree centrality to measure the influence of each node in the network as shown in Fig. 1. We then identified local community for each node based on its sphere of influence (a node and its neighbors) [4]. In the next step, to assess the node's connectedness, community clustering coefficient was measured. Figure 2 shows the average clustering coefficient values. The model is designed to construct all necessary parameters from the dataset such as the centrality and average clustering coefficient bounds (Upper, Lower), and it is not required to define the number of focal structures initially. The model filters communities based on the computed bounds.

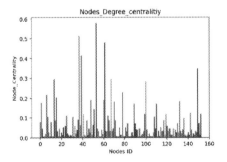

Fig. 1. Degree centrality measures

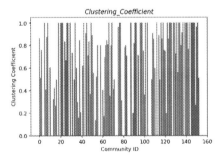

Fig. 2. Avg. clustering coefficient values

The next step, we have sorted the filtered communities based on the preferred centrality values, starting from the lowest central community to the highest central community. This step provides two advantages. First, it allows to study the powers of small communities in the network. And second, it overcomes a drawback of modularity and does not leave small clustered groups hidden within prominent groups [28]. In the final step of this level, filtered communities are converted into their vectors as shown in Eq. (1), where $\left(C_{v1}^{T}\right)$ is the vector $(v1)$ of the first local community, and (T) is the vector transpose. Also, 1 represents that node (1) is connected to node (8), otherwise its (0).

$$C_{v1}^{T} = [0,0,0,0,0,0,0,1,0,0,0,0,1,0,0,0,0,0,\ldots,0,0,0,0,0,0,0,0,0,0,1,0,0,0,0,0]$$

(1)

3.3   Global Analysis-Network Level

In this level, we use network level measures to study community structure by implementing the spectral modularity method [22, 23, 29]. In this level, communities that can maximize the modularity objective function as shown in Eq. (2) are identified. The number of required communities k is not predefined, but the algorithm will measure the modularity values for the communities created in the previous level considering only those communities that can maximize Q.

Let the modularity matrix be $B = A - dd^{T}/2m$, where $d \in R^{n \times 1}$ is the degree vector of all commenters. Then the modularity community detection can be reformulated based on the spectral modularity Eqs. (2) and (3) as explained by Wang et al. [18], where $X \in R^{n \times k}$ is the function of all partitions, and P is the k partitions $P = (P_1, P_2, P_3, \ldots, P_k)$. Where these method can explore more details and simplify the computations [30, 31], the goal of utilizing it is to finding the best community/nodes that maximize the graph modularity values in each iteration. Accordingly, Q is a function employed to maximize the modularity value. The vectors are considered to represent the node's community (node and its neighbors) not only nodes as presented by [18]. The overlapped nodes are omitted from second level solution structure. Finding the assignment matrix X that maximizes Q is an NP-hard problem too. At this level, the number of clusters of partitions of eigenvectors that are needed to have a clean separation of communities is not known in advance, but instead, we can calculate $(P > 1)$ of them and search for the highest modularity partition among those communities imported from the model's first level.

When the best joint community is found, it will be exported to the first level accordingly. So, the node level will reorganize the remaining local communities, and send them back to the second level until all communities checked or the modularity starts decreasing. The results of this section are the FSA candidates as illustrated in Fig. 3.

$$Q = \frac{1}{2m} Tr\left(X^{T} B X\right)$$

(2)

$$X = \begin{cases} 1 & v_i \in P_j \\ 0 & elsewhere \end{cases} \tag{3}$$

$$S = \begin{bmatrix} 0 & S_{12} & \cdots & S_{1n} \\ S_{21} & 0 & \cdots & S_{2n} \\ \vdots & \vdots & \vdots & \vdots \\ S_{n1} & S_{n2} & \cdots & 0 \end{bmatrix} \tag{4}$$

Once the model identifies the FSA candidates, we stitch them based on their node's similarity utilizing the Jaccard similarity method. The resulting similarity matrix S as shown in Eq. (4). A tunable threshold is considered to identify most similar FSA groups and groups with similarity values above the threshold are stitched together. Where S_{ij} is the similarity weight between F_i, and F_j, and the total weights of an FSA with all other pairs should be equal to 1.

The last part, we set quality control constraints inherited from the real-world network's criteria on the FSA. The Average clustering coefficient, the groups' density, average degree centrality, groups diameter, and average path length of all FSA groups are measured [32–34]. The candidates must not violate any of the mentioned constraints. The model constructs lower and upper bounds parameters based on utilized metrics from the original graph. Considering more than one metric helps to study each group from different aspects. However, measuring FSA sets based on a single constraint will not reveal enough information about them, but considering many judges for the explored sets could bring lots of information and increase the resolution of the selected FSA groups.

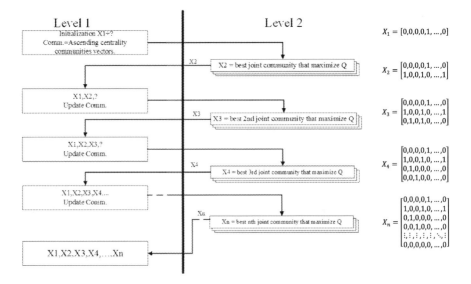

Fig. 3. Interactions between level 1 and level 2 for identifying FSA groups.

4   Experimental Results

In this section, we apply the above proposed algorithm to a YouTube disinformation dissemination dataset [2] to identify prominent actors. Using the obtained dataset, a commenter network was constructed by connecting two commenters if they commented on the same video which was fed to the above proposed algorithm. Table 1 provides some basic statistics.

The model identified fifty-four FSA sets from the network. These groups met the evaluation criteria and did not violate any applied constraints in Sect. 3. Also, Fig. 4 shows the FSA groups and their belonging commenters. The smallest FSA is a triad group such as (FSA14, FSA15, FSA16, etc.). Also, the identified FSA groups are not mutually exclusive. Commenters/nodes from one FSA group can also belong to other groups.

Table 1. YouTube network statistics

Comments	16,493
Commenters	9,661
Node in network	9,162
Edges in network	4,415,410

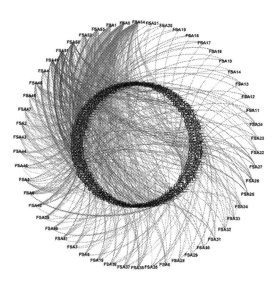

Fig. 4. YouTube commenters network focal structures

Figure 5 illustrates the communities' ability to make friends or linking friends in the network. Figure 6 shows power, opportunities, and resources for the identified FSA sets. For example, FSA5 has the highest centrality value and the highest friendships between the members at the same time.

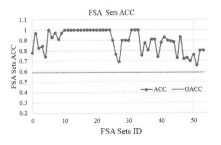

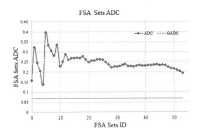

Fig. 5. The FSA's sets (ACC) vs. the graph's (GACC) clustering coefficient values.

Fig. 6. The FSA's Avg. centrality (ADC) vs. normalized graph's centrality (GADC).

In this paper, we have implemented a multi-criteria optimization problem to rank the explored FSA groups. The model considered the same constraints mentioned in the last part of Sect. 3. These judges are the best to use for ranking any FSA set.

The first set of equations in Eq. (3) are used to weight the groups' average clustering coefficient (AC_F), average degree centrality (ADC_F), and average density (DN_F). The numeric thresholds are tunable and can be estimated based on the user desires. Where W_{Fi}, is showing the weight of a metric when it is close to its max/min respectively and x is the actual value of each metric.

Since focal structure analysis requires groups to be associated with high clustering coefficient, density, and average degree centrality, the set of equations in (5) are used to assign high weights to any FSA associated with high $(AC_F, ADC_F, and\ DN_F)$, and the opposite is true. At the same time, the analysis needs to have sets with small diameters and average path length. The second set of equations in (6) are applied to give high weights to any FSA with small diameter $\{D_F\}$, and small average path length $\{Al_F\}$. The numeric thresholds are tunable, where W_{Fi}, shows the weights of a metric when it is close to its max/min respectively and y is the actual value of each FSA.

$$R_{AC} = \begin{cases} W_{Fi} = 5 & \mathcal{AC}_{Fi} \geq x \\ W_{Fi} = 4 & x > AC_{Fi} \geq 0.9\,x \\ W_{Fi} = 3 & 0.9x > \mathcal{AC}_{Fi} \geq 0.8x \\ W_{Fi} = 2 & 0.8x > \mathcal{AC}_{Fi} \geq 0.6\,x \\ W_{Fi} = 1 & otherwise \end{cases} \tag{5}$$

$$R_{A\ell} = \begin{cases} W_{Fi} = 5 & \mathcal{A\ell}_{Fi} \leq y \\ W_{Fi} = 4 & y < \mathcal{A\ell}_{Fi} \leq 1.5y \\ W_{Fi} = 3 & 1.5y < \mathcal{A\ell}_{Fi} \leq 2y \\ W_{Fi} = 2 & 2y < \mathcal{A\ell}_{Fi} \leq 3y \\ W_{Fi} = 1 & otherwise \end{cases} \tag{6}$$

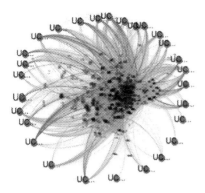

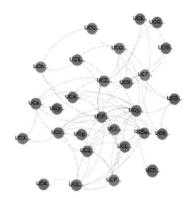

Fig. 7. Links between FSA sets and other nodes. (Color figure online)

Fig. 8. Links between top-ranked FSA sets.

The resultant network, after applying the above multi-criteria model to the identified FSA sets were prioritized into multiple levels as shown in Fig. 4. Also, Fig. 7, shows the top-ranked identified FSA groups from the YouTube commenter network. Although this network consists of 9,162 nodes and 4,415,410 edges, the identified nodes can influence almost all the nodes in the network. By allocating the top-ranked groups from different applied metrics used in this paper, we can monitor the links and resources used for information diffusion. Also, Fig. 8, shows the top-ranked nodes connection, where it monitors their abilities to communicate with each other, and at the same time they are creating unique structures.

These groups have enough resources to spread information to most parts of the network (red/green edges) and at the same time, they can receive information from other nodes using existing links (green/red edges) as shown in Fig. 7. Also, if these people agreed to coordinate with each other to spread the same information about videos, as shown in Fig. 8, they have the power, resources, opportunities in the network structure and enough links to influence the whole network fast and easy.

5 Conclusion and Discussion

In this paper, we have studied the focal structures analysis on a YouTube commenters network, where the commenters tried to spread fake news into the network by commenting on the posted videos many times. The proposed max-max model is a mix between the members-based community detection algorithm (degree centrality) and the group-based detection algorithms (modularity). Utilizing the real-world measures helped to increase the results' resolution, and a multi-criteria problem was employed to rank the extracted groups. These groups are unique in their structures, including commenters acting in different parts of the network. They can get/send their information quickly and deceive others by spreading false information at the same time. The proposed multi-criteria problem was utilized to rank the identified FSA groups. The model utilizes a decomposition and multi-criteria optimization procedure to identify the intensive groups, but it also validates the identified clusters using constraints and only valid FSA groups published on output. One of the benefits of the model is it does not require information (e.g., number of clusters, lower and upper bounds) from the input except the network dataset. Finally, The model was able to overcome drawbacks of current state of art FSA algorithm and finds smallest possible triads as intensive groups including members that span multiple FSA groups. Also, this algorithm was able to propose higher number of FSA groups and effectively ranks these groups.

Acknowledgment. This research is funded in part by the U.S. National Science Foundation (IIS-1636933, ACI-1429160, and IIS-1110868), U.S. Office of Naval Research (N00014-10-1-0091, N00014-14-1-0489, N00014-15-P-1187, N00014-16-1-2016, N00014-16-1-2412, N00014-17-1-2605, N00014-17-1-2675), U.S. Air Force Research Lab, U.S. Army Research Office (W911NF-16-1-0189), U.S. Defense Advanced Research Projects Agency (W31P4Q-17-C-0059), Arkansas Research Alliance, and the Jerry L. Maulden/Entergy Foundation at the University of Arkansas, Little Rock. Any opinions, findings, and conclusions or recommendations expressed in this material are those of the authors and do not necessarily reflect the views of the funding organizations.

References

1. Şen, F., Wigand, R., Agarwal, N., Tokdemir, S., Kasprzyk, R.: Focal structures analysis: identifying influential sets of individuals in a social network. Soc. Netw. Anal. Min. **6**(1), 17 (2016)
2. Hussain, M.N., Tokdemir, S., Agarwal, N., Al-Khateeb, S.: Analyzing disinformation and crowd manipulation tactics on YouTube. In: 2018 IEEE/ACM International Conference on Advances in Social Networks Analysis and Mining, pp. 1092–1095 (2018)
3. Leskovec, J., McGlohon, M., Faloutsos, C., Glance, N., Hurst, M.: Cascading behavior in large blog graphs. In: Proceedings of the 2007 SIAM International Conference on Data Mining, pp. 551–556 (2007)
4. Li, C., Wang, L., Sun, S., Xia, C.: Identification of influential spreaders based on classified neighbors in real-world complex networks. Appl. Math. Comput. **320**(11), 512–523 (2018)
5. Borgatti, S.P.: Centrality and network flow. Soc. Netw. **27**(1), 55–71 (2005)
6. Agarwal, N., Liu, H., Tang, L., Yu, P.S.: Modeling blogger influence in a community. Soc. Netw. Anal. Min. **2**(2), 139–162 (2012)
7. Agarwal, N., Liu, H., Tang, L., Yu, P.S.: Identifying the influential bloggers in a community. In: Proceedings of 2008 International Conference on Web Search and Data Mining, pp. 207–218 (2008)
8. Richardson, M., Domingos, P.: Mining knowledge-sharing sites for viral marketing. In: Proceedings of the Eighth ACM SIGKDD International Conference on Knowledge Discovery and Data Mining, pp. 61–70 (2002)
9. Kempe, D., Kleinberg, J.: Maximizing the spread of influence through a social network. In: Proceedings of the Ninth ACM SIGKDD International Conference on Knowledge Discovery and Data Mining, pp. 137–146 (2003)
10. Chen, W., Wang, Y.: Efficient influence maximization in social networks categories and subject descriptors. In: Proceedings of the 15th ACM SIGKDD International Conference on Knowledge Discovery and Data Mining, pp. 199–207 (2009)
11. Leskovec, J., Mcglohon, M., Faloutsos, C., Glance, N., Hurst, M.: Patterns of cascading behavior in large blog graphs. In: Proceedings of the 2007 SIAM International Conference on Data Mining, pp. 551–556 (2007)

12. Page, L., Brin, S., Motwani, R., Winograd, T.: The PageRank citation ranking: bringing order to the web. World Wide Web Internet Web Inf. Syst. **54**(1999–66), 1–17 (1998)
13. Kleinberg, J.O.N.M.: Authoritative sources in a hyperlinked environment. In: Proceedings of the ACM-SIAM Symposium on Discrete Algorithms, vol. 46, no. 5, pp. 604–632 (1999)
14. Hong, S., Park, N.: SENA: preserving social structure for network embedding. In: Proceedings of the 28th ACM Conference on Hypertext and Social Media, pp. 235–244 (2017)
15. Hagen, L., Kahng, A.B.: New spectral methods for ratio cut partitioning and clustering. IEEE Trans. Comput. Des. Integr. Circ. Syst. **11**(9), 1074–1085 (1992)
16. Von Luxburg, U.: A tutorial on spectral clustering. Stat. Comput. **17**(4), 395–416 (2007)
17. Newman, M.E.J.: Modularity and community structure in networks. Proc. Natl. Acad. Sci. **103**(23), 8577–8582 (2006)
18. Wang, G., Shen, Y., Luan, E.: Measure of centrality based on modularity matrix. Prog. Nat. Sci. **18**(8), 1043–1047 (2008)
19. Briscoe, E.J., Appling, D.S., Mappus, R.L., Hayes, H.: Determining credibility from social network structure. In: Proceedings of the 2013 IEEE/ACM International Conference on Advances in Social Networks Analysis and Mining, pp. 1418–1424 (2014)
20. Massen, C.P., Doye, J.P.K.: Identifying communities within energy landscapes. Phys. Rev. E - Stat. Nonlinear Soft Matter Phys. **71**(4), 1–13 (2005)
21. Kivran-Swaine, F., Govindan, P., Naaman, M.: The impact of network structure on breaking ties in online social networks. In: Proceedings of the SIGCHI Conference on Human Factors in Computing Systems, pp. 1101–1104 (2011)
22. Newman, M.E.J.: Detecting community structure in networks. Eur. Phys. J. B - Condens. Matter **38**(2), 321–330 (2004)
23. Newman, M.E.J., Girvan, M.: Finding and evaluating community structure in networks, pp. 1–16 (2003)
24. Zachary, W.W.: An information flow model for conflict and fission in small groups. J. Anthropol. Res. **33**(4), 452–473 (1977)
25. Hugo, V.: Les misérables. TY Crowell & Company (1887)
26. Lusseau, D., Schneider, K., Boisseau, O.J., Haase, P., Slooten, E., Dawson, S.M.: The bottlenose dolphin community of doubtful sound features a large proportion of long-lasting associations: can geographic isolation explain this unique trait? Behav. Ecol. Sociobiol. **54**(4), 396–405 (2003)
27. YouTube Data API. Google Developers
28. Sato, K., Izunaga, Y.: An enhanced MILP-based branch-and-price approach to modularity density maximization on graphs. Comput. Oper. Res. **106**, 1–25 (2018)
29. Girvan, M., Newman, M.: Community structure in social and biological networks. PNAS **99**(12), 7821–7826 (2002)
30. Tsung, C.K., Ho, H., Chou, S., Lin, J., Lee, S.: A spectral clustering approach based on modularity maximization for community detection problem. In: Proceedings - 2016 International Computer Symposium, ICS 2016, pp. 12–17 (2017)
31. Stone, E.A., Ayroles, J.F.: Modulated modularity clustering as an exploratory tool for functional genomic inference. PLoS Genet. **5**(5), e1000479 (2009)
32. Zafarani, R., Abbasi, M.A., Liu, H.: Social Media Mining: An Introduction. Cambridge University Press, Cambridge (2014)
33. Wilson, C., Boe, B., Sala, A., Puttaswamy, K.P.N., Zhao, B.Y.: User interactions in social networks and their implications. In: Proceedings of the 4th ACM European Conference on Computer Systems - EuroSys 2009, p. 205 (2009)
34. Zitnik, M., Sosič, R., Leskovec, J.: Prioritizing network communities. Nat. Commun. **9**(1), 1–9 (2018)

User Behavior Modelling for Fake Information Mitigation on Social Web

Zahra Rajabi[✉], Amarda Shehu[✉], and Hemant Purohit[✉]

Volgenau School of Engineering, George Mason University, Fairfax, VA 22030, USA
{zrajabi,ashehu,hpurohit}@gmu.edu

Abstract. The propagation of fake information on social networks is now a societal problem. Design of mitigation and intervention strategies for fake information has received less attention in social media research, mainly due to the challenge of designing relevant user behavior models. In this paper we lay the groundwork towards such models and present a novel, data-driven approach for user behavior analysis and characterization. We leverage unsupervised learning to define user behavioral categories over key behavior dimensions. We then relate these categories to content-based, user-based, and network-based features that can be extracted in near-real time and identify the most discriminative features. Finally, we build predictive models via supervised learning that leverage these features to determine a user's behavior category. Rigorous evaluation indicates that the constructed models can be valuable in predicting user behavior from recent activity. These models can be employed to rapidly identify users for intervention in mitigation strategies, crisis communication, and brand management.

Keywords: Disinformation · Fake news mitigation ·
User behavioral model · Social media mining · Unsupervised learning

1 Introduction

In 2017, 67% of Americans reported that they obtained at least some part of their news on social media[1]. This massive burst of data is naturally accompanied with the threat of disinformation, such as spam and fake news spread by malicious intent users, affecting different aspects of democracy, journalism, and freedom of expression [11]. Fake news dissemination was best highlighted during the 2016 presidential election, in which the spread of fake stories favoring each party gravely threatened trust in government [2]. The propagation of fake information on social networks is now a recognized societal problem [14]. Despite many efforts, social networking platforms have yet to effectively address this challenge In particular, mitigating the dissemination of fake content is now a

---

[1] http://www.journalism.org/2017/09/07/news-use-across-social-media-platforms-2017/.

R. Thomson et al. (Eds.): SBP-BRiMS 2019, LNCS 11549, pp. 234–244, 2019.
https://doi.org/10.1007/978-3-030-21741-9_24

critical challenge for researchers across academia leading to emerging research areas of social cyber-security [4] and social cyber forensics [1,3].

In this paper, we present a novel, data-driven approach for user behavioral analysis and characterization that enables us to identify vulnerable users for fake news mitigation. Our main contributions are: (a) Identification of key user behavior dimensions in reactions to the exposure of fake vs. fact information, specifically *initiation*, *propagation*, and *reception* behavior types. These dimensions allow us to organize users in behavior categories via unsupervised learning. (b) Validation of the hypothesis that behavior categories for users can be predicted by features extracted from shared content, user profile and activity, as well as the structural characteristics in the corresponding user interaction network. We also employ feature selection approaches to analyze the significance of our content-based, user-based, and network-based set of features to identify the most representative features of user behavior categories. These features are then used to build classification models to predict such categories. (c) Extensive experiments to evaluate state-of-the-art multiclass classification algorithms for user behavioral pattern prediction using a dataset collected from *Hoaxy* [8] platform. Our evaluations show that the predictive models demonstrate promising performance in categorizing users based on their reactions in response to fake/fact exposures, which consequently gives us an oversight to develop a solid baseline for designing an effective fake content mitigation strategy. In the remainder of this paper, we briefly present in Sect. 2 related works. Section 3 describes our proposed approach, followed by its evaluation and results in Sect. 4.

2 Related Work

In recent years, with increasing consumption of news over social media, the extreme consequences of fake information dissemination, from misleading election campaigns to inciting violence during crises, have led many researchers to focus on the problem of fake news detection [7,8,15]. Comprehensive reviews of this area of research in [14,17] show that existing studies mostly rely on static datasets to develop models based on supervised learning methods rather than online learning settings due to potential concept drifts. These approaches involve exploitation of user, content, and network-based information which inspired us in the feature design of our user behavioral modeling.

Users play a critical role as the creators and spreaders of fake content in social web. Therefore, assessing the credibility of users and modeling their behavior types could provide a valuable approach to design intervention strategies [6,12], which can optimize the dissemination of real news. For instance, [6] proposed an intervention framework using multivariate point process, however, authors did not consider the types of users and their behaviors. Given the uncertainty of user intent and activities, it is essential, although very challenging, to discriminate between malicious and naive users who unintentionally engage in fake content propagation. Therefore, modeling user behavior for identifying candidate vulnerable users for intervention strategies is an emerging research need.

While there exist extensive research on social media on user modeling and user credibility [9,10,16,17], the main goal of these studies has centered around content filtering for spam, improving user interest profiling for content and link recommendation systems, personalization in search, as well as influencer ranking. Our research instead complements such user modeling research by investigating user behavior types to inform the mitigation strategies for the propagation of malicious, fake content.

3 Methods

We first represent the key behavior dimensions relevant to the mitigation task and then describe the unsupervised learning setup that allows elucidating the organization of users in different behavior categories. Once such categories are identified, information theoretic measures expose characteristics/features that best relate with the identified categories. Supervised learning methods then yield predictive models of behavior categories from such features.

3.1 Key Dimensions of User Behavior

We have identified three major behavioral dimensions to capture user reactions to fake over fact cascade exposures, namely initiating, propagating, and receiving (but no further action) fake content. We define FoF_{prop}, $FoF_{received}$, and FoF_{init} to be log-ratio of fake over factual information respectively propagated, received or initiated by a user. As described in Sect. 4, these dimensions allow visualization of a three-dimensional semantic space as a baseline representation of user engagements within the network in response to different information cascade exposures. More importantly, they facilitate the application of unsupervised learning methods to identify behavioral categories based on user engagement in spread of fake versus fact cascades.

3.2 Identifying Behavior Categories via Unsupervised Learning

Clustering algorithms can group and categorize users within the three-dimensional space defined above. We consider several clustering algorithms such as kmeans, Agglomerative clustering, DBScan, and spectral clustering (c.f. survey on algorithms in [13]). In Sect. 4, we evaluate the performance of clustering algorithms (over different parameter values) along popular metrics, such as the Silhouette coefficient, the Calinski-Harabaz score, and the Davies-Bouldin score. These metrics do not rely on ground truth availability, which is our case.

Clustering Performance Evaluation. The Silhouette coefficient is calculated as $(b-a)/max(a,b)$, where a is the mean intra-cluster distance, and b is the mean nearest-cluster distance for each sample (user). This metric is computed as the mean Silhouette Coefficient of all samples ranging from -1 (worst) to 1 (best).

The Calinski-Harabaz score, also known as the variance ratio criterion, is defined as the ratio between the within-cluster dispersion and the between-cluster dispersion. A higher Calinski-Harabaz score indicates a model with better-defined clusters. Unlike the Silhouette score, the Calinski-Harabaz score is unbounded; the higher the score the better the cluster separation. The Davies-Bouldin is defined as the ratio of within-cluster distances to between-cluster distances and is bounded in $[0, 1]$. A lower score is better.

3.3   User Behavior Categories Representation

The above evaluation measures highlight the most effective clustering method and corresponding behavior categories, which can be used to label users. Our dataset has only ground truth for fake/fact content and no user labels for our analysis is provided. Therefore, our proposed approach of automatically labeling users with behavioral categories opens a way to supervised learning models that relate features to the discovered behavior categories/classes for users. Information-theoretic measures, such as Mutual Information (MI) measure is employed to evaluate characteristics/features of user nodes in the diffusion cascades that best relate with the identified behavior classes. A feature selection algorithm is employed to identify the most important features, and supervised learning methods are then utilized to build predictive models of behavior categories from the extracted features. In this section, we propose our data mining framework for user behavior category representation and prediction.

Features Extraction. The features representing each user are extracted using the propagated retweets' content, user profiles, and network structure.

1. *Content-based Features*:
   - sentiment: the average of sentiment intensity score of tweet texts shared by user is computed using Sentiment Analyzer tool in *nltk* (Natural Language Toolkit) Sentiment package.
   - Tweet text length: the average of length of tweet texts shared by user.
2. *User-based Features*:
   - followers_count: the number of users following a user; it shows a Twitter account's popularity;
   - friends_count: the number of users a user is following; it informs the user's interest-driven participation;
   - influence_score: the social reputation of each user based on follower and following counts that is computed by $\log((1 + \text{followers\_count})^2 + \log(\text{statuses\_count}) - \log(\text{friends\_count}))$;
   - listed_count: the number of public lists of which a user is a member;
   - statuses_count the number of tweets and retweets shared by a user;
   - has_url: a boolean feature showing if a user has a url or not;
   - sociability: the ratio of the number of friends_count to followers_count: $\log(1 + \frac{1+\text{friends\_count}}{1+\text{followers\_count}})$;

- `favorability`: ratio of the number of favorites to the total number of tweets; it informs higher engagement in contrast to just posting tweets. $\log(1 + \frac{(1+\text{favourites\_count})}{(1+\text{statuses\_count})})$;
- `survivability`: the potential active existence on the platform over time, and it is measured as the difference between current timestamp and the timestamp at which a tweet is created;
- `activeness`: the number of tweet statuses to the period of time since account creation; it determines the likelihood of a user to be active over a period of time on average: $\log(1 + \frac{(1+\text{statuses\_count})}{(1+\text{survivability}))})$;
- `favourites_count`: the number of tweets a user has favored over time as a measure of user engagement level.

3. **Network-based Features**:
  - `betweenness`: the normalized sum of the fraction of all-pairs' shortest paths that pass through a node/user. Betweenness values are normalized by $b = b\frac{(n-1)}{(n-2)}$ where n is the number of nodes in graph G.
  - `degree_centrality`: the fraction of nodes/users to which a particular user is connected.
  - `load_centrality`: the normalized fraction of all shortest paths that pass through a node.

Feature Ranking. Feature Selection is the process of identifying relevant features from a feature set that contribute most to the prediction variable and removing the irrelevant ones, in order to improve performance of predictive model. Features can be ranked according to different metrics, e.g. F-test statistic, Mutual Information (MI) measure, and p-value.

- F-value: ANOVA F-test statistic captures linear dependency of two random variables and computes F-value for each feature. This test measures the ratio of between-groups to within-groups variances; we note that the groups here are user behavior categories. When F-values are near 1, the null hypothesis is true (establishing independence).
- p-value: This allows determining whether the null hypothesis can be rejected with 95% confidence level (corresponding to p-values of 0.05). The smaller the p-value, the stronger the evidence to reject the null hypothesis.
- MI measure: This measure captures mutual dependency between random variables. MI quantifies the amount of information obtained about one random variable through observing the other random variable, with zero value showing two random variables are independent, whereas higher values meaning higher dependency. In Sect. 4 we provide the F-value, p-value, and MI measure for each of the features.

F-value along with the p-value enables us in deciding whether results are significant enough to reject the null hypothesis. We investigated both univariate feature selection and recursive feature elimination (RFE) and in both methods, the top 1/3 of the features (5 of the 16 initial features) are very similar. In particular, RFE selects smaller sets of features recursively with the least important features pruned at each iteration. We employed an SVM classifier with linear kernel in the RFE estimator for this purpose.

User Behaviors Estimation: In previous sections, we described how we group similar users based on their associated behaviors into three user behavioral clusters (classes). In this section, we focus on building a behavioral model that predicts a user behavior class by incorporating extracted features into supervised classification models.

We consider ten different classifiers such as Nearest Neighbors (k-NN), SVM with RBF kernel, Random Forest, a multilayer perceptron classifier (MLP), AdaBoost, XGBoost, Naive Bayes, Decision Tree, and Quadratic Discriminant Analysis (QDA). Each classifier is used with recommended default parameter settings and not optimized for performance. Specifically, the number of neighbors in k-NN is 3, for SVM with RBF kernel γ is chosen automatically, the maximum depth of decision tree is set to 5 both in the Decision Tree and the Random Forest classifier (`max_depth = 5`). In the latter, the number of estimators is set to 10 (`n_estimators = 10`), and the maximum number of features is set to 1 (`max_features = 1`). In the MLP classifier, settings include L2 penalty (regularization term) parameter $\alpha = 1$. In XGBoost classifier, the parameters are set as follows: `n_estimators = 100`, `learning_rate = 1.0`, `max_depth = 1`, `random_state = 0`.

Each classifier is trained on a balanced version of the training dataset to effectively compare performance while addressing the class imbalance. Two options are considered for this: Balanced bagging versus SMOTE. We note that the balanced bagging effectively provides an ensemble method with each of the ten classifiers acting as the base classifier. While balanced bagging undersamples, SMOTE (Synthetic Minority Over-sampling Technique) oversamples [5].

The classification performance is evaluated using accuracy and F1 score. Accuracy evaluates the number of correct predictions over the total number of predictions, whereas the F1 score = 2·(precision·recall)/(precision+recall), where $precision = \frac{TP}{TP+FP}$ and recall/sensitivity = $\frac{TP}{TP+FN}$; TP, FP, and FN refer to the number of true positives, false positives, and false negatives, respectively. We note that in this multi-class setting, the F1 score is a micro-average; that is, contributions of all classes are aggregated to compute an average metric.

4 Experiments and Results

4.1 Experimental Setup

We describe three sets of experiments. First, we evaluate the performance of various clustering algorithms that allow us to learn user behavior groups (categories) in an unsupervised manner. Comparative analysis shows the most effective approach that we employ for getting the associated behavior category user labels for training. Second, we conduct a detailed selective analysis on user feature sets to choose the most relevant set for the identified behavior classes. Third, we show multi-class classification models that learn the relationship between features and the behavior classes. A principled comparison of these models along several performance metrics is presented.

Dataset: Our dataset contains records of retweets between May 16th 2016 and Dec 31st 2017 provided by [8]. Each record is a retweet of a tweet that contains at least one link to an article, which can be either a claim or a fact-checking source. The dataset consists of $20,987,210$ retweets, with $19,917,712$ (95%) linking to claim articles (fake) and $1,069,498$ (5%) to fact-checking articles (fact). We randomly sample $5,000$ users participating in retweet cascades over which we identify behavior categories via clustering to consider as the labels (inferred ground truth) for users.

4.2   Visualization of User Behavior Categories

The three key behavior dimensions introduced in Sect. 3 are used to visualize the user behavior space shown in Fig. 1(a). We observe groups of users who receive but block major fake over fact cascades, users that propagate more fakes than facts cascades, and users that act somewhere in between. These observations can be quantified via automatic groupings of users, as done via clustering algorithms. Figure 1(b)–(c) shows the three user behavior clusters detected by Agglomerative and kmeans clustering algorithms with the best performances and Table 1 shows their clustering performance metrics as described in Sect. 3).

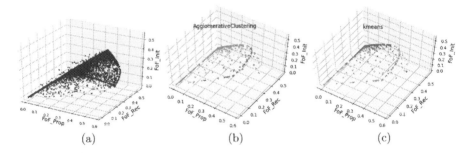

Fig. 1. (a) Visualization of 3D user behavior space. Each dimension represents the log of ratio of user reactions (initiation, propagation, or reception) to fake over fact cascade exposures. The results of the Agglomerative clustering and kmeans clustering are shown in (b) and (c), respectively. Different colors show the emergent user clusters. (Color figure online)

We call the three user categories detected in the semantic space of user behaviors as `malicious`, `good`, and `vulnerable/naive users`, based on their locality: `good user` has insignificant participation in the spread of fakes, `malicious user` who participate significantly in spreading or initiating fake over facts; and vulnerable user who mostly have lower rate of fakes to facts propagation and higher fake to fact reception. These are users who have volatile reactions in terms of behavior of reception, initiation, and propagation of fakes through the network.

Table 1. Comparing the performance of clustering algorithms using evaluation metrics.

Clustering algorithm	Silhouette	Calinski-Harabaz	Davies-Bouldin
kmeans	0.88	22913.55	0.30
Agglomerative clustering	0.87	20951.39	0.32

4.3   Feature Ranking and Selection

In this section, we evaluate the significance of each feature using F-value, p-value, and MI measure calculated as described in Sect. 3 for the user behavior categories obtained via clustering. Table 2 shows results computed over user behavior categories obtained via Agglomerative clustering. Features with MI measure above 0.5 (important ones) are highlighted in bold. (Results computed over categories obtained via kmeans are similar and omitted due to space limit).

Table 2. The F-value, p-value, and MI measure for features computed over user behavior categories obtained via Agglomerative clustering.

Feature	F-value	MI	p-value
influence score	0.13	**0.68**	0
betweenness	0.00	0.04	0.45
deg centrality	0.07	0.06	0
clustering coefficient	0.00	0.00	0.1
load centrality	0.00	0.00	0.45
followers_count	0.01	**0.55**	0
friends_count	0.02	0.38	0
listed_count	0.00	0.39	0.02
statuses_count	0.01	**0.60**	0
has_url	0.03	0.16	0
tweet character length	0.16	**1.00**	0
sentiment	0.01	0.44	0
sociability	0.08	**0.65**	0
favorability	0.04	**0.66**	0
survivability	1.00	**0.67**	0
activeness	0.01	**0.62**	0

Feature selection by recursive feature elimination (RFE algorithm) ranks the following top five features as the most significant within each user behavior category obtained by Agglomerative clustering: *followers_count, friends_count, statuses_count, survivability* and *tweet_character_length*; And kmeans clustering: *followers_count, friends_count, listed_count, statuses_count, survivability*. We note great agreement between these two sets and the features with MI measure higher than 0.5 shown in Table 2. Further, besides belonging to all feature types - content, user profile, and interaction network, visual comparisons of feature distributions within each group can be found here: https://drive.google.com/drive/folders/1B5-xVFMK9y6yW20GmpCY8BvcltFGMGMV.

4.4 Comparative Analysis of Predictive User Behavior Models

We examined the performance of various multi-class prediction models learned using aforementioned top features for classification of three user behavior categories obtained via both kmeans and agglomerative clusterings. A representative dataset of 5, 000 sampled users over which clustering is performed to obtain labels is subjected to both 10-fold cross validation (CV) and a $60 - 40$ split strategy for train-test sets. Table 3 shows the performance of 10 different classifiers on the test set (and average over 10 folds); the classifiers are trained over a balanced version of the training set, where we address class imbalance using balanced bagging and SMOTE sampling methods. As Table 3 shows, the majority of the classifiers saturate in performance around 0.80 in both accuracy and F1 score, highlighting the scope of further improvement in predicting user behavior.

Overall, the good performance belongs to MLP and SVM with balanced bagging and also k-NN classifier with balanced training set using SMOTE (n-nearest = 3). We also had additional experiments (results omitted due to space

Table 3. Comparison of performance of classification approaches in terms of Accuracy and F1 score for kmeans clustering labels for both 10-fold CV and split-strategy (in brackets). Notations: *BB-Acc(10-CV (Split)): Balanced Bagging Accuracy, BB-F1: Balanced Bagging F1, SMOTE-Acc: Synthetic Minority Oversampling Technique (SMOTE) Accuracy, SMOTE-F1: SMOTE F1-score.*

Classifier	BB-Acc	BB-F1	SMOTE Acc	SMOTE F1
3-NN	0.52 (0.52)	0.75 (0.51)	**0.76** (0.74)	0.76 (0.74)
RBF SVM	**0.80** (0.52)	0.81 (0.79)	0.53 (0.54)	0.53 (0.54)
Decision tree	0.50 (0.55)	0.79 (0.52)	0.59 (0.58)	0.59 (0.58)
Random forest	0.60 (0.54)	0.81 (0.61)	0.57 (0.56)	0.58 (0.56)
MLP	**0.80** (0.52)	0.81 (0.79)	0.53 (0.51)	0.53 (0.51)
AdaBoost	0.50 (0.36)	0.79 (0.49)	0.58 (0.57)	0.58 (0.57)
XGBoost	0.53 (0.54)	0.80 (0.54)	0.60 (0.59)	0.60 (0.59)
Gaussian NB	0.77 (0.38)	0.79 (0.76)	0.53 (0.54)	0.53 (0.54)
QDA	0.53 (0.42)	0.61 (0.62)	0.45 (0.35)	0.45 (0.35)

limit) for imbalanced settings and found both accuracy and F1 score reaching up to 0.80 for 10-fold CV. These results provide a preliminary evidence that it is possible to build user behavior predictive models that can be further improved with larger datasets or more features, by exploiting information available in near real-time for mitigation strategies.

5   Conclusion and Future Work

In this paper we have presented a novel, data-driven approach for user behavior analysis on social web for assisting fake content mitigation strategies. The identification of key behavior dimensions allows leveraging unsupervised learning to organize users along behavior categories. We identified diverse features from user information that is available in near realtime to validate predictability of user behavior categories. Supervised learning models show that user behavior categories can be predicted from such features. Although, we acknowledge the limitation of the experiments, in particular, the approach to data sampling and extracted features. Given this preliminary foundation work for user modeling to serve user intervention strategies, we will address these limitations in our future work. Furthermore, behavioral psychologists can contribute detailed models of user behavior that can inform or refine the presented data-driven modeling approach. This research provides the groundwork for advanced user modeling toward mitigation-focused social cyber-security research.

References

1. Al-khateeb, S., Hussain, M.N., Agarwal, N.: Social cyber forensics approach to study Twitter's and Blogs' influence on propaganda campaigns. In: Lee, D., Lin, Y.-R., Osgood, N., Thomson, R. (eds.) SBP-BRiMS 2017. LNCS, vol. 10354, pp. 108–113. Springer, Cham (2017). https://doi.org/10.1007/978-3-319-60240-0_13
2. Allcott, H., Gentzkow, M.: Social media and fake news in the 2016 election. J. Econ. Perspect. **3**(2), 211–236 (2017)
3. Bock, K., Shannon, S., Movahedi, Y., Cukier, M.: Application of routine activity theory to cyber intrusion location and time. In: 2017 13th European Dependable Computing Conference (EDCC), pp. 139–146 (2017)
4. Carley, K.M., Cervone, G., Agarwal, N., Liu, H.: Social cyber-security. In: Thomson, R., Dancy, C., Hyder, A., Bisgin, H. (eds.) SBP-BRiMS 2018. LNCS, vol. 10899, pp. 389–394. Springer, Cham (2018). https://doi.org/10.1007/978-3-319-93372-6_42
5. Chawla, N.V., Bowyer, K.W., Hall, L.O., Kegelmeyer, W.P.: SMOTE: synthetic minority over-sampling technique. J. Artif. Intell. Res. **16**, 321–357 (2002)
6. Farajtabar, M., et al.: Fake news mitigation via point process based intervention. In: 34th International Conference on Machine Learning, pp. 1097–1106. JMLR.org (2017)
7. Ferrara, E., Varol, O., Davis, C., Menczer, F., Flammini, A.: The rise of social bots. Commun. ACM **59**(7), 96–104 (2016)
8. Hui, P.M., Shao, C., Flammini, A., Menczer, F., Ciampaglia, G.L.: The Hoaxy misinformation and fact-checking diffusion network. In: ICWSM (2018)

9. Jiang, M., Cui, P., Faloutsos, C.: Suspicious behavior detection: current trends and future directions. IEEE Intell. Syst. **31**(1), 31–39 (2016)

10. Jin, L., Chen, Y., Wang, T., Hui, P., Vasilakos, A.V.: Understanding user behavior in online social networks: a survey. IEEE Commun. Mag. **51**(9), 144–150 (2013)

11. Purohit, H., Pandey, R.: Intent mining for the good, bad, and ugly use of social web: concepts, methods, and challenges. In: Agarwal, N., Dokoohaki, N., Tokdemir, S. (eds.) Emerging Research Challenges and Opportunities in Computational Social Network Analysis and Mining. LNSN, pp. 3–18. Springer, Cham (2019). https://doi.org/10.1007/978-3-319-94105-9_1

12. Sameki, M., Zhang, T., Ding, L., Betke, M., Gurari, D.: Crowd-o-meter: predicting if a person is vulnerable to believe political claims. In: HCOMP, pp. 157–166 (2017)

13. Saxena, A., et al.: A review of clustering techniques and developments. Neurocomputing **267**, 664–681 (2017)

14. Shu, K., Sliva, A., Wang, S., Tang, J., Liu, H.: Fake news detection on social media: a data mining perspective. ACM SIGKDD Explor. Newsl. **19**(1), 22–36 (2017)

15. Starbird, K.: Examining the alternative media ecosystem through the production of alternative narratives of mass shooting events on Twitter. In: ICWSM, pp. 230–239 (2017)

16. Varol, O., Ferrara, E., Davis, C.A., Menczer, F., Flammini, A.: Online human-bot interactions: detection, estimation, and characterization. In: ICWSM, pp. 280–289 (2017)

17. Viviani, M., Pasi, G.: Credibility in social media: opinions, news, and health information–a survey. Wiley Interdisc. Rev.: Data Min. Knowl. Disc. **7**(5), e1209 (2017)

Effect of E-Cigarette Use and Social Network on Smoking Behavior Change: An Agent-Based Model of E-Cigarette and Cigarette Interaction

Yang Qin[1]([✉]), Rojiemiahd Edjoc[2], and Nathaniel D. Osgood[1]

[1] University of Saskatchewan, Saskatoon, Canada
{yang.qin,nathaniel.osgood}@usask.ca
[2] Statistics Canada, Ottawa, Canada
rojiemiahd.edjoc@canada.ca

Abstract. Despite a general reduction in smoking in many areas of the developed world, it remains one of the biggest public health threats. As an alternative to tobacco, the use of electronic cigarettes (ECig) has been increased dramatically over the last decade. ECig use is hypothesized to impact smoking behavior through several pathways, not only as a means of quitting cigarettes and lowering risk of relapse, but also as both an alternative nicotine delivery device to cigarettes, as a visible use of nicotine that can lead to imitative behavior in the form of smoking, and as a gateway nicotine delivery technology that can build high levels of nicotine tolerance and pave the way for initiation of smoking. Evidence regarding the effect of ECig use on smoking behavior change remains inconclusive. To address these challenges, we built an agent-based model (ABM) of smoking and ECig use to examine the effects of ECig use on smoking behavior change. The impact of social network (SN) on the initiation of smoking and ECig use were also explored. Findings from the simulation suggest that the use of ECig generates substantially lower prevalence of current smoker (PCS), which demonstrates the potential for reducing smoking and lowering the risk of relapse. The effects of proximity-based influences within SN increases the prevalence of current ECig user (PCEU). The model also suggests the importance of improved understanding of drivers in cessation and relapse in ECig use, in light of findings that such aspects of behavior change may notably influence smoking behavior change and burden.

Keywords: E-cigarette · Smoking · Agent-based modeling · Distance-based network

1 Introduction

Smoking and secondhand smoke harm nearly every organ of the body and contribute to many preventable diseases, including lung cancer, coronary heart disease, chronic obstructive pulmonary disease, and other cardiovascular diseases

© Springer Nature Switzerland AG 2019
R. Thomson et al. (Eds.): SBP-BRiMS 2019, LNCS 11549, pp. 245–255, 2019.
https://doi.org/10.1007/978-3-030-21741-9_25

[6,17]. Nicotine products come in various forms, e.g., cigarettes, nicotine gum, patch, and ECig [7]. ECigs, vaporizing a liquid mixture which is used as a substitute for tobacco leaves and stored inside cartridges [13,16], were introduced to the market in 2003, promoted and marketed by major tobacco companies in the last decade [9,13]. The use of ECig as a cigarette alternative has increased dramatically. The PCEU among US adults increased from 0.3% in 2010 to 6.8% in 2013 [8]. Within recent years, there has been a particularly dramatic and alarming rise in the use of ECig amongst youth.

The health behaviors associated with smoking have been studied in detail. The majority of smokers attempt to quit smoking, but fewer than 5% of them remain quit for more than three months [5]. Effective tools for smoking cessation (SC) may help current smoker (CS) quit, and forestall an individual at risk of smoking, e.g., former smoker (FS), struggling with avoiding relapse. ECigs also allow never smoker (NS) seeking to experiment with nicotine as an alternative to cigarettes. The rise of ECig use is associated with a perception that ECig is safer than cigarettes and a useful SC device. However, there remains little solid scientific evidence confirming the effectiveness and safeness of ECig as a SC tool [8,20]. By surveying 2028 US smokers in 2012 and 2014 and two-years of follow-up, Zhuang et al. [20] concluded that long-term ECig users had a higher rate of SC of 42.4% than short-term Ecig users and non-users (14.2% and 15.6%, respectively). Zhu et al. [19] concluded that ECig users have a higher rate of SC, and are more likely to remain quit than non-ECig users. Cherng et al. [8] proposed an ABM to exmaine the effect of ECig on the smoking prevalence of US adults, and concluded that the simulated effects of ECig on SC largely changed smoking behavior. The ABM simulated the influences of smoking behavior on ECig use initiation and cessation, and how ECig reversely affected SC and smoking initiation SI.

While promising, previous studies have predominantly relied upon self reported surveys, cohort studies and clinical trials. Such larger studies are expensive, are associated with high delay until they show effect, and can be difficult to plan and execute given the wide variety of patterns of behavior possible (e.g., initiation of exclusive smoking following ECig use, initiation of exclusive ECig use following tobacco, dual use, start of ECig use following quitting tobacco, etc.). Clinical trials often regulate or exclude factors that play a key role in shaping outcomes in society, such as switching of nicotine delivery modality, varying rates of compliance, and peer influence effects.

In this paper, extending the preliminary model structure introduced by Cherng et al. [8], we build an ABM of smoking and ECig use with modalities of initiation, cessation, and relapse to examine the effects of ECig use on individual-level smoking behavior change and population-level smoking patterns according to the aggregation of individual outcomes. Our model incorporates strong SN effects involving both selection of networks and influence over networks, age, sex and history-dependent effects regarding the rate of initiation, cessation, and relapse for both smoking and ECig use, and individual decision-making effects based on characteristics of social contacts. In particular, we use the model to investigate whether the ECig is an effective SC device and the impact of ECigs on non-smokers with regards to SI.

2   Methods

Model Overview. ABM can simulate complex social dynamics and behaviors with considerably high resolution, and generate population-level results by aggregating individual outcomes in different scenarios [8]. Equally notable, ABM is widely applied to probe the impacts of counter-factual interventions, as well as to help prioritize data collection in a complex milieu of complex interactions of behaviors and product types. In this study, a high level of heterogeneity characterizing both exogenous and endogenous components, specific traits at individual level and modularity also strongly motivated the use of ABM.

Our model was built in AnyLogic (version 8.3.3), and used four interacting statecharts for each agent, featuring smoking states, ECig use states, birth, and mortality. The parameters, transition rates and statecharts in the Person class serve as influences from within an agent on smoking and ECig use behavior. The model further incorporates a distance-based network to simulate social contacts between agents.

The model simulates a population of 100,000 agents with age distribution based on population pyramid of Canada [1]. The model time unit is 1yr, and the length of the time horizon is 70yrs. The initial states may misestimate the prevalence of each smoking and ECig use state, so a period of burn time (52 years) is used for the model to achieve equilibrium. Over the continuous time of the simulation, agents either maintain their current state of smoking and ECig use or transit to other status based the (hazard) rates discussed in the next section.

Model Formulation. Smoking statechart describes three smoking states: never smoker (NS), CS and FS. An individual can switch its presence in each of the three states of statechart according to specified transition rates, namely the rate of SI, the rate of SC, and the rate of smoking relapse (SR).

ECig use statechart separates the states of ECig use as never ECig user (NEU), CEU and former ECig user (FEU). The transition of ECig use initiation (ECigUI) is fired with a hazard rate, transferring an agent from NEU to CEU. Other transitions are message triggered transitions, which will be activated only under scenarios when we consider: A CS who never used ECig may possibly initiate ECig use after quitting smoking, transiting from CS∧NEU to FS∧CEU by chance; An FS who is CEU may possibly quitting ECig after relapse to smoking, transferring from FS∧CEU to CS∧FEU by chance; And a CS who is FEU may possibly relapse to CEU after quitting smoking, transiting from CS∧FEU to FS∧CEU by chance. For the two statecharts, agents can occupy a specific, concrete state of one statechart at any one time, while being in any state of the other statechart.

Rate of SI, SC and SR denoted as r_{si}, r_{sc} and r_{sr}, respectively, are each the product of its corresponding hazard rate (α_{si}, α_{sc} and α_{sr} for the calculation of

r_{si}, r_{sc} and r_{sr}, respectively), a multiplier (m_{si}, m_{sc} and m_{sr} for the calculation of r_{si}, r_{sc} and r_{sr}, respectively) and a coefficient (e_{si}, e_{sc} and e_{sr} for the calculation of r_{si}, r_{sc} and r_{sr}, respectively).

The hazard rates reflect the magnitude of the effect of age, gender and smoking history on r_{si}, r_{sc} and r_{sr}. We transformed the annual probabilities of SI and SC (p_{si} and p_{sc}, respectively) of male and female of 1970 birth cohort, reported by Holford et al. [10], into their corresponding α_{si} and α_{sc} as table functions in AnyLogic by using $p = 1 - e^{-\alpha}$. The model assumed that α_{sr} declines with growing time since quit; thus, individuals who only recently quit have far higher relapse risk than an agent who has remained as FS for a prolonged period. The value of multipliers is driven by the state of ECig use. Wills et al. [18] suggested that NS who tried ECig is three times more likely to start smoking. Leventhal et al. [11] reported that ECig users were four times likely to uptake cigarettes. McRobbie et al. [12] suggested that the rate for SC was significantly higher in the presence of ECig use (RR 2.3; 95% CI: 1.05–4.96). Based on the linkages between ECig use and r_{si} and r_{sc} mentioned above. As ECig use can help relieve the symptoms of nicotine withdrawal to some degree and might provide an additional avenue towards continued socialization with companions who remain tobacco users, CEUs are less likely to relapse in smoking, compared to non-ECig users [14]. Therefore the model assumes m_{si} is 4.0 for agents who are CS∧CEU [11], or is 2.87 for agents who are FEU [18], m_{sc} is 2.3 for agents who are CS∧CEU [12], m_{sr} is 0.5 for agents who are CS∧CEU [14]. If each rate is only the product of its hazard rate and a multiplier, the rate may misestimate the projection of smoking. Therefore, the coefficients e_{si}, e_{sc} and e_{sr} were calibrated to match simulation outcomes against historical data.

For the rate of ECigUI, the model adapted the time-based sigmoid function and divisors introduced by Cherng et al. [8], to characterize the increasing use of ECig after its introduction into the market and the influence of smoking status on ECigUI. Additionally, the rate of ECigUI is strongly related to the agent's smoking status and demographic factors [15], suggesting that ECig is popular in smokers and young people; thus, we assumed an hazard of ECigUI of male agents using a table function, which has a x-axis of age of the agents and y-axis of the hazard rate and follows same pattern as for the hazard rate of SI for male. If an individual is female, the hazard of ECigUI of this agent is given by the corresponding point on the table function divided by the variable divECigFemale with a value of 1.5. The overall rate of ECigUI is the product of the hazard of ECigUI given by the time-based sigmoid function [8] and a coefficient (e_{ECig}), which was calibrated by matching model generated incidence of ECig use against corresponding historical data.

The model assumes that the transition of ECig use cessation (ECigUC) and ECig use relapse (ECigUR) are affected only by the smoking behavior, that is, the model assumed that individuals who are CEUs or FEUs would remain so unless changes occurred in their smoking behavior. Specifically, in the absence of identified evidence with respect to the fraction of individuals whose state of ECig use will be affected by smoking behavior, the model posited that 85% of CEUs∧FSs will quit ECig if they relapse to smoking, since their nicotine cravings were satisfied by smoking, and 80% of agents who are FEU∧CS will transit to CEU if they quit smoking. As ECigs may be used as cessation tools, the model further assumed that 50% of smoking quitters would uptake ECig immediately after quitting smoking. Therefore, message dichotomous branching transitions were built for ECigUI and ECigUC under these assumptions in addition to the rate of ECigUI discussed above.

Age-specific birth and mortality rates drawn from Statistics Canada of 2016 [3,4] are used in the model. The total fertility rate of Canada in 2016 is 1.54 per woman. To maintain population replacement (with a total fertility rate of 2.1) for successive years of the model running, we thus multiplied a coefficient (with a value of 1.357) by the fertility rate of each age group.

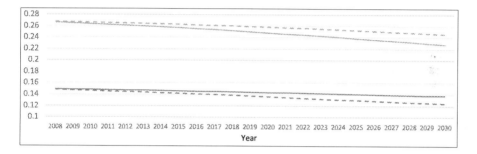

Fig. 1. PCS and PFS of Scn1 and Scn2. Orange and blue solid line represent the PFS and PCS in Scn1, respectively. Orange and blue dashed line represents the PFS and PCS in Scn2, respectively. (Color figure online)

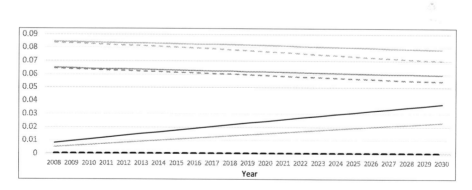

Fig. 2. PFCS, PMCS, PFCEU and PMCEU of Scn1 and Scn2. Blue, orange, green, and black solid line represent the PFCS and PMCS of Scn1, PFCEU and PMCEU of Scn2, respectively. Blue, orange, green, and black dashed line represent the PFCS and PMCS of Scn2, PFCEU and PMCEU of Scn1. (Color figure online)

Smoking is well recognized as both an individual habit and a social phenomenon [5]. The baseline model was extended with a distance based network to simulate the effect of social connection and peer pressure on the SI and ECigUI. To build a localized SN for each agent, connecting with its nearby agents, the model assumed that an agent establishes the network with the agents in proximity (50 m). The SN was implemented as a dynamic network driven by agent mobility in continuous space with width and height both equal to 250,000 m. Specifically, the agent moves to a new location within the space, and disconnects from the current network then re-establish a network based on agents layout by using a cyclic timeout event with an interval of 2 yrs. As dynamic network, the fraction of CS and CEUs among its connected agents are modified with the change of the SN, therefore, influence the effect of SN on SI and ECigUI.

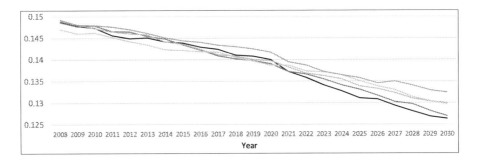

Fig. 3. SA of rate of ECigUC on PCS. Black, red, grey, yellow, and blue line represent a successively larger rate of ECigUC of 0.2, 0.4, 0.6, 0.8, and 1.0, respectively. (Color figure online)

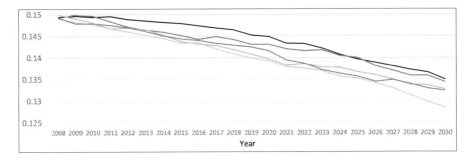

Fig. 4. SA of rate of ECigUR on PCS. Black, red, grey, yellow, blue line represent a successively larger rate of ECigUR of 0.2, 0.4, 0.6, 0.8, 1.0, respectively. (Color figure online)

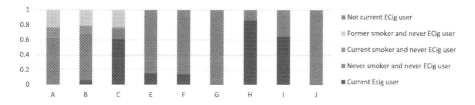

Fig. 5. Panels A, B and C depict the population breakdown by smoking category in Scn1, Scn2, and Scn3, respectively. Panels E, F, G, H, I and J illustrate fraction of CEU among CS, among FS and among NS in Scn2 and those fractions in Scn3, respectively.

The effect of SN was modeled using multipliers (m_{net}), and applied them to the baseline r_{si} and r_{ECig}, respectively. The overall rate of SI and ECigUI of a particular agent were increased by m_{net}, relative to the rates in the baseline scenario. Without a specific mathematical model to quantify the effects of connected neighbors of a particular agent, the model employed a sigmoid function to describe the progression of the influence from the connected neighbors, which increases small at the beginning then accelerates fast and reaches the plateau. We, therefore, assumed m_{net} follows a sigmoid function (Eq. 1), where f is the fraction of CS (or, correspondingly, CEUs) among its connected agents if m_{net} is used to calculate the rate of SI or ECigUI, respectively, and f_0, α and γ in Eq. 1 are 0.25, 2.0 and 1.0, respectively. Similarly, in Eq. 1, r_{average} represents the average rate of SI or rate of ECigUI of population for the calculation of rate of SI or ECigUI of this agent, respectively. The r_{average}s are re-calculated every year based on the smoking and ECig use status of the population at the beginning of each year.

$$m_{\text{net}} = \frac{\alpha + e^{-\gamma \times (f - f_0)}}{(1 + e^{-\gamma \times (f - f_0)}) \times r_{\text{average}}} \tag{1}$$

Model Calibration. e_{si}, e_{sc}, e_{sr} and e_{ECig} were calibrated to match the estimated PCS, prevalence of former smoker (PFS) and PCEU generated by the rates (r_{si}, r_{sc}, r_{sr} and r_{ECig}) in the baseline model, against historical data of 2013–2017 from CTADS [2]. The calibrated result of e_{si}, e_{sc}, e_{sr} and e_{ECig} is 1.088, 2.435, 1.51 and 7.898, respectively.

Model Scenarios. We examine here simulated population-level smoking behaviour change and ECig use under following three scenarios: smoking behavior in scenario one (Scn1) which is in absence of ECig use and the SN, smoking behavior in scenario two (Scn2) which is under the use of ECig, and smoking behavior in scenario three (Scn3) which SN exists and supports the SI and ECigUI (Scn3). The outputs from these scenarios examined the difference in prevalence and incidence of smoking arising from considering ECigs as well as SN both separately and in combination. The simulation of Scn1 and Scn2 were run for 100 realizations, and simulation under Scn3 were run for 40 realizations with respect to the considerably large computation of SN in AnyLogic, with random seeds making each simulation run unique, then the means of the outputs of all runs were calculated for the comparison. Furthermore, to examine the statistical significance between the results from Scn1 and Scn2, we performed a Mann-Whitney-U test on the per-realization output (PCS), from the two scenarios.

Sensitivity Analysis. To assess the sensitivity of model parameters on model outputs, we performed sensitivity analysis (SA) on the parameters such as the rate of ECigUC and the rate of ECigUR. The message transitions for ECigUC and ECigUR were replaced by the rate transitions. In the SA of the rate of ECigUC, the model assumed the rate of ECigUR is 1.0, and the range of the rate of ECigUC was 0.2 to 1.0 with a step of 0.2 for each iteration. Similarly, for the SA of the rate of ECigUR, the model assumed the rate of ECigUC is 1.0 and the rate of ECigUR had the same range and step with the rate of ECigUC in its SA experiment. The SA experiments examined the potential change of PCS resulting from changes in the value of the rate of ECigUC and the rate of ECigUR.

3   Results

Comparison between Scn1 and Scn2. Mean, median and standard deviation of the results for PCS, generated by the model realizations in Scn1, are 0.1438, 0.1440 and 0.0037, respectively, and those from the model realizations in Scn2 are 0.1369, 0.1374 and 0.0074, respectively. The results of a two-sided Mann-Whitney-U test for the results of two scenarios, $p < 2.2e^{-16}$, demonstrates that the distributions in the results of two scenarios differed significantly.

The message transitions in ECig use statechart were disabled in Scn1 and Scn2, therefore, in the stacked column chart showing the breakdown by smoking category (Figs. 5A, B and C), the agents were divided into four categories: CEU regardless of their smoking status, NS∧NEU, CS∧NEU, and FS∧NEU, respectively. The portion of FS and NEU (23%) in Scn1 is slightly higher than that (21%) in Scn2, due to a large portion (6%) of CEU, as shown in Figs. 5A and B. This reflects the fact that the FS in Scn1 is located within the FS∧NEU category, whereas in Scn2 some of those individuals are located within the CEU category.

Comparison between Scn2 and Scn3. In Scn3, at the end of simulation, the maximum and minimum degree centrality of a given agent is 2 and 1, respectively. With the presence of SN (in Scn3), as shown in Fig. 5C, the fraction of CEU in the population increased dramatically – rising from 6% in Scn2 to 61% in Scn3. With the exposure to ECig use from connected individuals or neighbors, people tend to initiate ECig use. The increased portion of CEU are mostly from the agents who were NS∧NEU. In Scn2, the fraction of CEU among CS and that among FS are similar, with value of 15% and 14% in Figs. 5E and F, respectively, which are considerably larger than fraction of CEU among NS, as shown in Fig. 5G. In Scn3, the SN significantly increased the fraction of CEU among CS and FS, with the value of 86% and 65%, respectively in Figs. 5H and I, while the fraction of CEU among NS does not show obvious increase due to SN, compared with that of Scn2.

Sensitivity Analysis. Results from the SA on the rate of ECigUC and the rate of ECigUR suggests that the PCEU and prevalence of former ECig user can substantially change the PCS, as shown in Figs. 3 and 4. Figure 3 demonstrates that when ECR is increased from 0.2 to 1.0 – holding invariant the value of ERR – PCS are gradually increased, and PCEU decreases. The results in Fig. 3 suggest that although incidence of SI is reduced by the lower PCEU, the decreased rate of SC and elevation in SR due to the decreased PCEU compensates for the decrease in the rate of SI. Similarly, the change in the rate of ECigUR also influences PCEU. Holding constant the rate of ECigUC, an increase in the rate of ECigUR generally increases the PCEU, but lowers the PCS, with a possible exception at the lowest levels of the rate of ECigUR. Results in Fig. 4, the line from the rate of ECigUR of 0.2 having the lowest PCEU, reflect that agents were more likely to remain as CS.

4 Discussion

From the results in the three scenarios, the model demonstrates that ECig use and SN encourage agents to uptake ECig, therefore, shape population-level smoking behavior. Although the use of ECig increases the rate of SI, the combined effect of the increase in the rate of SC and the decrease in the rate of SR results a considerably large decline in PCS and increase in PFS. The results of SA further shows the PCS is sensitive to the ECig use behavior change. The outputs of the model largely depend on the feedback between smoking and ECig use, and interactions between agents. First, we assumed the rate of ECigUI of CS, FS and NS are in a declining order, specifically, the CS has highest rate of ECigUI compared with other smoking category. Second, if an individual is CS, being a CEU increases the probability of quitting smoking and staying in FS state, which means they have a relatively higher probability of using ECig as a SC tool. We assumed the ECig use helps greatly in SI for NS. Given the model results, fraction of CEU among NS is considerably lower, compared with CS and FS. Accordingly, as a combined result of the rate of SI, the rate of SC and rate of SR, the PCS is decreased due to ECig use. Furthermore, we assumed gender effect as divisors in the rate of ECigUI. Thus, the model behaves a relatively stronger influence from ECig use on smoking behavior. The effect of SN is modeled as a multiplier to the rate of ECigUI, which generates more CEU during simulation.

Despite fine resolution of the model, there are some limitations. First, the model is highly sensitive with the use of ECig, however, the model has no good assumption on the rate of ECigUC and the rate of ECigUR. Second, at this resolution, the model cannot capture the smoking episodes, dynamics of nicotine metabolism, allowing model to analyze whether ECig use helps in relieving nicotine cravings at fine-grained level as SC tool. Finally, the model assumes the effect of SN in a relatively simple way.

Although with some limitations, the model outcomes can provide some straightforward understanding of the complex feedback between smoking and ECig use at individual level, then allow us to analyze population-level smoking behaviour. Additionally, the model is also a useful tool for examining how SN influences smoking and ECig use, particularly among adolescents.

References

1. Population pyramids of the world from 1950 to 2100 (2017). https://www.populationpyramid.net/canada/2017/
2. Data about canadians' use of tobacco, alcohol and drugs, 30 October 2018. https://www.canada.ca/en/health-canada/services/canadian-tobacco-alcohol-drugs-survey.html
3. Geography: Canada, province or territory, 26 January 2019. https://www150.statcan.gc.ca/t1/tbl1/en/tv.action?pid=1310071001
4. Fertility: Overview, 2012 to 2016, 5 June 2018. https://www150.statcan.gc.ca/n1/pub/91-209-x/2018001/article/54956-eng.htm
5. Axtell, R., et al.: Social influences and smoking behaviour: Final report to the American Legacy Foundation (2006)
6. Center for Disease COntrol and Prevention. In: How Tobacco Smoke Causes Disease: The Biology and Behavioral Basis for Smoking-Attributable Disease: A Report of the Surgeon General (2010)
7. Chaturvedi, P., Mishra, A., Datta, S., Sinukumar, S., Joshi, P., Garg, A.: Harmful effects of nicotine. Indian J. Med. Paediatr. Oncol. **36**(1), 24 (2015)
8. Cherng, S.T., Tam, J., Christine, P.J., Meza, R.: Modeling the effects of e-cigarettes on smoking behavior. Epidemiology **27**(6), 819–826 (2016)
9. Demick, B.: A high-tech approach to getting a nicotine fix, April 2009. http://articles.latimes.com/2009/apr/25/world/fg-china-cigarettes25
10. Holford, T.R., et al.: Patterns of birth cohort-specific smoking histories, 1965–2009. Am. J. Prev. Med. **46**(2), e31–e37 (2014)
11. Leventhal, A.M., et al.: Association of electronic cigarette use with initiation of combustible tobacco product smoking in early adolescence. JAMA **314**(7), 700–707 (2015)
12. McRobbie, H., Bullen, C., Hartmann-Boyce, J., Hajek, P.: Electronic cigarettes for smoking cessation and reduction. Cochrane Database of Syst. Rev. (2014)
13. Pisinger, C., Døssing, M.: A systematic review of health effects of electronic cigarettes. Prev. Med. **69**, 248–260 (2014)
14. Polosa, R., Caponnetto, P., Morjaria, J.B., Papale, G., Campagna, D., Russo, C.: Effect of an electronic nicotine delivery device (e-cigarette) on smoking reduction and cessation: a prospective 6-month pilot study. BMC Pub. Health **11**(1), 786 (2011)
15. Reid, J.L., Rynard, V.L., Czoli, C.D., Hammond, D.: Who is using e-cigarettes in Canada? nationally representative data on the prevalence of e-cigarette use among Canadians. Prev. Med. **81**, 180–183 (2015)
16. Rom, O., Pecorelli, A., Valacchi, G., Reznick, A.Z.: Are e-cigarettes a safe and good alternative to cigarette smoking? Ann. N. Y. Acad. Sci. **1340**(1), 65–74 (2014)
17. US Department of Health and Human Services: The Health Consequences of Smoking—50 Years of Progress: A Report of the Surgeon General. US Department of Health and Human Services, Centers for Disease Control and Prevention National Center for Chronic Disease Prevention and Health Promotion Health, Office on Smoking and Health, pp. 1–36 (2014). NBK179276. ISBN 24455788
18. Wills, T.A., Sargent, J.D., Gibbons, F.X., Pagano, I., Schweitzer, R.: E-cigarette use is differentially related to smoking onset among lower risk adolescents. Tob. Control **26**(5), 534–539 (2016)

19. Zhu, S.H., Zhuang, Y.L., Wong, S., Cummins, S., Tedeschi, G.: E-cigarette use and associated changes in population smoking cessation: evidence from US current population surveys. BMJ **358**, j3262 (2017)
20. Zhuang, Y.L., Cummins, S.E., Sun, J.Y., Zhu, S.H.: Long-term e-cigarette use and smoking cessation: a longitudinal study with us population. Tobacco Control **25**(1), i90–i95 (2016)

Multi-scale Simulation Modeling for Prevention and Public Health Management of Diabetes in Pregnancy and Sequelae

Yang Qin[1]([✉]), Louise Freebairn[2,3,4], Jo-An Atkinson[5,6,7], Weicheng Qian[1], Anahita Safarishahrbijari[1], and Nathaniel D. Osgood[1]

[1] University of Saskatchewan, Saskatoon, Canada
{yang.qin,weicheng.qian,anahita.safarishahrbijari,
nathaniel.osgood}@usask.ca
[2] ACT Health, Canberra, Australia
Louise.Freebairn@act.gov.au
[3] The Australian Prevention Partnership Centre, Sydney, Australia
[4] University of Notre Dame, Sydney, Australia
[5] Decision Analytics, The Sax Institute, Ultimo, Australia
jo-an.atkinson@saxinstitute.org.au
[6] Sydney Medical School, University of Sydney, Sydney, Australia
[7] Translational Health Research Institute, Western Sydney University,
Sydney, Australia

Abstract. Diabetes in pregnancy (DIP) is an increasing public health priority in the Australian Capital Territory, particularly due to its impact on risk for developing Type 2 diabetes. While earlier diagnostic screening results in greater capacity for early detection and treatment, such benefits must be balanced with the greater demands this imposes on public health services. To address such planning challenges, a multi-scale hybrid simulation model of DIP was built to explore the interaction of risk factors and capture the dynamics underlying the development of DIP. The impact of interventions on health outcomes at the physiological, health service and population level is measured. Of particular central significance in the model is a compartmental model representing the underlying physiological regulation of glycemic status based on beta-cell dynamics and insulin resistance. The model also simulated the dynamics of continuous BMI evolution, glycemic status change during pregnancy and diabetes classification driven by the individual-level physiological model. We further modeled public health service pathways providing diagnosis and care for DIP to explore the optimization of resource use during service delivery. The model was extensively calibrated against empirical data.

Keywords: Gestational diabetes mellitus · Agent based model · System dynamic model · Discrete event model

© Springer Nature Switzerland AG 2019
R. Thomson et al. (Eds.): SBP-BRiMS 2019, LNCS 11549, pp. 256–265, 2019.
https://doi.org/10.1007/978-3-030-21741-9_26

1   Introduction

Gestational diabetes mellitus (GDM) is an increasing public health priority in the Australian Capital Territory (ACT), particularly on account of its impact on the risk of Type 2 Diabetes (T2DM) across the population [11, 19]. The increase of GDM is associated with increasing prevalence of risk factors including advanced maternal age [18], obesity [6], and sedentary behavior, growing GDM risk factors in those with family history of diabetes, and a growing number of residents whose ethnic background has traditionally been subject to elevated rates [11].

Mathematical models characterizing diabetes progression, glucose hemostasis, pancreatic physiology and complications related to diabetes have been built by many researchers [5, 27]. De Gaetano et al. [8] formulated a model representing the pancreatic islet compensation process, related to insulin resistance, beta-cell mass and glycemia (G) of a diabetic individual. Hardy et al. [13] proposed a model, characterizing mechanisms of anti-diabetic intervention and the corresponding impact on glucose homeostasis. Lehmann and Deutsch [20] modeled the physiology underlying the interaction between insulin sensitivity (K_{xgI}) and G of an individual with Type 1 diabetes (T1DM).

Health simulation models commonly apply one of three types of modeling techniques: system dynamics modeling (SDM), agent based modeling (ABM) and discrete event simulation (DES). SDM captures and describes complex patterns of feedback and accumulation by solving sets of differential equations. While SDM can be applied at different scales [25], it is most commonly applied at the aggregated level, and its core components include the accumulation of elements (stocks), rate (flows), causal loops involving stocks (feedback), and delays [15, 17]. By contrast, ABM simulates complex social dynamics by characterizing emergent system behavior as the result of within-environment interactions between individual elements in a system that are referred to as agents. ABM readily captures heterogeneous characteristics of agents, including agent history, situated decision making, structured interaction between agents typically evolving along multiple aspects of states and transitions and aggregation of individual outcomes [21, 25]. DES characterizes individual-level, resource-limited progression through structured workflows which often associated with service delivery, queuing processes, waiting times and lists and resource utilization [22].

Previous studies examining the health burden of GDM and its risk factors have predominantly relied upon cohort studies, administrative data or clinical trials [10, 16, 29]. While filling a key set of research needs, given the dynamically complex nature of the interactions including feedback, accumulations, delays, heterogeneity, and interacting factors across many levels, it is difficult to use such studies to answer "what-if" question related with the risk factors and effects of interventions, particularly counter-factual whose outcomes have not yet been observed. Given the long time scales involved, cost, logistics, and ethical concerns, clinical trial studies may not be feasible for providing timely evaluation of novel portfolios of clinical-level and population-level interventions (PLI).

In this work, we built a multi-scale hybrid model in AnyLogic (version 8.3.3) including SDM, ABM, and DES, to describe the dynamics of glycemic regulation (DGR), weight status and pregnancy, and to evaluate impacts of the interventions on DGR. While leaving most aspects of examination of model health findings to other forthcoming contributions, this paper introduces the design and structure of the model, provides illustrations of some of the types of interventions that the model can capture and simulation outputs.

The structure of the remainder of the paper is as follows: Model overview section describes the model structure and the simulation description. The next section briefly discusses model calibration and assumption. The model formulation section then describes the statecharts, DGR, weight dynamics, interventions, service delivery and offspring outcomes by hyperglycemia. Part 3 and 4 illustrate and discuss some of the model outputs and limitations of the model.

2   Methods

Model Overview. The Person class of the ABM includes the individual level characteristics such as evolving states, actions that change them, and the rules to trigger those actions (all captured in statecharts), parameters and functions. The ABM further represents family structure, weight at birth and evolution over the adult life course, individual history, inter-generational family context, pregnancy and diabetes classification, and implementation of the PLI. The SDM describing the DGR forms a sub-model encapsulated in the Person class. By encapsulating this SDM in the ABM structure, the model can capture side-by-side both individual characteristics and their evolution and continuous dynamics of the glucose-insulin system. The clinical service pathway for pregnant women in the ACT is described by a shared (global) DES, building on top of the ABM. The model will be discussed in its essentials in the following sections. Added elements of detail, the technical description associated with the model, are listed in the supplementary material, https://www.cs.usask.ca/faculty/ndo885/GDM-ACT, other material will be available later.

The model simulates a population of 200,000 female agents, each an instance of Person class. During the simulation, the agents can become pregnant, thereby experiencing the risk of GDM, and subsequently give birth, influencing the weight status and DGR of their descendants. The second generation agents also have their life-course shaped thoroughly by model dynamics. The information available for the descendants is, therefore, richer than in the initial population. Thus, the simulation requires a burn time of 60 years.

Model Calibration. To estimate poorly- or non-measured parameters and to support the projection of status quo future incidence of DIP using model outputs, we calibrated a baseline model without interventions against the following historical data: the incidence of DIP of each ethnicity in ACT from 2008–2016, the prevalence of macrosomia by DIP status of in ACT from 2010–2016. To further capture the effects of inter-generational transfer of risk for GDM and T2DM which has been recognized from multi-generational epidemiological studies [7,12], and occurrence of later-life diabetes, we drew on data related to developing diabetes by age 30 of offspring by their mother's G status (e.g., GDM, T2DM) and birth weight from population-wide administrative data from Saskatchewan, Canada. Birth weight can serve as an important marker of control of G in utero, and epidemiological studies suggest that it further influences the tendency towards GDM. The details of model calibration, e.g., objective function, will be available in other contributions.

Model Formulation. We discuss here several aspects of model formulation, particularly concentrating on statecharts, which encapsulate a discrete set of collectively exhaustive and mutually exclusive (lowest-level) states with respect to particular concerns, the actions by which the individual transitions between such states, and the rules under which such actions take place.

Pregnancy statechart (Fig. 1) indicates whether an agent is pregnant or not, and their transitions through different stages of pregnancy. Female agents with ages between 15 and 50 can transit between the notPregnat, planPregnant, and pregnant states. Agents in the pregnant hierarchical state will be in one of three substates, corresponding to trimesters of pregnancy. notPregnant agents will either be in the fertile or PostPartum state. PostPartum agent will either be in the breastFeeding or notBreastFeeding state. Of these, two of seven state transitions are memoryless transitions driven by a hazard rate (henceforth known as rate transitions), becomePregnant and leaveBreastFeeding; the hazard rate for becomePregnant is an age-ethnicity-specific fertility rate [3]. While the others are timeout transitions triggered after a specified residence time. The timeout transition birthTransition is particularly notable, as it introduces a new agent into the model.

Population statechart separates the population into three categories, initial female population, female descendant and male descendant. Agents are initialized with different ages [1], and assigned their ethnicity according to ACT demographic information taken from the 2011 Australian Census and National Health Survey. Type of ethnicity includes Australian Born, Australasian Diabetes in Pregnancy Society at risk group (ADIPS) [24], Aboriginal and Torres Strait Islander (ATSI) and Other. All male descendants are excluded during simulation, and female agents leave the model upon reaching age 50. The female agents aged less than 50 leave the model by the age-specific death rate [4].

Dysglycemia classification statechart (Fig. 2) divides the G of an agent into four categories: T1DM, NormoglycemicAndIGR, T2DM and GDM states, according to clinical classification categories. Agents can occupy one of four states, and switch states by checking whether the G of agents exceed the threshold of each state (known as condition transition). Reflecting the fact that residence in the GDM state is only an option during pregnancy, pregnancy status is also considered. The GDM agents will either be in NormoglycemicAndIGR or T2DM state after pregnancy. Thresholds for T2Dm and GDM state are denoted as G_{T2DM} and G_{gdm}, which are calculated by $C_{T2DM} \times G_t$ and $G_t \times C_{T2DM} \times C_{gdm}$, respectively, where C_{T2DM}, C_{gdm} and G_t are calibrated and equal to 1.636, 0.642 and 5.504, respectively.

DGR, the interaction between beta-cells, G and K_{xgI}, is represented as an SDM based on the ordinary differential equation models of diabetes progression

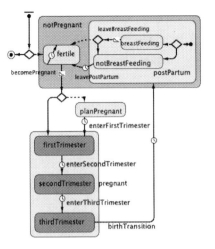

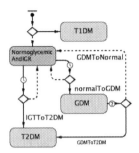

Fig. 1. Pregnancy statechart   **Fig. 2.** Dysglycemia classification statechart

by De Gaetano et al. [8,9] and Hardy et al. [13]. To improve model scalability, a cyclic timeout event with time interval (dt) is used to solve the compartmental equations in SDM [8,9,13]. Another cyclic timeout event with a time interval of 5 and 30 days during pregnant and non-pregnant periods, respectively, updates G and K_{xgI} using the Newton-Raphson method and other components in SDM. Parameter and function details are listed in the supplementary material.

To capture dynamics of K_{xgI} in different trimesters of pregnancy, postpartum and different weight status, respectively, we modified the (exogenous) equations giving K_{xgI} over time introduced by De Gaetano et al. [8,9], The model assumes the diminished K_{xgI} in pregnancy will gradually recover during the postpartum period to the value it would have held absent the pregnancy. The model further assumes the K_{xgI} of overweight and obese agents would decline faster over age than that of agents with normal weight. In addition to the modification of equations giving K_{xgI}, we modified the equation giving the spontaneous recovery rate of pancreas (T_η) for ADIPS. While ADIPS represents agents from a recognized high risk group with respect to GDM, the empirical data revealed that the ADIPS group actually had a higher proportion of healthy weight agents than that were present in the other groups, indicating that weight as a risk factor did not fully account for the higher risk levels. Therefore, to capture the high incidence of GDM of ADIPS, the model assumes the T_η of ADIPS declines faster than that of other ethnic groups. Furthermore, to investigate effects on various types of intervention on the DGR, we incorporated the mechanism introduced by Hardy et al. [13] for the impact of lifestyle change (LC), metformin treatment (MT) and insulin treatment (IT) on K_{xgI}. Elements of interventions making use of the LC, IT and MT are discussed in the next sections.

Weight dynamics are characterized as a continuous variable of BMI value, and a variable of Z-score of a BMI distribution (BMID), representing the position of

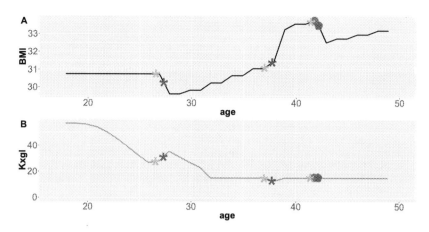

Fig. 3. Illustration of the individual trajectory of BMI change (A) and K_{xgI} change (B) over age without PLIs and Services. Green and red dots are the start and end of the GDM period, respectively. Orange and blue stars are the beginning and the end of the pregnancy, respectively. (Color figure online)

BMI within the age group (AG) specific BMID. Upon entry to adulthood, agents are assigned a BMI value based on an AG specific BMID introduced by Hayes et al. [14], and its corresponding Z-score calculated by the BMI and mean of the BMID. Hayes et al. [14] reported that the BMID of the population within AG move toward higher BMI value through their life course. Applying an identical Z-score into the BMID of different AGs may position the agents into different weight categories. Therefore, for simplicity, the Z-score of agents are assumed to stay the same as they age, unless intervention or pregnancy [16] changes their BMI value and assigns a new Z-score to them. When an agent transfers from one AG to another, the BMI value of next AG of the agent will be calculated by applying the Z-score to the BMID of next AG in an event with a cyclic timeout of 10 years. As a continuous variable, another cyclic timeout event with an interval of 1 year is used to make the BMI of current AG change towards the BMI of next AG gradually. With this BMI-Z-score mechanism, we also captured BMI change following pregnancy; due to space considerations, the interested reader is referred to the supplemental material. Time-Varying weight distribution is required in light of the simulation burn time. We, therefore, employed importance sampling using an alternative BMID of female adults aged 25 to 64 years in 1980 and 2000 [2] and AG specific BMID in 1995 and 2008 [14], to estimate the AG specific BMID in 1980.

ACT clinical service pathway, Services, is modeled using DES. And in Person, a statechart reflects type of health care that an agent is currently being delivered, which is separated as the InPrimaryCare state reflecting that a non-pregnant woman is receiving usual health care services through a general practitioner, and the InACTHealthService state reflecting that a pregnant woman is moving through the Services. The DES and statechart not only models the

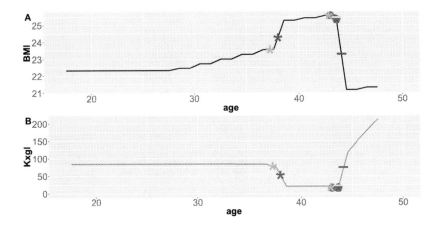

Fig. 4. Illustration of the individual trajectory of BMI change (A) and K_{xgI} change (B) over age with DRI. Color Labels of GDM and pregnancy are same with Fig. 3. Yellow and brown bars are the start and the end of the DRI, respectively. (Color figure online)

effects of LC and IT in reducing the risk of progression to T2DM after delivery and implemented PLIs in the InPrimaryCare, but also leave room for investigating resource use and costs associated with service provision. The blocks (i.e. dipAssessment, antenatalCare, dieticianReview and lifestyleOrInsulinTreatment) in the Services form a sequence of operations, providing pregnant agents the DIP assessment test, education of LC, IT for the agents with DIP, and further deliver postpartum checks to agents. Details of the DES are described in the supplemental material.

PLI includes consideration of a public health messaging and mobile app support intervention (PHMMASI), health professional support intervention (HPSI), diet review intervention (DRI), and public health messaging and support intervention (PHMSI). The difference between the Services and the PLI is that PLIs are initiated during the non-pregnant period, while the Services is triggered during pregnancy. All PLIs share a similar mechanism of taking optional LC and BMI reducing. Specifically, overweight and obese agents reduce their BMI, drawing the extent of that reduction from a normal distribution, while the normal weight agents keep their BMI invariant. The interventions are variants of each other with respect to who they are target, the intervention triggering time, and the length and strength of adherence of the LC. In detail, the agents with age between 20–35 take the PHMMASI and retake it according to certain probability [23]. The HPSI takes place at the planPregnant state, and works on women with risk factors, e.g., BMI > 28, age > 30, ADIPS ethnicity [28]. DRI and PHMSI both take place between pregnancies, and target on women who had DIP in previous pregnancies and on women who have given birth, respectively. For the HPSI and PHMSI, the adherence and length of LC are flexible, whereas the agents who are subject to DRI take mandatory, lifelong, strongly adherent LC.

The outcomes for baby and mother including birth weight (e.g., macrosomia), type of birth (e.g., Caesarean section), NICU admission, and shoulder injury, are triggered in birthTransition of the pregnancy statechart. The probability of occurrence of baby outcomes was calculated based on the study by The HAPO Study Cooperative Research Group [26]. Furthermore, information of the mother is passed on to the new child, including DIP status, age, weight status and ethnicity. The mother's DIP status influences the K_{xgI} of the child by multiplying a coefficient to the K_{xgI} calculated by the modified equations giving K_{xgI} over time.

3   Results

We show here several scenarios that demonstrate the functioning of the model at the level of an individual's health history. Figure 3A and B show the individual trajectories of BMI changes and K_{xgI} over age without the PLIs and the Services, respectively. Figure 3A illustrates the agent entered adulthood with a BMI of 30.75, and her BMI reduced over one BMI unit after the first and third pregnancy and three BMI units after the second pregnancy. Other than pregnancy, Fig. 3A also demonstrates continuous BMI change over age. The K_{xgI} remained constant under the age of 18 but declined over time according to the value of BMI after the age of 18. The agent developed GDM at the third pregnancy, as shown by the dots in Fig. 3. Furthermore, Fig. 3B reflects the decreasing K_{xgI} during pregnancy and recovery in postpartum. We can see from Fig. 3 that K_{xgI} is declining in parallel to, and, in fact, in response to the increase in BMI.

From Fig. 4, we can see that the agent increased over one BMI unit after the first pregnancy, and K_{xgI} was decreased in response to this BMI increase. At the second pregnancy, the agent developed GDM but retained their BMI and corresponding Z-score after pregnancy. But the DRI reduced that agent's BMI and significantly increased K_{xgI} from 20 to 116 at the end of BMI reduction period (6 months), following which the K_{xgI} continued to increase due to strong adherence in LC, as shown by the bar labels in Fig. 4A and B.

4   Discussion

This paper has described a novel multi-scale model that utilizes three types of system simulation methods to provide a versatile, powerful and general platform for examining interventions to address the growing epidemic of GDM and T2DM in the ACT. The model achieves such versatility by virtue of maintaining a core underlying physiological representation that captures the common generative pathways mediating diverse needs in the model, to capture effects of lifestyle and clinical interventions, to capture clinical categorization, to represent the effects of each of pregnancy, aging and BMI change, and the longer term-effects of one pregnancy (via beta-cell mass and function) on later pregnancies and subsequent material risk of T2DM, and outcomes of interest. Such a representation can also flexibly capture the impacts of maternal status on the offspring.

A high level of heterogeneity at the individual level, e.g., family context, risk factors for diabetes and life course trajectories motivated the use of ABM as the core component of this hybrid model. The ABM permits a high-resolution representation of relevant dynamics of individual objects and further allows the implementation of finely targeted interventions. Compared to ABM, SDM simulates a system in a more abstract and general way. The high level of abstraction of DGR makes it a suitable candidate for SDM. The Services can be described as a sequence of operations, DES, therefore, was selected to model the Services, and to study the resource allocation and effect of clinical interventions.

While empirical models of necessity represent simplifications of processes in the world, the model here includes a requisite degree of detail to capture a remarkably broad set of factors. Nonetheless, they remain important limitations in the model that are ripe for addressing. These notably include a lack of detail with regards to childhood dynamics (including weight change), neglect to social network effects on behavior, and an overly simple representation of changes in K_{xgI} and behavior change. Extensions of the model to capture such effects, and to capture cost and resource components of scenarios, remain an important priority.

References

1. Population pyramid 1978. https://www.populationpyramid.net/australia/1978/
2. Walls, H.L., et al.: Trends in BMI of urban Australian adults, 1980–2000. Public Health Nutr. **13**(5), 631–638 (2010)
3. ACT Maternal Perinatal Data Collection, Australian Capital Territory. Health Directorate & 2011 Census of Population and Housing. Australian Bureau of Statistics (2011)
4. ABS.Stat: Deaths, year of occurrence, age at death, age-specific death rates, sex, states, territories and Australia. http://stat.data.abs.gov.au/Index.aspx?DataSetCode=DEATHS_AGESPECIFIC_OCCURENCEYEAR
5. Ajmera, I., Swat, M., Laibe, C., Le Novère, V.C.: The impact of mathematical modeling on the understanding of diabetes and related complications. Pharmacomet. Syst. Pharmacol. **2**, 1–14 (2013)
6. Athukorala, C., Rumbold, A.R., Willson, K.J., Crowther, C.A.: The risk of adverse pregnancy outcomes in women who are overweight or obese. BMC Pregnancy Childbirth **10**(1), 56 (2010)
7. Dabelea, D., et al.: Intrauterine exposure to diabetes conveys risks for type 2 diabetes and obesity: a study of discordant sibships. Diabetes **49**(12), 2208–2211 (2000)
8. De Gaetano, A., et al.: Mathematical models of diabetes progression. Am. J. Physiol.-Endocrinol. Metab. **295**(6), E1462–E1479 (2008)
9. DeGaetano, A., Panunzi, S., Palumbo, P., Gaz, C., Hardy, T.: Data-driven modeling of diabetes progression. In: Marmarelis, V., Mitsis, G. (eds.) Data-driven Modeling for Diabetes. LNB, pp. 165–186. Springer, Heidelberg (2014). https://doi.org/10.1007/978-3-642-54464-4_8
10. Feig, D.S., Zinman, B., Wang, X., Hux, J.E.: Risk of development of diabetes mellitus after diagnosis of gestational diabetes. CMAJ **179**(3), 229–234 (2008)
11. Ferrara, A.: Increasing prevalence of gestational diabetes mellitus. Diabetes Care **30**(Supplement 2), S141–S146 (2007)

12. Franks, P.W., et al.: Gestational glucose tolerance and risk of type 2 diabetes in young pima indian offspring. Diabetes **55**(2), 460–465 (2006)
13. Hardy, T., Raddad, E., Pørksen, N., De Gaetano, A.: Evaluation of a mathematical model of diabetes progression against observations in the diabetes prevention program. Am. J. Physiol. Endocrinol. Metab. **303**, E200–E212 (2012)
14. Hayes, A., Gearon, E., Backholer, K., Bauman, A., Peeters, A.: Age-specific changes in BMI and BMI distribution among Australian adults using cross-sectional surveys from 1980 to 2008. Int. J. Obes. **39**(8), 1209–1216 (2015)
15. Homer, J.B., Hirsch, G.B.: System dynamics modeling for public health: background and opportunities. Am. J. Public Health **96**(3), 452–458 (2006)
16. Knight-Agarwal, C.R., et al.: Association of BMI and interpregnancy BMI change with birth outcomes in an australian obstetric population: a retrospective cohort study. BMJ Open **6**(5), e010667 (2016)
17. Kreuger, L.K., Osgood, N., Choi, K.: Agile design meets hybrid models: using modularity to enhance hybrid model design and use. In: 2016 Winter Simulation Conference (WSC), pp. 1428–1438, December 2016
18. Lao, T., Ho, L.F., Chan, B., Leung, W.C.: Maternal age and prevalence of gestational diabetes mellitus. Diabetes Care **29**(4), 948–949 (2006)
19. Lee, H., Jang, H.C., Park, H.K., Metzger, B.E., Cho, N.H.: Prevalence of type 2 diabetes among women with a previous history of gestational diabetes mellitus. Diabetes Res. Clin. Pract. **81**(1), 124–129 (2008)
20. Lehmann, E., Deutsch, T.: A physiological model of glucose-insulin interaction in type 1 diabetes mellitus. J. Biomed. Eng. **14**(3), 235–242 (1992). Annual Scientific Meeting
21. Luke, D.A., Stamatakis, K.A.: Systems science methods in public health: dynamics, networks, and agents. Ann. Rev. Public Health **33**(1), 357–376 (2012)
22. Marshall, D.A., et al.: Applying dynamic simulation modeling methods in health care delivery research-the simulate checklist: report of the ISPOR simulation modeling emerging good practices task force. Value Health **18**(1), 5–16 (2015)
23. Mateo, G.F., Granado-Font, E., Ferré-Grau, C., Montaña-Carreras, X.: Mobile phone apps to promote weight loss and increase physical activity: a systematic review and meta-analysis. J. Med. Internet Res. **17**(11), e253 (2015)
24. Nankervis A, et al.: ADIPS consensus guidelines for the testing and diagnosis of gestational diabetes mellitus in Australia, June 2014
25. Osgood, N.: Using traditional and agent based toolset for system dynamics: present tradeoffs and future evolution, January 2007
26. The HAPO Study Cooperative Research Group: Hyperglycemia and adverse pregnancy outcomes. Obstet. Gynecol. Surv. **63**(10), 615–616 (2008)
27. Topp, B., Promislow, K., Devries, G., Miura, R.M., Finegood, D.T.: A model of β-cell mass, insulin, and glucose kinetics: pathways to diabetes. J. Theoret. Biol. **206**(4), 605–619 (2000)
28. Weisman, C.S., et al.: Improving women's preconceptional health: long-term effects of the strong healthy women behavior change intervention in the central pennsylvania women's health study. Women's Health Issues **21**(4), 265–271 (2011)
29. Xiong, X., Saunders, L., Wang, F., Demianczuk, N.: Gestational diabetes mellitus: prevalence, risk factors, maternal and infant outcomes. Int. J. Gynecol. Obstet. **75**(3), 221–228 (2001)

Cough Detection Using Hidden Markov Models

Aydin Teyhouee$^{(\boxtimes)}$ and Nathaniel D. Osgood

University of Saskatchewan, Saskatchewan, Canada
ayt227@usask.ca, osgood@cs.usask.ca

Abstract. Respiratory infections and chronic respiratory diseases impose a heavy health burden worldwide. Coughing is one of the most common symptoms of many such infections, and can be indicative of flare-ups of chronic respiratory diseases. Whether at a clinical or public health level, the capacity to identify bouts of coughing can aid understanding of population and individual health status. Developing health monitoring models in the context of respiratory diseases and also seasonal diseases with symptoms such as cough has the potential to improve quality of life, help clinicians and public health authorities with their decisions and decrease the cost of health services. In this paper, we investigated the ability to which a simple machine learning approach in the form of Hidden Markov Models (HMMs) could be used to classify different states of coughing using univariate (with a single energy band as the input feature) and multivariate (with a multiple energy band as the input features) binned time series using both of cough data. We further used the model to distinguish cough events from other events and environmental noise. Our Hidden Markov algorithm achieved 92% AUR (Area Under Receiver Operating Characteristic Curve) in classifying coughing events in noisy environments. Moreover, comparison of univariate with multivariate HMMs suggest a high accuracy of multivariate HMMs for cough event classifications.

Keywords: Cough detection · Machine learning ·
Hidden Markov Model · Frequency domain · Time domain ·
Pattern recognition · Spectrogram · Health care · Public health

1 Introduction

Symptoms such as cough are important clinical signs. Coughing is the most common symptom in respiratory diseases, and awareness of the occurrence or persistent presence of a cough can provide valuable information to physicians. Detailed awareness of coughing can aid physicians with their treatment on the basis of quantitative assessments such as frequency or intensity as well as qualitative assessments such as dry or wet coughs [1]. Moreover, cough detection analysis has the potential to reduce the cost of health services by – for example – detecting the early signs of diseases and making preemptive diagnosis possible

© Springer Nature Switzerland AG 2019
R. Thomson et al. (Eds.): SBP-BRiMS 2019, LNCS 11549, pp. 266–276, 2019.
https://doi.org/10.1007/978-3-030-21741-9_27

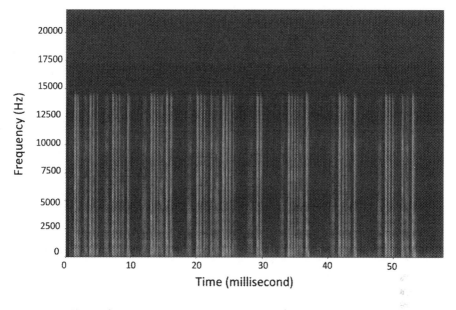

Fig. 1. A spectrogram of a sample cough (Color figure online)

and prescribing basic treatments while they are still effective [2]. However, the benefits of securing reliable, and timely quantification of coughing behavior can also offer benefits beyond the physician's office. Collecting cough data using monitoring devices such as mobile sensors or other devices and analyzing the audio signals of coughs can support remote monitoring of patients with chronic respiratory illnesses or restricted mobility. For such diseases, awareness of flare-ups of coughing can motivate the need to present for care, and can inspire changes to treatment recommendations. A final and important advantage of cough recognition resides in its potential to provide health authorities with timely surveillance information about emergence of high-burden respiratory conditions, thereby supporting earlier outbreak identification in particular geographic areas, thereby better supporting public health decision making, including the design of public health interventions.

The duration of a cough sound typically varies between 0.2 and 1 second [3], and exhibits a sequence of distinct acoustic patterns. The origin of these patterns is airway narrowing and bifurcation. The airway narrowing is due to a change in the thickness of the airflow walls (inflammation, mucus collection, bronchoconstriction and fibrosis). A typical cough sound usually is composed of three stages: an explosive expiration due to the abrupt opening of glottis, the intermediate stage in which cough sounds are reduced, and the voiced stage due to the closing of the vocal cord. There are a variety of patterns of coughing based on the presence or absence of each of these stages [4].

A visual representation of the spectrum of frequencies of a cough signal as it varies over time is shown in the spectrogram of Fig. 1, which is depicted as a heat map, with the lowest and highest intensities being represented by dark and light green, respectively.

Several studies have described methods to analyze cough characteristics, considering the subjective interpretation of cough sound recordings and the analysis of spectrograms [5–10]. There are two main research streams for cough recognition. One stream investigates audio signals frame-by-frame and combines consecutive cough frames as a cough event [11]. The second stream consists of event detection and cough classification steps. Event detection identifies cough event candidates; each candidate is then classified as a cough or non-cough event [12]. Our work follows the first stream, by seeking to detect cough signals in continuous audio recording using a Hidden Markov Model (HMM).

This paper investigated the performance of an HMM, where each state of that model corresponds to a portion of a typical cough, and where observables represent summaries of information from sound profiles. We further investigated the performance of the model in detecting each state and thus distinguishing a period of time in which a cough was occurring from when it was not. The HMM could further be used to distinguish coughing from non–coughing behaviour when considering a longer period of time, and when the main focus is to identify bouts of cough present in an sound recording events. To achieve this, the acoustic energy was selected as the observable and measurable feature to feed into a univariate HMM. In another attempt, using the frequency or pitch of the sound, the energy spectrum as the observation input was split into a vector of three sub-features as low, mid and high energy bands. Finally, we compared the performance of these two scenarios.

2   Materials and Methods

2.1   Data Collection and Labeling

The cough data used in this article is collected from recordings of cough sounds from individuals in Computational Epidemiology and Public Health Informatics Laboratory in the Department of Computer Science at the University of Saskatchewan. A duration of 20 min of such cough sounds were manually annotated by the authors.

We divided each audio signal into 25 ms time slots (bins) and extracted the following information from each bin: the time corresponding to the mid–point of each bin, the sum of the energy density of frequencies under 2 KHz (low-band energy), the sum of the energy density of frequencies between 2 KHz and 4 KHz (mid-band energy) and – finally – the sum of the energy density of frequencies between 4 KHz and 22 KHz (high-band energy). In light of the limited span of the audio frequency range, no frequencies above 22 KHz were considered. We considered the sum of energy densities as our training features for the Hidden Markov model. In this work, each cough recording was divided into five distinct states/stages, and each 25 ms time bin was labeled as to the state with which it was associated. Specifically, we considered three states inside a single cough (states A, B and C), a brief state of silence between each cough inside a bout of coughs (D) and a longer state of silence between bouts of coughs for cough-prone cases (E). Bouts of coughing were considered to trigger additional coughing (thus returning from state D to state A) with higher probability than in a general non-coughing state (state E); alternatively, a bout of coughing could then end, via a transition to state E. Figure 2 depicts different coughing states in the time domain. B contrast, a schematic diagram showing posited transitions between different coughing states is demonstrated in Fig. 3.

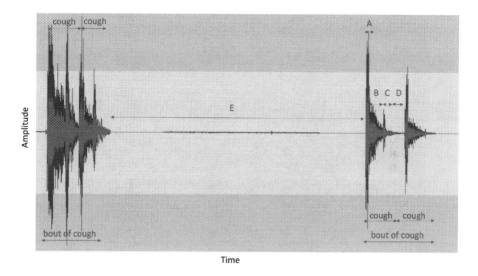

Fig. 2. Different states of coughing in an acoustic signal of four cough epochs

The length of the cough sounds vary from cough-to-cough, and the distinctions between the successive stages are not always clear – leading to imprecision in human classification of such stages. The beginning of the cough sound was used as the starting point of state A, the start of state B was selected when the sound amplitude was significantly lower than the initial peak and the start of state C was chosen when there was a rise in the sound amplitude after state B.

This work sought to investigate the effectiveness of an HMM in predicting the underlying state of a given time interval of a cough–recording by feeding our model with low, mid and high band energy-density values. Given the characteristics of a single 25 ms bin and the energy density values, we investigated the capacity of that model to predict with which state of coughing this bin was associated.

2.2   Model Training

The calculated probability for each hidden state is obtained by multiplying two values; one inferred from the observation i.e., the likelihood of observing that

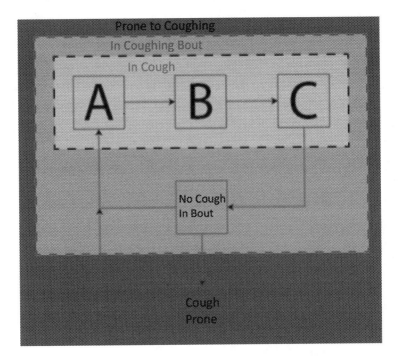

Fig. 3. Cough transitions captured in the HMM

hidden state given the current observation vector and the other one derived from the transition matrix – i.e., the probability of being in that specific state according to the probability of having been in different states in the previous time bin. The initial states' values, i.e. the probability of being in any of the hidden states in the initialization step was set assuming that the model starts in state A with the probability of 1.

2.3   Model Evaluation

We employed a two-fold cross validation approach in a repeated manner for training our model and used the average AUC – Area Under the Receiver Operating Characteristic [ROC] Curve – of the cross validation steps as the primary evaluation metrics. The confusion matrix, sensitivity, and specificity were considered to further evaluate model performance.

Since the ultimate goal is of this work to classify Cough from Non-Cough (correctly identifying an epoch of cough in a bout of coughs) or Coughing from Non-Coughing (correctly identifying a bout of coughs), we further investigated the capacity to classify audio signals according to two dichotomous categories: Cough vs. Non-Cough, and Coughing vs. Non-Coughing. To accomplish this, we grouped the states in binary format as follows:

- **Cough vs. Non-Cough**: states A, B and C were grouped in a single state of Cough and states D and E as a single state of Non-Cough
- **Coughing vs. Non-Coughing**: States A, B, C and D were grouped as sate of Coughing and E as the state of Non-Coughing.

The details of the preferred classifier will differ depending on our goals. Here, we applied the Youden's index [13] (by applying the "best" argument of the "coords" method from pROC package [14]) to maximize the sum of sensitivity and specificity. The Confusion matrix and the optimal accuracy, sensitivity and specificity are demonstrated in Sect. 3.

Transition and Emission Matrices. Table 1 shows a sample of data points extracted from cough signals. The HMM states and transitions captured the posited structure of transition ins between cough stages as shown in Fig. 3.

Table 1. Training data sample

Ground truth label	Low–band energy	Mid–band energy	High–band energy
A	31855.85	1155.99	678.39
B	5630.51	895.47	1704.09
B	9672.26	1891.19	1126.83
C	371.24	8.47	2.07
D	189.62	6.65	1.22
E	3.12	0.39	0.06
E	2.16	0.15	0.05
E	1.13	0.10	0.02

At any given time bin, the HMM can be in one of the five (hidden) states of A, B, C, D or E, resulting in the transition matrix shown as Table 2. It bears emphasis that there are no transitions between some pairs of states – for example, from A to C, or A to D; the probability of such transitions was treated as zero.

Table 2. Transition table for sample data

	A	B	C	D	E
A	$P_{A|A}$	$P_{A|B}$	0.0	0.0	0.0
B	0.0	$P_{B|B}$	$P_{B|C}$	0.0	0.0
C	0.0	0.0	$P_{C|C}$	$P_{C|D}$	0.0
D	$P_{D|A}$	0.0	0.0	$P_{D|D}$	$P_{D|E}$
E	$P_{E|A}$	0.0	0.0	0.0	$P_{E|E}$

Table 3. Performance statistics of the testing set for univariate HMM

		Observed				
		Class: A	Class: B	Class: C	Class: D	Class: E
Predicted	Class: A	31	6	1	1	7
	Class: B	3	45	19	6	25
	Class: C	3	17	29	4	5
	Class: D	3	2	31	21	19
	Class: E	13	9	65	84	714
Sensitivity		0.585	0.570	0.200	0.181	0.927
Specificity		0.986	0.951	0.971	0.947	0.565
Accuracy		0.722				

To calculate the probability P_{xy} of transition from a current state x to any of the probable states y, we first found the probability of leaving a given state to any destination. Based on the HMM assumption of memoryless transition processes, this is given by the reciprocal of the mean residence time (in time bins) within that state. For states exhibiting a single outgoing transition (states A, B, C and E), that probability was employed directly. For state D (which can be followed by either state A and state E), to arrive at the probability of making the transition to each of states A and E, we further multiplied the probability of leaving the state by the empirically observed proportion of transitions from state D to states A and E, respectively.

Since the model in this work makes use of continuous observations, instead of having an emission matrix, we used density functions extracted from and fitted to empirical observations, where the observations are assumed to be independent from each other, conditional on being in a given state. As a simplifying assumption, the joint likelihood of observing a given vector of low-band, mid-band and high-band energy quantities was approximated as the product of independent likelihood functions (each associated with a univariate probability density function). For a case of univariate HMM where a single observation (i.e, the total energy inside each bin), for any given state, only one empirical density function was defined.

3   Results

Two experiments were conducted using the HMM. Experiment A trained and evaluated a univariate HMM considering just a single feature: the total energy in a time-binned audio signal. By contrast, in Experiment B, all the three band of energies were considered as a vector of observations, and a multivariate HMM was trained. Both experiments used the "mhsmm" package in the statistical software R. Both Experiments evaluated the HMMs according to ability to classify, for a given time bin, the particular coughing state as well as dichotomous classification regarding the presence of absence of coughing.

Table 4. Performance statistics of the testing set for the univariate HMM in cough/no_cough and coughing/no_coughing classification mode

	Identifying a cough epoch in bout of coughs			Identifying a bout of coughs		
	Observed			Observed		
	Cough		No–cough	Coughing		No–coughing
Predicted cough(ing)	247		230	371		214
Predicted no-cough(ing)	30		656	22		556
		Accuracy: 78%			Accuracy: 80%	
		Sensitivity: 89%			Sensitivity: 94%	
		Specificity: 74%			Specificity: 72%	

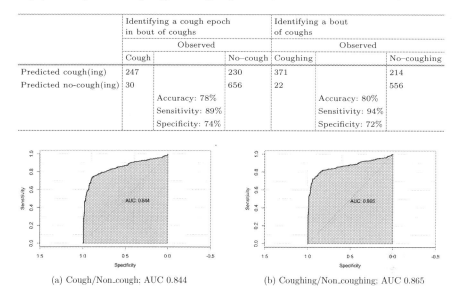

(a) Cough/Non_cough: AUC 0.844   (b) Coughing/Non_coughing: AUC 0.865

Fig. 4. ROC curve for uni-variate HMM after grouping

3.1 Results of the Univariate HMM: Experiment A

Using the total energy in bins as the single feature, an AUC value of 0.751 and 0.744 was obtained for training and testing sets, respectively. The performance statistics of the model over the testing set – including a confusion matrix, sensitivity, specificity, and accuracy – is shown in Table 3. Performance statistics of the testing set for the univariate HMM in cough/nocough and coughing/nocoughing classification mode is shown in Table 4.

To investigate the obtained models performance in classifying Cough from Non-Cough or Coughing from Non-Coughing, the identified states were grouped as per the process discussed in Sect. 2.3 resulting in the following ROC curves shown in Fig. 4 for Cough/Non-Cough and Coughing/Non-Coughing classifications.

3.2 Multivariate HMM Results: Experiment B

The multivariate HMM trained with a vector of three features containing the acoustic energy in low, medium and high bands improved by 6% the performance of the AUC for the testing set, increasing it from 0.744 to 0.789. The AUC for training set was almost the same as for the univariate case, reaching 0.752. The performance statistics of the chosen by Youden's-index-selected multivariate model over the testing set is demonstrated at Table 5. Also, the results of the Cough/Non-Cough and Coughing/Non-Coughing classifications as the results

Table 5. Performance statistics of the testing set for multivariate HMM

		Observed				
		Class: A	Class: B	Class: C	Class: D	Class: E
Predicted	Class: A	41	12	0	1	5
	Class: B	4	41	6	0	8
	Class: C	0	21	33	2	10
	Class: D	2	4	43	26	3
	Class: E	6	1	63	87	744
Sensitivity		0.774	0.519	0.223	0.224	0.966
Specificity		0.984	0.984	0.968	0.950	0.600
Accuracy		0.761				

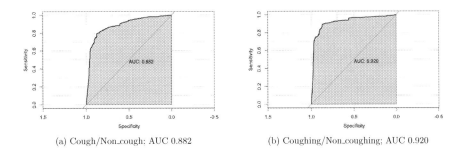

(a) Cough/Non_cough: AUC 0.882   (b) Coughing/Non_coughing; AUC 0.920

Fig. 5. ROC curve for multivariate HMM after grouping

Table 6. Performance statistics of the testing set for multi-variate HMM in cough/no_cough and coughing/no_coughing classification mode

	Identifying a cough epoch in bout of coughs			Identifying a bout of coughs		
	Observed			Observed		
	Cough		No–cough	Coughing		No–coughing
Predicted cough(ing)	243		181	342		82
Predicted no-cough(ing)	34		705	51		668
		Accuracy: 82%			Accuracy: 89%	
		Sensitivity: 88%			Sensitivity: 87%	
		Specificity: 80%			Specificity: 90%	

of dichotomously grouping the cough states are depicted in Fig. 5. The AUC for the cases of Cough/Non-Cough and Coughing/Non-Coughing classification were increased by 4.5% and 6.4% when compared to their univariate HMM counterparts.

Using the curves demonstrated in Fig. 5 and to maximize both the sensitivity and specificity, the best cut-off point was calculated on which the confusion matrix and the optimal accuracy, sensitivity and specificity were obtained, according to Youden's index. Results of using the best threshold in terms of balancing the sensitivity and specificity are shown in Table 6.

4   Conclusion

The HMMs evaluated here demonstrated favorable results, especially when the obtained results were interpreted as a dichotomously problem of distinguishing Coughs from Non_Coughs, or Coughing from Non_Coughing periods. Moreover, the multivariate HMM performed slightly more favourably than did a univariate HMM.

Unsurprisingly, the results presented in this work further suggest that the multivariate HMM demonstrates classification and detection of cough events with higher accuracy than does a to univariate HMM. Splitting the energy of cough sounds into three separate bands lead to density functions corresponding to each band which can provide more detailed information to the HMM.

While the results presented here demonstrate much promise, the approach applied exhibits significant limitations and room for improvements. The added accuracy associated with multivariate analysis invites investigation into both alternative bands, but also classification according to a larger number of such bands. The library of cough sounds examined here were greatly limited in their sourcing; results presented here may differ significantly for alternative coughing etiologies, and according to the pulmonary and upper-respiratory character and physical shape of the individual coughing, and potentially according to cultural norms involved. Greater variety in sourcing of cough source remains a high priority. Moreover, the classification accuracy exhibited in this study needs to be considered in light of the limited library of recordings employed here; other audio recordings containing a variety of background noise or other respiratory-related sounds may exhibit marked difference in the accuracy of classification that they support using similar HMMs. Finally, it will be important to consider examining other classifiers, that provide additional avenues for predictive accuracy, including classifiers that are less theory-based, such as artificial recurrent neural networks or deep learning networks employing recurrent network structures.

Despite its limitations, the cough analysis presented approach can provide a foundation towards support both clinical research on pulmonary distress at a clinical level and for capturing patient outcomes. It further offers intriguing potential for early-warning outbreak detection in public areas using mobile sensor data – such as from wearable devices and smartphones, particularly when coupled with transmission modeling and tools such as particle filtering. Another potential application of this study can be symptomatically-triggered treatment of patients suffering from respiratory diseases, particularly in patients that lack ready capacity to communicate their distress, such as in infants and young children, and among adults suffering from dementia or verbal limitations. The technique also offers potential for recognizing animal vocalization and diagnosing animal health status.

References

1. Swarnkar, V., Abeyratne, U., Chang, A., Amrulloh, Y., Setyati, A., Triasih, R.: Automatic identification of wet and dry cough in pediatric patients with respiratory diseases. Ann. Biomed. Eng. **41**(5), 1016–1028 (2013)
2. Larson, E.C., Lee, T., Liu, S., Rosenfeld, M., Patel, S.N.: Accurate and privacy preserving cough sensing using a low-cost microphone. In: Proceedings of the 13th International Conference on Ubiquitous Computing, pp. 375–384. ACM (2011)
3. Korpas, J., Sadlonova, J., Vrabec, M.: Analysis of the cough sound: an overview. Pulm. Pharmacol. **9**, 261–268 (1996)
4. Morice, A.H., et al.: ERS guidelines on the assessment of cough. Eur. Respir. J. **29**, 1256–1276 (2007)
5. Korpas, J., Vrabec, M., Sadlonova, J., Salat, D., Debreczeni, L.A.: Analysis of the cough sound frequency in adults and children with bronchial asthma. Acta Physiol. Hung. **90**, 27–34 (2003)
6. Day, J., Goldsmith, T., Barkley, J., Afshari, A., Frazer, D.: Identification of individuals using voluntary cough characteristics. In: Biomedical Engineering Society Meeting, vol. 1, pp. 97–107 (2004)
7. Doherty, M.J., et al.: The acoustic properties of capsaicin-induced cough in healthy subjects. Eur. Respir. J. **10**, 202–207 (1997)
8. Murata, A., Taniguchi, Y., Hashimoto, Y., Kaneko, Y., Takasaki, Y., Kudoh, S.: Discrimination of productive and non-productive cough by sound analysis. Internal Med. **37**, 732–735 (1998)
9. Thorpe, C.W., Toop, L.J., Dawson, K.P.: Towards a quantitative description of asthmatic cough sounds. Eur. Respir. J. **5**, 685–692 (1992)
10. Toop, L.J., Dawson, K.P., Thorpe, C.W.: A portable system for the spectral-analysis of cough sounds in asthma. J. Asthma **27**, 393–397 (1990)
11. Matos, S., Birring, S.S., Pavord, I.D.: Detection of cough signals in continuous audio recording using hidden Markov models. IEEE Trans. Biomed. Eng. **53**(6), 1078–1083 (2006)
12. Matos, S., Birring, S.S., Pavord, I.D., Evans, D.H.: An automated system for 24-h monitoring of cough frequency: the leicester cough monitor. IEEE Trans. Biomed. Eng. **54**(8), 1472–1478 (2007)
13. Youden, W.J.: Index for rating diagnostic tests. Cancer **3**(1), 32–35 (1950)
14. Robin, X., et al.: pROC: an open-source package for R and S+ to analyze and compare ROC curves. BMC Bioinform. **12**(1), 77 (2011)

Modeling Belief Divergence and Opinion Polarization with Bayesian Networks and Agent-Based Simulation
A Study on Traditional Healing Use in South Africa

Kamwoo Lee[✉] and Jeanine Braithwaite

University of Virginia, Charlottesville, VA 22904, USA
{kl9ch,jeaninebraithwaite}@virginia.edu

Abstract. This study uses agent-based simulation with human settlement patterns to model belief revision and information exchange about health care options. We adopt two recent microeconomic theories based on Bayesian Network formulations for individual belief update then examine the macro-level effects of the belief revision process. This model tries to explain traditional healing usage at the village and regional level while providing a causal mechanism with a single conceptual factor, mobility, at the individual level. The resulting simulation estimates the dependency on traditional healing in villages in Limpopo, South Africa, and the estimates are validated with empirical data.

Keywords: Traditional healing · Belief divergence ·
Settlement patterns · Bayesian Networks · Agent-based modeling

1 Introduction

Traditional healing (TH) and its practitioners have long been considered an important supplement to the health care system in South Africa [1,7,14]. However, it is unclear to what extent individuals depend on TH and why some groups of people are more likely to use it than others. Even in the scientific literature, the estimated prevalence of TH use has wide discrepancies between studies. While it is frequently stated that 80% of people in African countries use traditional medicine practitioners for some part of their primary health care [2,10], other studies argue that the prevalence is negligibly low, anywhere between 0.1% and 2% [11,13].

Opinions towards TH are as polarized as the estimation itself. Some people vehemently advocate that modern or Western medicine is firmly rooted in a scientific paradigm and biomedical science explains the cause of disease while others believe that traditional medicine has long been successfully practiced, operating within a spiritual realm [15]. Although attitude polarization is a common, social phenomenon, it is puzzling why some groups of people are consistently inclined towards one side in the first place.

R. Thomson et al. (Eds.): SBP-BRiMS 2019, LNCS 11549, pp. 277–287, 2019.
https://doi.org/10.1007/978-3-030-21741-9_28

This paper is an attempt to demonstrate how beliefs towards TH can diverge between groups of the general public in a systematic fashion across age, income, and geography. We hypothesize that people's mobility, which is closely related to all three factors, causes divergence when the information is inherently ambiguous like opinions about medical services. Since people with different mobility have different information flows, they give different credence to the same pieces of information they encounter. This discrepancy can eventually result in differing, or even opposing, opinions through polarization. We test this hypothesis by simulating continuous opinion exchanges in several geographic regions with people who approximate probabilistic inference. Starting with random levels of belief regarding traditional healers and hospitals, people revise their beliefs after listening to others and develop fairly regular and consistent opinions over time, based on age, income, and settlement patterns. The result of the simulation matches both our survey results and national statistics.

2   Background and Related Work

This paper draws together three separately researched topics and suggests a new combination of methods and subject matter. This section summarizes selected findings in these topics that comprise the foundation for the model of this study.

2.1   Traditional Healing in South Africa

The World Health Organization (WHO) defined traditional medicine (TM) as "the sum total of the knowledge, skill, and practices based on the theories, beliefs, and experiences indigenous to different cultures", while suggesting that TM is an important but underestimated form of health care [19]. It is arguable that TM can encompass all TH in South Africa because TH sometimes encompass supernatural counselling or even fortune telling. It is true that South African traditional healers play many different roles such as health care provider, spiritual advisor, and cultural heritage keeper. This paper particularly focuses on the primary health care function of TH and uses research findings and data of TM and TH interchangeably in this narrow functionality.

2.2   Belief Revision, Divergence, and Polarization

Belief revision is the process of changing beliefs in response to new information. The logic of belief revision is a relatively young field of research originated from philosophy and computer science [8]. Basically, a belief-revision rule defines how a person accepts and internalizes outside information when she or he encounters it. There are many logical representations for this process [6]. When assumed that believing is a probabilistic degree instead of a dichotomous conviction, one of the widely used methodologies to conceptualize belief revision is Bayesian frameworks [12].

It is not uncommon that two people change their beliefs in the opposite direction even when they observe the same set of information. Jern et al. [9] showed that the belief divergence can be expressed with a Bayesian network for a family of cases and that probabilistic inference of rational people can lead to divergence. Attitude polarization, a similar but different concept, is a phenomenon in which a disagreement becomes more extreme as people consider evidence about the issue. This is especially true when people encounter ambiguous evidence and interpret it to align with their existing beliefs. Through this process, people reinforce their previously-held beliefs, which further widens disagreement. Fryer Jr. et al. [5] introduced a model that represents polarization with Bayes' rule when information is ambiguous and open to interpretation.

When modeling belief on traditional healing, we combine and extend the above-mentioned microeconomic models to incorporate both belief divergence and attitude polarization into macroeconomic dynamics. Our model extension allows us to follow the macroscopic effects of a sequence of belief revision provided by the microbehavioral insights in our real-world application.

2.3   Agent-Based Modeling of Spatial Dynamics

As outlined in Sect. 1, the main hypothesis of this research is that mobility and information flows drive polarization of opinion. Thus, it is necessary to incorporate spatial dynamics into the model. It is extremely difficult, if not impossible, to model the spatial pattern formation using purely mathematical abstractions. Agent-based models (ABM) can simulate autonomous agents with spatial interactions to allow macroscopic patterns to emerge from microscopic rules and make it possible to understand human behaviors and interactions that give rise to complex patterns [20]. Although, to our knowledge, ABM has not been applied in TH studies specifically, it has been widely used in health behavior and health care policy research [18]. One of the closest studies is the overview of possibilities of applying ABM to a wide range of complementary and alternative medicine research [4].

3   Study Area and Associated Data

The study villages are located in Vhembe district, Limpopo province, South Africa and are approximately 500 km Northeast of Johannesburg (Fig. 1). Vhembe district, and more broadly Limpopo province, is a suitable study area for traditional healing in the rural context since it is one of the most rural and poorest provinces in the country.

3.1   General Household Survey in South Africa

According to the General Household Survey (GHS) conducted by the national statistical office of South Africa [16], the percentage of households using TH as their primary health aid is quite low. When asked "If anyone in this household

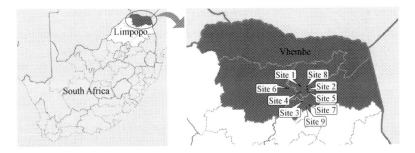

Fig. 1. Map of study area and nine survey sites in Vhembe district, Limpopo province, Limpopo province, South Africa (Source: Wikimedia Commons licensed under CC BY-SA)

becomes ill and decides to seek medical help, where do they usually go first?", less than 1% of household heads answered they would go to a "traditional healer" or a "spiritual healer" first. The Chi-Square test of independence on GHS 2017 data shows that income and geography related variables have significant relationships with the choice of health aid, such as "means of transport to nearest health facility" $(p < 10^{-7})$, "distance to the nearest transport" $(p < 10^{-4})$, "net household income" $(p < 10^{-4})$, and "roof material" $(p < 10^{-6})$.

3.2   Field Study in Limpopo

Our local utilization study took place in nine, mostly rural, traditional villages in Vhembe district during May 23–31, 2018 conducted by a joint research team from the University of Virginia, USA, and the University of Venda, South Africa (IRB protocol UVA IRB-SBS #2018-0156-00). We surveyed 112 people, whose ages were quite evenly distributed from 20 to 90, and the elderly formed a considerably large portion of our respondents compared to the national population-age composition. Since the percentage of people using TH is so low in the GHS, it is advantageous to focus on this admittedly biased sample to study TH user groups as it likely reflects a more sensible depiction of the prevalence of TH use in South Africa.

The main questions of the questionnaire asked for the choice of primary medical assistance. The two-step questions were "Do you primarily use traditional healers, western health centers, or both in conjunction?" and "If both, which would you visit first?". In summary, 8% of respondents answered "primarily traditional healers", 45% answered "primarily western health centers", and 44% answered "both". Among the respondents whose answer was "both", 37% and 51% responded that they would go first to traditional healers and western health centers, respectively. One noticeable finding of our study was that the dependency on TH was highly variable even among rural, traditional villages. When counted people who either primarily use TH or who use both but go to TH first, TH prevalence in each village ranges from 0% to 50%.

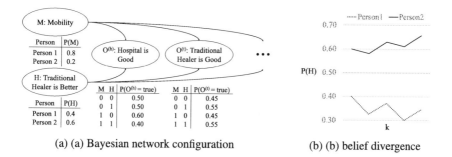

(a) (a) Bayesian network configuration   (b) (b) belief divergence

Fig. 2. Illustration of suggested Bayesian Network (a) and an example of belief divergence after encountering two $O^{(h)}$ and two $O^{(t)}$ opinions (b).

3.3 Human Settlement Patterns

Including the settlement pattern is necessary to differentiate people in our model. Since it is almost impossible to obtain such settlement pattern data in rural areas in developing countries, an alternative source of information is utilized. A synthetic population density data source – the high resolution settlement layer (HRSL) – provides estimates of human settlement distribution at a resolution of 1 arc-second (approximately 30 m) for the year 2015. This dataset was created using computer vision techniques to detect objects from satellite images and was validated with the national census data and the World Bank Living Standards Measurement Study (LSMS) surveys [3].

4 Proposed Model

The simulation model proposed in this paper formulates attitude polarization as a result of belief divergence and reinforcement of such belief. Divergence and reinforcement are treated as separate mechanisms but in the same Bayesian network formulation. After introducing the Bayesian network formulation, this section delineates parameters and interactions of agents in the model.

4.1 Bayesian Network Formulation for Individual Belief Revision

At the individual level, a Bayesian network is represented by conditional probability distributions of random variables with causal relationships. The three types of variables in our proposed model are:

1. Hidden fact in question (H): Traditional healer is better than hospital
2. Mobility of an agent (M): Mobility of an agent is high
3. Opinion (O_k^{type}): At time k, an agent hears from another agent either that a hospital is good (type h) or that a traditional healer is good (type t)

Figure 2 shows the causal relationships and conditional probabilities of the variables. Here, the fact that an agent hears an opinion about a hospital or a traditional healer obviously depends on the effectiveness of the hospital or the healer. Also, one's mobility affects the chance of listening to other agents' opinions about hospitals, which are usually more remote than traditional healers. Intuitively, agents with high mobility think that they can judge the reality about remote hospitals better than agents with low mobility because they have more chances to listen to many opinions. However, everyone has the same chance to listen to opinions about traditional healers regardless of their mobility because the healers are close to them. The opinions about traditional healers are less clear due to the nature of healing procedure. All conditional probability on observation nodes are close to 0.5, which means that opinions about medical aid are inherently not clear and the advantage of agents with high mobility is only slightly higher.

Once the building blocks of the Bayesian network are constructed, it can be expanded for a recursive update as follows:

$$P(H|O_{1:k}) = \frac{P(O_k|H)}{P(O_k|O_{1:k-1})} \times P(H|O_{1:k-1})$$

$$= \frac{\left\{ P(O_k|M,H) \cdot P(M|H,O_{1:k-1}) + P(O_k|M^c,H) \cdot P(M|H^c,O_{1:k-1}) \right\}}{P(O_k|O_{1:k-1})} \times P(H|O_{1:k-1})$$

where the denominator $P(O_k|O_{1:k-1})$ is constant relative to H, and the numerator can be calculated and then normalized. $P(H|O_{1:k-1})$ is a previous belief on the hidden fact before encountering O_k and, the base case is $P(H)$ which we give normally distributed random value with mean 0.5 and standard deviation 0.1. $P(M|H,O_{1:k-1})$ and $P(M|H^c,O_{1:k-1})$ can be updated recursively, too as follows:

$$P(M|H,O_{1:k-1}) = \frac{P(O_{k-1}|M,H)}{P(O_{k-1}|M,O_{1:k-2})} \times P(M|H,O_{1:k-2}),$$

$$P(M|H^c,O_{1:k-1}) = \frac{P(O_{k-1}|M,H^c)}{P(O_{k-1}|M,O_{1:k-2})} \times P(M|H^c,O_{1:k-2}).$$

With this recursion, agents fine-tune their beliefs. It is worth noting that the model does not assume any human bias at this point. Since perfectly rational probabilistic inference can cause divergence due to the nature of the hidden fact and the mobility, the simulation takes advantage of this structure. As shown in Fig. 2, two perfectly rational people can reach different conclusions after encountering the same set of opinions. The beliefs of person 1 and person 2 have changed from 0.40 to 0.34 and from 0.60 to 0.66, respectively, after observing two $O^{(h)}$ and two $O^{(t)}$ opinions.

It is assumed that people interpret each ambiguous opinion and then make inferences about the interpreted opinions. Even though people lose some information since only the interpretation of the opinions is retained, it has been shown that this is optimal if agents sufficiently discount the value of time as shown by [5].

4.2   Parameters of Agents

Individual mobility for the simulation is parameterized as the following formula:

$$\text{mobility} = a \cdot \frac{1}{\text{age}} + b \cdot log(\text{family income}) + c \cdot \sqrt{\text{available cars}} + \varepsilon,$$

where a, b and c are constants that are calibrated during the initial simulation test-runs. The reasoning behind this formulation is that mobility of a person decreases with age and increases with income or the number of available cars around. We take the log of income and the square root of the number of available cars to better differentiate population by utilizing the fact that income and density of cars are more normally distributed on logarithmic scale and square-root scale, respectively, while the original distribution is highly skewed. ε is a small random value with a fraction of the standard normal distribution which reflects individual peculiarity and simulation randomness.

The agents in the simulation are populated according to the HRSL. The distributions of age and family income are from the Limpopo data in the GHS 2017, and the demographic characteristics are randomly assigned in each run of simulation. It is assumed that the decisions for people under 20 are made by adult members of the population group. The number of vehicles around is estimated from the population density and statistics of the South African Department of Transport [17] assuming that the number of vehicles in a village is proportional to the number of residents.

4.3   Interactions of Agents

In this model, agents are simulated to exchange their opinions about hospitals and traditional healers. Each person regards other people's opinion as evidence about the quality of the medical practitioners and makes probabilistic inferences about the question whether traditional healers are better than hospitals. If the probability is greater than 0.5, one concludes that the traditional healers are better and if it is less than 0.5, one reaches the other conclusion. Over time, agents keep revising their beliefs while they hear opinions of others. Since expression can be ambiguous, it is assumed that 50% of opinions needs to be interpreted and agents retain the interpreted opinions in their mind.

5   Results

This section presents the analysis of simulation runs and validation. The parameters of agents are calibrated with the responses in three villages of our field study described in Sect. 3.2. The study responses from the remaining 6 villages are used to validate the model. The national statistics is also used for the validation.

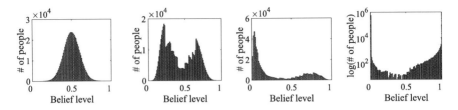

Fig. 3. Beliefs on traditional healing are diverging gradually from initial normal distribution. Note that the last figure is displayed in logarithm scale for a better depiction.

5.1 Simulation

The simulation demonstrates how the initial random beliefs about traditional healing changes over time and how the prevalence of TH use asymptotically approach the current real-world distributions, which range from 0% to 50% of the biased population in our study areas (Sect. 3.2) and 1% of the total population in Limpopo province (Sect. 3.1). Our results show that the opinions of agents diverge and eventually polarize. Each simulation runs until opinions of people do not change for sufficient time. Since each run begins with random beliefs and random demographic characteristics, the end result is different for every run. The final outputs of the simulation are the average and confidence interval out of 100 simulation runs. The simulation parameters are calibrated using study responses from village 3, 4, and 5 to capture a wide range of responses for TH use. The fixed parameters are used for simulations for the remaining of villages as well as Limpopo province.

Figure 3 illustrates the divergence of beliefs. The initial bell curve splits right after the start of the simulation, and then ends up with polarized beliefs on both sides. Even though the numbers are much higher on the left side, neither side of the curve moves after some point in time. One could casually conjecture that more rural, isolated areas use more TH. But this may not necessarily be true. The variances of TH use are high in very rural, isolated areas depending on the initial demographic distribution and interactions among them. For example, village 7 is one of the most isolated areas, but the usage of TH is found to be relatively low in our field study. Our simulation shows that very isolated areas have high variance of TH use depending on the initial demographic characteristics and initial beliefs of small population.

5.2 Validation

The model results are validated with our study responses and the GHS 2017 results. It is not reasonable to compare the percentages of TH use between our study responses and the simulation since the sample of our study is biased as discussed in Sect. 3.2 while the simulation reflects the whole population distribution of Limpopo. Thus, in the simulation, we keep track of the same sub-population group divided by age and income in each village (Fig. 4). We, then, compare the

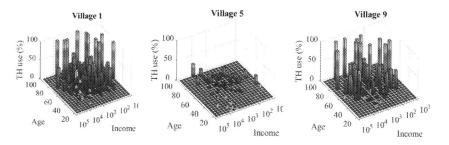

Fig. 4. Average TH use of sub-population groups divided by age and income in each village out of 10 simulation runs. For instance, in Village 1, the percentage of TH use for the sub-group with age 80–83 and income 2500–3900 Rand/month is 12% from this set of simulation runs. The red bars represent the sub-groups of which we keep track for the comparison between the simulation result and the field study. (Color figure online)

Table 1. Comparison of simulation results with field study and GHS

Category	Name	Field study and GHS		Simulation	
		Avg TH use	95% CI	Avg TH use	95% CI
Village 1	Vhutanda	50.0%	(25.4%–74.6%)	26.9%	(21.6%–32.2%)
Village 2	Khwevha	7.1%	(1.3%–31.5%)	11.9%	(6.5%–17.3%)
Village 3	Tshivhulana	36.4%	(15.2%–64.6%)	27.9%	(18.3%–37.4%)
Village 4	Ha-Gelebe	11.8%	(3.3%–34.3%)	8.0%	(3.4%–12.6%)
Village 5	Muledane	0.0%	(0.0%–29.9%)	0.8%	(0.0%–1.9%)
Village 6	Vuvha	28.6%	(11.7%–54.6%)	6.4%	(3.2%–9.5%)
Village 7	Thandani	6.7%	(1.2%–29.8%)	20.9%	(13.3%–28.4%)
Village 8	Univen	0.0%	(0.0%–56.2%)	0.1%	(0.0%–0.2%)
Village 9	Davhana	31.3%	(14.2%–55.6%)	30.8%	(23.3%–38.4%)
Province	Limpopo	1.05%	(0.30%–1.95%)	1.77%	(0.89%–2.64%)

average TH use of that sub-population group with our study responses. For the comparison with GHS, we run the same model on Limpopo province HRSL data and compare the TH use of the whole population.

Since both survey data and simulation result are estimation, we check the average and confidence interval (CI) of both estimates. Table 1 shows the validation results. The average TH use in the simulation falls within the 95% CI of the average TH use in most villages of our field study. It is notable that the simulation average of Limpopo province is also within the tight CI of GHS average. Conversely, the average TH use in the field study and GHS is within the CI of simulation average in the majority of villages and Limpopo province. Village 8 does not represent this geographic location because it is a university campus. It is included for only presentation purpose. The average TH use in Village 6 is significantly different between our field study and simulation. We will discuss this matter in the discussion section.

6   Discussion and Conclusion

This paper proposed a causal mechanism that explained the distribution and variance of TH use in Limpopo region, South Africa. It is surprising that mobility alone can explain much of the variance in this abstract micro-macro model since there could be many other factors that affect TH use such as cultural backgrounds, education, and types of illness [13]. This is in part because people in our study areas shared much of the same cultural experience, and we regarded many cultural variables as constant. Another reason could be that the focus was on the health care function of traditional healing. If people had been asked whether they use TH for other reasons, the distribution could have been much different. Other roles of TH such as supplementary health care, spiritual mentoring, and forecasting the future are beyond of the scope of this research. This research only emphasizes the primary health care function of TH.

Several limitations remain, and many extensions are possible. In validation, the simulation estimate was much lower than the average study responses in Village 6. Our conjecture is that the ratio of vehicles to population is different from region to region, especially in very isolated villages. To solve this, we need more accurate data. One very effective and efficient way is to use an object detection methods on satellite images and count the number of vehicles in each village. On the macro level, the result may seem similar to a regression analysis. But this research study is more than just identifying factors that are related TH usage. We tried to explain the belief changing process, which could be applicable to other aspects of development as well. By explaining the causal link between the individual level and the macroscopic phenomenon, it enables policy makers to develop individual level interventions for systematically marginalized populations since the model can serve as a test bed to conduct dynamic experiments.

Acknowledgments. This research was supported by UVA's Center for Global Health, the Harrison Research Award, and the Frank Batten School of Leadership and Public Policy. We would also like to show our deepest gratitude to the University of Venda for its partnership, hospitality and support. We especially thank the coordinators/interpreters/translators for our field study in Limpopo: Livhuwani Daphney, Mphatheleni Makaulule, Faith Musvipwa, Rendani Nematswerani, Phathutshedzo Nevhutalu, and Wisani Nwankoti.

References

1. Babb, D.A., Pemba, L., Seatlanyane, P., Charalambous, S., Churchyard, G.J., Grant, A.D.: Use of traditional medicine by hiv-infected individuals in south africa in the era of antiretroviral therapy. Psychol. Health Med. **12**(3), 314–320 (2007)
2. Ekor, M.: The growing use of herbal medicines: issues relating to adverse reactions and challenges in monitoring safety. Front. Pharmacol. **4**, 177 (2014)
3. Facebook Connectivity Lab and Center for International Earth Science Information Network - CIESIN - Columbia University: High Resolution Settlement Layer (HRSL) (2016)

4. Frantz, T.L.: Advancing complementary and alternative medicine through social network analysis and agent-based modeling. Complement. Med. Res. **19**(Suppl. 1), 36–41 (2012)
5. Fryer Jr., R.G., Harms, P., Jackson, M.O.: Updating beliefs with ambiguous evidence: implications for polarization. Technical report, National Bureau of Economic Research (2013)
6. Gärdenfors, P.: Belief Revision, vol. 29. Cambridge University Press, Cambridge (2003)
7. Gqaleni, N., Moodley, I., Kruger, H., Ntuli, A., McLeod, H.: Traditional and complementary medicine: health care delivery. S. Afr. Health Rev. **2007**(1), 175–188 (2007)
8. Hansson, S.O.: Logic of belief revision. In: Zalta, E.N. (ed.) The Stanford Encyclopedia of Philosophy. Metaphysics Research Lab, Stanford University, winter 2017 edn. (2017)
9. Jern, A., Chang, K., Kemp, C.: Bayesian belief polarization. In: Advances in Neural Information Processing Systems, pp. 853–861 (2009)
10. World Health Organization: Promoting the role of traditional medicine in health systems: a strategy for the African region (2000)
11. Oyebode, O., Kandala, N.B., Chilton, P.J., Lilford, R.J.: Use of traditional medicine in middle-income countries: a who-sage study. Health Policy Plan. **31**(8), 984–991 (2016)
12. Pearl, J.: Belief networks revisited. In: Artificial Intelligence in Perspective, pp. 49–56 (1994)
13. Peltzer, K.: Utilization and practice of traditional/complementary/alternative medicine (TM/CAM) in south africa. Afr. J. Tradit. Complement. Altern. Med. **6**(2), 175 (2009)
14. Richter, M.: Traditional medicines and traditional healers in South Africa. Treat. Action Campaign AIDS Law Proj. **17**, 4–29 (2003)
15. Sobiecki, J.: The intersection of culture and science in South African traditional medicine. Indo-Pac. J. Phenomenol. **14**(1) (2014)
16. Stats SA: General household survey (revised 2017) (2017). http://nesstar.statssa.gov.za:8282/webview/
17. The electronic National Administration Traffic Information System (eNATIS): Number of registered vehicles by province in South Africa (2017). http://www.enatis.com/index.php/statistics/
18. Tracy, M., Cerdá, M., Keyes, K.M.: Agent-based modeling in public health: current applications and future directions. Ann. Rev. Public Health **39**, 77–94 (2018)
19. World Health Organization: WHO traditional medicine strategy, 2014–2023. World Health Organization (2013)
20. Ye, X., Mansury, Y.: Behavior-driven agent-based models of spatial systems. Ann. Reg. Sci. **57**(2–3), 271–274 (2016)

Author Index

Printed in the United States
By Bookmasters